THEORY OF LINEAR OPERATIONS

North-Holland Mathematical Library

VOLUME 38

NORTH-HOLLAND
AMSTERDAM · NEW YORK · OXFORD · TOKYO

Theory of Linear Operations

S. BANACH †

English translation by

F. JELLETT
London, United Kingdom

1987

NORTH-HOLLAND
AMSTERDAM · NEW YORK · OXFORD · TOKYO

ISBN: 0 444 70184 2

This volume is a translation of:

Théorie des Operations Linéaires
©PWN-Państwowe Wydawnictwo Naukowe, Warsaw, Poland, 1979

and includes 'comments' by

A. Pełczyński and Cz. Bessaga
under the title
'Some Aspects of the Present Theory of Banach Spaces'

Published by:

ELSEVIER SCIENCE PUBLISHERS B.V.
P.O. Box 1991
1000 BZ Amsterdam
The Netherlands

Sole distributors for the U.S.A. and Canada:

ELSEVIER SCIENCE PUBLISHING COMPANY, INC.
52 Vanderbilt Avenue
New York, N.Y. 10017
U.S.A.

PRINTED IN THE NETHERLANDS

Preface

The theory of operators, created by V. Volterra, has as its object the study of functions defined on infinite-dimensional spaces. This theory has penetrated several highly important areas of mathematics in an essential way: suffice it to recall that the theory of integral equations and the calculus of variations are included as special cases within the main areas of the general theory of operators. In this theory the methods of classical mathematics are seen to combine with modern methods in a remarkably effective and quite harmonious way. The theory often makes possible altogether unforeseen interpretations of the theorems of set theory or topology. Thus, for example, the topological theorem on fixed points may be translated, thanks to the theory of operators (as has been shown by Birkhoff and Kellogg) into the classical theorem on the existence of solutions of differential equations. There are important parts of mathematics which cannot be understood in depth without the help of the theory of operators. Contemporary examples are: the theory of functions of a real variable, integral equations, the calculus of variations, etc.

This theory, therefore, well deserves, for its aesthetic value as much as for the scope of its arguments (even ignoring its numerous applications) the interest that it is attracting from more and more mathematicians. The opinion of J. Hadamard, who considers the theory of operators one of the most powerful methods of contemporary research in mathematics, should come as no surprise.

The present book contains the basics of the algebra of operators. It is devoted to the study of so-called *linear* operators, which corresponds to that of the linear forms $a_1x_1 + a_2x_2 + \ldots + a_nx_n$ of algebra.

The notion of linear operator can be defined as follows. Let E and E_1 be two abstract sets, each endowed with an associative addition operation as well as a zero element. Let $y = U(x)$ be a function (operator, transformation) under which an element y of E_1 corresponds to each element x of E (in the special case where E_1 is the space of real numbers, this function is also known as a *functional*). If, for any x_1 and x_2 of E, we have $U(x_1 + x_2) = U(x_1) + U(x_2)$, the operator U is said to be *additive*. If, in addition, E and E_1 are *metric* spaces, that is to say that in each space the *distance* between pairs of elements is defined, one can consider *continuous* operators U. Now operators which are both additive and continuous are called *linear*.

In this book, I have elected, above all, to gather together results concerning linear operators defined in general spaces of a certain kind, principally in the so-called *B-spaces* (i.e. *Banach spaces* [trans.]), examples of which are: the space of continuous functions, that of the pth-power-summable functions, Hilbert space, etc.

I also give the interpretation of the general theorems in various mathematical areas, namely, group theory, differential equations, integral equations, equations with infinitely many unknowns, functions of a real variable, summation methods, orthogonal series, etc. It is interesting to see certain theorems giving results in such widely varying fields. Thus, for example, the theorem on the extension of an additive functional settles simultaneously the general problem of measure, the moment problem and the question of the existence of solutions of a system of linear equations in infinitely many unknowns.

Along with algebraic tools, the methods are principally those of general set theory, which in this book are to the fore in gaining, for this theory, several new applications. Also to be found in various chapters of this book are some new general theorems. In particular, in the last two chapters and the appendix: no part of the results included therein has been published before. They constitute an outline of the study of invariants with respect to linear transformations (of B-spaces). In particular, Chapter XII includes the definition and analysis of the properties of *linear dimension*, which in these spaces plays a rôle analogous to that of dimension in the usual sense in euclidean spaces.

Results and problems, which, for want of space, have not been considered, are discussed briefly in the Remarks at the end of the book. Some further references are also to be found there. In general, except in the Introduction or, rather, its accompanying Remarks at the end of the book, I do not indicate the origin of theorems which either I consider too elementary or else are proved here for the first time.

Some more recent work has appeared and continues to appear in the periodical *Studia Mathematica*, whose primary purpose is to present research in the area of functional analysis and its applications.

I intend to devote a second book, which will be the sequel to the present work, to the theory of other kinds of functional operators, using topological methods extensively.

In conclusion, I would like to express my sincere gratitude to all those who have assisted me in my work, in undertaking the translation of my Polish manuscript, or helping me in my labours with their valuable advice. Most particularly, I thank H. Auerbach for his collaboration in the writing of the Introduction and S. Mazur for his general assistance as well as for his part in the drafting of the final remarks.

Stefan Banach

Lwów, July 1932

Contents

Some aspects of the present theory of Banach spaces

Introduction

A. THE LEBESGUE - STIELTJES INTEGRAL

We assume the reader is familiar with measure theory and the Lebesgue integral.

§1. Some theorems in the theory of the Lebesgue integral.

If the measurable functions $x_n(t)$ form a (uniformly) bounded sequence and the sequence $(x_n(t))$ converges almost everywhere in a closed interval $[a,b]$ to the function $x(t)$, then

$$\lim_{n\to\infty} \int_a^b x_n(t)\,dt = \int_a^b x(t)\,dt. \tag{1}$$

More generally, if there exists a summable function $\phi(t) \geq 0$ such that $|x_n(t)| \leq \phi(t)$ for $n=1,2,\ldots,$ the limit function is also summable and (1) is still satisfied.

If the functions $x_n(t)$ are summable in $[a,b]$ and form a non-decreasing sequence which converges to the function $x(t)$, then (1) holds, when the function $x(t)$ is summable, and

$$\lim_{n\to+\infty} \int_a^b x_n(t)\,dt = +\infty$$

otherwise.

If the sequence $(x_n(t))$ of pth-power summable functions $(p \geq 1)$ converges almost everywhere to the function $x(t)$ and if

$$\int_a^b |x_n(t)|^p dt < K \quad \text{for } n=1,2,\ldots,$$

the function $x(t)$ is also p^{th}-power summable.

§2. Some inequalities for p^{th}-power summable functions.

The class of functions which are p^{th}-power summable $(p > 1)$ in $[a,b]$ will be denoted by L^p. To the number p, there corresponds the number q, connected with p by the equation $\frac{1}{p} + \frac{1}{q} = 1$, and known as the *conjugate exponent of* p. For $p = 2$, we have equally $q = 2$.

If $x(t) \in L^p$ and $y(t) \in L^q$, the function $x(t)y(t)$ is summable and its integral obeys the inequality

$$\left|\int_a^b xy\,dt\right| \leq \left(\int_a^b |x|^p dt\right)^{\frac{1}{p}} \left(\int_a^b |y|^q dt\right)^{\frac{1}{q}}.$$

In particular, we therefore have for $p = 2$:

$$\left|\int_a^b xy\,dt\right| \leq \left(\int_a^b x^2 dt\right)^{\frac{1}{2}} \cdot \left(\int_a^b y^2 dt\right)^{\frac{1}{2}}.$$

If the functions $x(t)$ and $y(t)$ belong to L^p, so does the function $x(t) + y(t)$ and we have:

$$\left(\int_a^b |x+y|^p dt\right)^{\frac{1}{p}} \leq \left(\int_a^b |x|^p dt\right)^{\frac{1}{p}} + \left(\int_a^b |y|^p dt\right)^{\frac{1}{p}}$$

These inequalities are analogues of the following arithmetic inequalities:

$$\left|\sum_{i=1}^n a_i b_i\right| \leq \left(\sum_{i=1}^n |a_i|^p\right)^{\frac{1}{p}} \cdot \left(\sum_{i=1}^n |b_i|^q\right)^{\frac{1}{q}},$$

$$\left(\sum_{i=1}^n |a_i + b_i|^p\right)^{\frac{1}{p}} \leq \left(\sum_{i=1}^n |a_i|^p\right)^{\frac{1}{p}} + \left(\sum_{i=1}^n |b_i|^p\right)^{\frac{1}{p}},$$

of which the first yields, for $p = 2$, the well-known Schwarz inequality:

$$\left|\sum_{i=1}^n a_i b_i\right| \leq \left(\sum_{i=1}^n a_i^2\right)^{\frac{1}{2}} \cdot \left(\sum_{i=1}^n b_i^2\right)^{\frac{1}{2}}.$$

For every p^{th}-power summable function ($p \geq 1$) and every $\varepsilon > 0$ there exists a continuous function $\phi(t)$ such that

$$\int_a^b |x - \phi|^p < \varepsilon.$$

§3. Asymptotic convergence.

The sequence $(x_n(t))$ of measurable functions defined on some set is said to be *asymptotically convergent* (or *convergent in measure*) to the function $x(t)$ defined on the same set, if for each $\varepsilon > 0$

$$\lim_{n\to\infty} m(\{t : |x_n(t) - x(t)| > \varepsilon\}) = 0,$$

where $m(A)$ stands for the (Lebesgue) measure of the set A.

A sequence $(x_n(t))$ which is asymptotically convergent to the function $x(t)$ always has a subsequence which converges pointwise to this function almost everywhere.

For a sequence $(x_n(t))$ to be asymptotically convergent, it is necessary and sufficient that, for each $\varepsilon > 0$,

$$\lim_{i,k\to\infty} m(\{t : |x_i(t) - x_k(t)| > \varepsilon\}) = 0.$$

§4. Mean convergence.

A sequence $(x_n(t))$ of pth- power summable functions ($p \geq 1$) in $[a,b]$ is said to be p^{th} - power *mean convergent* to the pth - power summable function $x(t)$ if

$$\lim_{n\to\infty} \int_a^b |x_n(t) - x(t)|^p dt = 0.$$

A necessary and sufficient condition for such a function $x(t)$ to exist is that

$$\lim_{i,k\to\infty} \int_a^b |x_i(t) - x_k(t)|^p dt = 0.$$

The function $x(t)$ is then uniquely defined in $[a,b]$, up to a set of measure zero.

A sequence of functions which converges in mean to a function $x(t)$ is also asymptotically convergent to this function and therefore (c.f. §3) has a subsequence which converges pointwise to the same function almost everywhere.

§5. The Stieltjes Integral.

Let $x(t)$ be a continuous function and $\alpha(t)$ a function of bounded variation in $[a,b]$. By taking a partition of the interval $[a,b]$ into subintervals, using the numbers

$$a = t_0 < t_1 < t_2 < \ldots < t_n = b$$

and choosing an arbitrary number θ_i in each of these subintervals, we can, by analogy with the definition of the Riemann integral, form the sum

$$S = \sum_{i=1}^{n} x(\theta_i)[\alpha(t_i) - \alpha(t_{i-1})] \qquad \text{where } t_i \geq \theta_i \geq t_{i-1}.$$

One shows that for every sequence of subdivisions, for which the length of the largest subinterval tends to 0, the sums S converge to a limit which is the same for all such sequences; this limit is denoted by

$$\int_a^b x(t)\,d\alpha(t)$$

and is called a *Stieltjes integral*.

This integral has the following properties:

$$\int_a^b x(t)\,d\alpha(t) = -\int_b^a x(t)\,d\alpha(t),$$

$$\int_a^b x(t)\,d\alpha(t) + \int_b^c x(t)\,d\alpha(t) = \int_a^c x(t)\,d\alpha(t),$$

$$\int_a^b [x_1(t) + x_2(t)]\,d\alpha(t) = \int_a^b x_1(t)\,d\alpha(t) + \int_a^b x_2(t)\,d\alpha(t).$$

The first mean value theorem here takes the form of the inequality

$$\left|\int_a^b x(t)\,d\alpha(t)\right| \leq MV,$$

where M denotes the supremum of the absolute value $|x(t)|$ and V the total variation of the function $\alpha(t)$ in $[a,b]$.

If the function $\alpha(t)$ is absolutely continuous, the Stieltjes integral can be expressed as a Lebesgue integral as follows:

$$\int_a^b x(t)\,d\alpha(t) = \int_a^b x(t)\alpha'(t)\,dt.$$

If $\alpha(t)$ is an increasing function (i.e. $\alpha(t') < \alpha(t'')$ whenever $a \leq t' < t'' \leq b$) and if, for each number $s \in [\alpha(a), \alpha(b)]$, one puts

$$\beta(s) = \sup(\{t : s \geq \alpha(t)\}),$$

one obtains:

$$(2)\qquad \int_a^b x(t)\,d\alpha(t) = \int_{\alpha(a)}^{\alpha(b)} x[\beta(s)]\,ds.$$

Proof. We have, by definition of $\beta(s)$:

$$(3)\qquad \beta[\alpha(t)] = t \quad \text{for } a \leqq t \leqq b.$$

Since $\beta(s)$ is increasing, by hypothesis, and takes all values in the interval $[a,b]$ where, by (3), $a = \beta[\alpha(a)]$ and $b = \beta[\alpha(b)]$, it is a continuous function. It follows that the function $x[\beta(s)]$ is continuous as well.

Consider a subdivision δ of $[a,b]$ given by the numbers $a = t_0 < t_1 < \ldots < t_n = b$ and put $\alpha(t_i) = \theta_i$ for $i=1,2,\ldots,n$. We have

$$I_i = \int_{\theta_{i-1}}^{\theta_i} x[\beta(s)]\,ds = (\theta_i - \theta_{i-1})\,x(\theta_i'),$$

where $\theta_i' = \beta(s_i')$ and $\theta_{i-1} \leqq \theta_i' \leqq \theta_i$. Clearly $\beta(\theta_{i-1}) \leqq \beta(s_i') = \theta_i' \leqq \beta(\theta_i)$. By (3) we have $\beta(\theta_{i-1}) = \beta[\alpha(t_{i-1})] = t_{i-1}$ and similarly $\beta(\theta_i) = t_i$. Consequently

$$t_{i-1} \leqq \theta_i' \leqq t_i,$$

so that

$$I_i = x(\theta_i')\,[\alpha(t_i) - \alpha(t_{i-1})],$$

whence

$$(4)\qquad \int_{\alpha(a)}^{\alpha(b)} x[\beta(s)]\,ds = \sum_{i=1}^{n} I_i = \sum_{i=1}^{n} x(\theta_i')\,[\alpha(t_i) - \alpha(t_{i-1})].$$

Now, since this last sum tends to $\int_a^b x(t)\,d\alpha(t)$ when the maximum length of the intervals of the subdivision δ tends to 0, the equality (4) yields (2), q.e.d.

This established, we now allow $\alpha(t)$ to be any function of bounded variation. Such a function $\alpha(t)$ can always be written as a difference $\alpha_1(t) - \alpha_2(t)$ of two increasing functions $\alpha_1(t)$ and $\alpha_2(t)$; denoting as before the corresponding functions by $\beta_1(s)$ and $\beta_2(s)$, we obtain

$$\int_a^b x(t)\,d\alpha(t) = \int_a^b x(t)\,d\alpha_1(t) - \int_a^b x(t)\,d\alpha_2(t) = \int_{\alpha_2(a)}^{\alpha_1(b)} x[\beta_1(s)]\,ds - \int_{\alpha_2(a)}^{\alpha_2(b)} x[\beta_2(s)]\,ds.$$

If the functions $x_n(t)$ are continuous and uniformly bounded and if the sequence $\{x_n(t)\}$ converges everywhere (pointwise) to a continuous function $x(t)$, we have, for every function $\alpha(t)$ of bounded variation

$$\lim_{n\to\infty} \int_a^b x_n(t)\,d\alpha(t) = \int_a^b x(t)\,d\alpha(t),$$

because

$$\lim_{n\to\infty} \int_{\alpha_1(a)}^{\alpha_1(b)} x_n[\beta_1(s)]\,ds = \int_{\alpha_2(a)}^{\alpha_1(b)} x[\beta_1(s)]\,ds,$$

and

$$\lim_{n\to\infty} \int_{\alpha_2(a)}^{\alpha_2(b)} x_n[\beta_2(s)]\,ds = \int_{\alpha_2(a)}^{\alpha_2(b)} x[\beta_2(s)]\,ds.$$

§6. Lebesgue's theorem.

Let us note the following theorem, due to H. Lebesgue (*Annales de Toulouse* 1909).

For a sequence $(x_n(t))$ *of summable functions over* $[0,1]$ *to satisfy*

$$\lim_{n\to\infty} \int_0^1 \alpha(t)x_n(t)\,dt = 0$$

for every bounded measurable function $\alpha(t)$ *on* $[0,1]$, *it is necessary and sufficient that the following three conditions be simultaneously satisfied:*

1° *the sequence* $\left(\int_0^1 |x_n(t)|\,dt\right)$ *is bounded,*

2° *for every* $\varepsilon > 0$ *there exists an* $\eta > 0$ *such that for every subset* H *of* $[0,1]$ *of measure* $<\eta$, *the inequality* $\left|\int_H x_n(t)\,dt\right| \leq \varepsilon$ *holds for* $n=1,2,\ldots,$

3° $\lim_{n\to\infty} \int_0^u x_n(t)\,dt = 0$ *for every* $0 \leq u \leq 1$.

We shall become acquainted with other theorems of this kind later in the book.

B. (B)-MEASURABLE SETS AND OPERATORS IN METRIC SPACES.

§7. Metric spaces

A non-empty set E is called a *metric space* or *D*-space when to each ordered pair (x,y) of its elements there corresponds a number $d(x,y)$ satisfying the conditions:

1) $d(x,x) = 0$, $d(x,y) > 0$ when $x \neq y$,

2) $d(x,y) = d(y,x)$,

3) $d(x,z) \leq d(x,y) + d(y,z)$.

The function d is called a *metric* and the number $d(x,y)$ is called the *distance* between the *points* (elements) x,y. A sequence of points (x_n) is said to be *convergent*, when

$$\lim_{p,q\to\infty} d(x_p,x_q) = 0; \tag{5}$$

the sequence (x_n) is said to be *convergent to the point* x_0, and we write $\lim_{n\to\infty} x_n = x_0$, when

$$\lim_{n\to\infty} d(x_n,x_0) = 0. \tag{6}$$

The point x_0 is then known as the *limit* of the sequence (x_n).

Remark. Sequences which are *convergent* in this sense are more usually known as *Cauchy* sequences. [Trans.]

It is easy to see that (6) implies (5), since we always have

$$d(x_p,x_q) \leq d(x_p,x_0) + d(x_0,x_q).$$

Consequently, a sequence convergent to a point is convergent for this reason; of course, the converse is not always true.

A metric space with the property that every convergent sequence in it converges to some point is said to be *complete*.

A metric space with the property that every (infinite) sequence of its points has a subsequence convergent to some point is said to be *compact*.

The euclidean spaces constitute examples of complete metric spaces. We shall now describe some other important examples.

1. *The set S of measurable functions* in the interval [0,1]. For each ordered pair (x,y) of elements of this set, put

$$d(x,y) = \int_0^1 \frac{|x(t) - y(t)|}{1 + |x(t) - y(t)|}\, dt.$$

It is easily verified that conditions 1) - 3) above are satisfied. In fact, it is clear that conditions 1) and 2) are satisfied, (we do not distinguish between functions which only differ on a set of measure zero) and to see that condition 3) also holds, it is enough to remark that for every pair of real numbers a,b one has:

$$\frac{|a+b|}{1+|a+b|} \leqq \frac{|a|}{1+|a|} + \frac{|b|}{1+|b|}.$$

Thus "metrised", the set S therefore becomes a metric space; this space is complete, since convergence of a sequence (x_n) of its points (to a point x_0) means convergence in measure of the sequence of functions $(x_n(t))$ (to the function $x_0(t)$) in [0,1].

2. *The set s of all sequences of numbers.* For each ordered pair (x,y) of its elements, put

$$d(x,y) = \sum_{n=1}^{\infty} \frac{1}{2^n} \cdot \frac{|\xi_n - \eta_n|}{(1 + |\xi_n - \eta_n|)},$$

where, as in all the examples of sequence spaces, $x = (\xi_n)$ and $y = (\eta_n)$.

The set s then becomes a complete metric space. In fact, convergence of a sequence of points (x_m) and its convergence to a point x_0 here mean (putting $x_m = (\xi_n^{(m)})$ and $x_0 = (\xi_n)$) that for each natural number n, each of the sequences $(\xi_n^{(m)})$ is convergent, and is convergent to ξ_n, respectively, as m tends to infinity.

3. *The set M of bounded measurable functions* in [0,1]. If one puts, for each pair x,y of its elements

$$d(x,y) = \operatorname*{ess\,sup}_{0 \leqq t \leqq 1} |x(t) - y(t)|,$$

one obtains a complete metric space. Convergence of a sequence of points (x_n) (to a point x_0, respectively) here means uniform convergence almost everywhere in [0,1] of the sequence of functions $(x_n(t))$ (to the function $x_0(t)$).

4. *The set m of bounded sequences of numbers.* Putting

$$d(x,y) = \sup_{1 \leqq n} |\xi_n - \eta_n|$$

one clearly obtains from m a complete metric space.

5. *The set C of continuous functions* in [0,1]. For each pair x,y of its elements put

$$d(x,y) = \max_{0 \leqq t \leqq 1} |x(t) - y(t)|.$$

The set C then forms a complete metric space; convergence of a sequence of its points (x_n) (to a point x_0, respectively) here becomes uniform convergence in [0,1] of the sequence of functions $(x_n(t))$ (to the function $x_0(t)$).

6. *The set c of convergent sequences* of numbers. We define, for each pair x,y of its elements, the distance $d(x,y)$ exactly as we did in the space m. It is then easily seen that c also forms a complete metric space.

7. *The set $C^{(p)}$ of functions with continuous p^{th} derivative* in [0,1]. Putting

$$d(x,y) = \max_{0\leq t\leq 1} |x(t) - y(t)| + \max_{0\leq t\leq 1} |x^{(p)}(t) - y^{(p)}(t)|,$$

we obtain a complete metric space. A necessary and sufficient condition for a sequence of points (x_n) to be convergent (to a point x_0, respectively) in this space is that both the sequences $(x_n(t))$ and $(x_n^{(p)}(t))$ of functions be uniformly convergent in [0,1] (the first to the function $x_0(t)$ and the second to the function $x_0^{(p)}(t)$).

8. *The set L^p, where $p\geq 1$, of pth- power summable functions* in [0,1]. Putting

$$d(x,y) = \left[\int_0^1 |x(t) - y(t)|^p dt\right]^{\frac{1}{p}},$$

we see that the set L^p becomes a complete metric space. For a sequence (x_n) of its points to be convergent (to the point x_0 respectively) it is necessary and sufficient that the sequence of functions $(x_n(t))$ be p^{th}- power mean convergent in [0,1] (to the function $x_0(t)$).

9. *The set l^p, where $p\geq 1$, of sequences of numbers such that the series $\sum_{n=1}^{\infty} |\xi_n|^p$ is convergent.* Putting, for elements x,y of l^p

$$d(x,y) = \left[\sum_{n=1}^{\infty} |\xi_n - \eta_n|^p\right]^{\frac{1}{p}}$$

one obtains a complete metric space.

10. *The set of analytic functions $f(z)$ which are uniformly continuous* in the circle $|z|\leq 1$ forms a complete metric space when one defines the distance between two functions $f(z)$ and $g(z)$ as

$$\max_{|z|\leq 1} |f(z) - g(z)|.$$

It should be noted that one can define *sets of functions of n variables corresponding to examples* 3,5,7 and 8.

§8. Sets in metric spaces.

Let E be any metric space and G an arbitrary set of elements (points) of E.

A point x_0 is said to be an *accumulation point* of the set G if there exists a sequence of points (x_n) such that $x_0 \neq x_n \in G$ for each n and $\lim_{n\to\infty} x_n = x_0$. The set of all accumulation points of G is called its *derived set* and is denoted by G'. The set

$$\overline{G} = G \cup G'$$

is called the *closure* of the set G; the set G is said to be *closed* when $G' \subseteq G$ and is called *perfect* when $G' = G$. One says that a set G is *open* when its complement, i.e. the set $E\setminus G$, is a closed set.

Every open set is also called an *entourage* or *neighbourhood* of each of its points.

Given a point $x_0 \in E$ and a number $r_0 > 0$, the set of all points x such that $d(x,x_0) \leq r_0$ is called a *sphere* and that of the points such that $d(x,x_0) < r_0$ is called an *open sphere*; the point x_0 is called the *centre* and the number r_0 the *radius* of this sphere or open sphere respectively. A set G is said to be *dense* when $\overline{G} = E$ and *nowhere dense* when $\overline{G}$ contains no sphere.

The space E is said to be *separable* if it contains a countable dense subset. It is easy to see that every *compact metric space*, i.e. a metric space such that every sequence of its points has a convergent subsequence, cf. p. 5, is separable.

A set G is said to be *of the first category or of category I* if it can be written as the union of a countable family of nowhere dense subsets; otherwise, it is said to be *of the second category or category II*. A set G is *of the first category at a point* x_0 when there exists a neighbourhood V of x_0 such that the set $G \cap V$ is of the first category; if no neighbourhood of the point x_0 has this property, one says that the set G is *of the second category at the point* x_0.

One can prove the following

THEOREM 1. *If a set G in an arbitrary metric space E is of the second category, there exists in E a sphere K such that the set G is of the second category at each point of $G \cap K$.*

For the time being, let E be a *complete* metric space. We shall prove the

LEMMA. *If (K_n) is a sequence of spheres of radius r_n in E such that $K_{n+1} \subseteq K_n$ for $n=1,2,\ldots$ and $\lim_{n\to\infty} r_n = 0$, there exists a point lying in all these spheres.*

Proof. Let x_n be the centre of the sphere K_n. By hypothesis, if $p < q$ we have $x_q \in K_q \subseteq K_p$, whence

$$d(x_p,x_q) \leq r_p. \tag{7}$$

It follows from this that the sequence of points (x_n) is convergent. Putting, as E is complete, $\lim_{n\to\infty} x_n = x_0$, we have for $p < q$, in view of (7), $d(x_p,x_0) \leq d(x_p,x_q) + d(x_q,x_0) \leq r_p + d(x_q,x_0)$, whence $d(x_p,x_0) \leq r_p$. Now, as p is arbitrary, the point x_0 belongs to all the spheres K_n, q.e.d.

A simple consequence of this lemma is the

THEOREM 2. *Every complete metric space E is of the second category.*

Proof. Suppose, on the contrary that

$$E = \bigcup_{n=1}^{\infty} G_n, \tag{8}$$

where each of the sets G_n is nowhere dense. There then exists a sequence of spheres (K_n) of radii (r_n) with the following properties:

$$K_1 \cap G_1 = \emptyset,\ r_1 < 1 \text{ and } K_{n+1} \subseteq K_n,\ K_{n+1} \cap G_{n+1} = \emptyset,\ r_{n+1} < \frac{1}{n+1}$$

By the lemma, there exists a point x_0 which belongs to all these spheres. Now, as $K_n \cap G_n = \emptyset$ for each $n=1,2,\ldots$, this point cannot belong to any G_n, which contradicts (8).

Now let E be any metric space and F an arbitrary subset of E. If one uses the same definition of distance for elements of F as that employed in the space E, the set F is itself a certain metric space.

Consider a set $G \subseteq F$. If it is, e.g., nowhere dense when regarded as a subset of the metric space F, we say that it is nowhere dense *relative to (the set) F*; only when $F = E$ do we usually omit the words "relative to (the set) F". The same applies to the other definitions introduced at the beginning of this section.

Theorem 1 implies that if the set G is of category I at each of its points relative to F, it is of category I relative to F. Similarly, theorem 2 implies that if the metric space E is complete and the set F is closed, then this set is of category II relative to itself.

Consider in an arbitrary metric space E the smallest class $\mathcal{B}$ of subsets of this space satisfying the following conditions:

1) each closed set belongs to $\mathcal{B}$,

2) each countable union of sets belonging to $\mathcal{B}$ belongs to $\mathcal{B}$,

3) each complement of a set belonging to $\mathcal{B}$ belongs to $\mathcal{B}$.

The sets of the class $\mathcal{B}$ are called the "*B-measurable sets*". A set G is said to *satisfy the Baire condition* if every non-empty perfect set P contains a point x_0 such that at least one of the sets $P \cap G$ and $P \setminus G$ is of category I at the point x_0 relative to P.

One has the following

THEOREM 3. *Every B-measurable set satisfies the Baire condition.*

§9. Mappings in metric spaces.

Let E and E_1 be arbitrary non-empty sets. If to each element $x \in E$ there corresponds a certain element of E_1 one says that a *mapping* or *operator* is defined in the set E. The element corresponding to x is called the *value* of this mapping at x; the set E is known as the *domain* and the set of values the *codomain* or *range* of the mapping concerned. In the special case where the values of the given mapping are numbers, it is called a *functional*.

Now let E be a metric space and let U be a mapping with E as domain and some metric space as codomain. The mapping U is said to be *continuous at the point x_0* if, for every sequence of points (x_n) converging to x_0, one has $\lim_{n\to\infty} U(x_n) = U(x_0)$; the mapping U is said to be *continuous in E* when it is continuous at each point of this space. If a sequence of mappings (U_n) and a further mapping U_0, all defined in E and all with codomain lying in the same metric space, are given, the sequence of mappings is said to *converge at the point x_0* to the mapping U_0 when the sequence of values $(U_n(x_0))$ converges to $U_0(x_0)$; the sequence of mappings (U_n) is *convergent in E to the mapping U_0*, when it is convergent at each point of E. If the sequence of mappings (U_n) is convergent in E to the mapping U_0 this last mapping is called the *limit of (U_n) in E*. Instead of saying "continuous mapping in E", one says, briefly, "continuous mapping", when it is understood which space is concerned; the same applies to the other terms.

Let $\mathcal{F}$ be the smallest class of mappings, all having the same given metric space E as domain and with codomains all lying in some other metric space, which satisfies the conditions:

1) every continuous mapping belongs to $\mathcal{F}$,

2) every limit of a convergent sequence of mappings belonging to $\mathcal{F}$, belongs to $\mathcal{F}$.

The mappings of this class are known as "*B-measurable mappings*".

A mapping U with domain E and codomain also a metric space is said to *satisfy the Baire condition* if, in each non-empty perfect set $P \subseteq E$ there exists a set G of category I relative to P such that the mapping U restricted to the space $P \setminus G$, is continuous in this space.

We have the following

THEOREM 4. *Every B-measurable mapping satisfies the Baire condition.*

Equally, one can prove

THEOREM 5. *If the mapping U defined in the space E is a limit of continuous mappings, there exists in E a set G of category I such that the mapping U is continuous at each point of the set $E \setminus G$.*

The following theorem establishes a relationship between the B-measurable sets and the B-measurable mappings; let E be the metric space where they are defined and E_1 the space where they take their values.

THEOREM 6. *If the mapping U is B-measurable, then for every B-measurable set $G_1 \subseteq E_1$, the set G of points x such that $U(x) \in G_1$ is B-measurable.*

THEOREM 7. *If the spaces E and E_1 are separable and the mapping U is continuous in E, then the images of the B-measurable sets $G \subseteq E$ satisfy the Baire condition. If, further, $x \neq x'$ always implies $U(x) \neq U(x')$, the images of the B-measurable sets are also B-measurable.*

The first part of the theorem follows from the fact that the continuous image of a B-measurable set is always a so-called "analytic" set and every analytic set satisfies the Baire condition. The proof of the second part of the theorem as well as that of theorem 6, is also to be found in *Set theory* by F. Hausdorff.

THEOREM 8. *If the mappings U' and U'' are B-measurable, so is the functional $d\big(U'(x), U''(x)\big)$.*

The proof follows from the fact that if the mappings U' and U'' are continuous, so is the functional $d\big(U'(x), U''(x)\big)$ and for each point $y_0 \in E_1$, the functional $d(y, y_0) = d(y_0, y)$ is continuous in E_1.

THEOREM 9. *If (U_n) is a sequence of B-measurable mappings, the set of points where this sequence is convergent is a B-measurable set.*

Proof. For natural numbers p, q, r let $G_{p,q,r}$ be the set of points x such that $\big(U_p(x), U_q(x)\big) \leqslant \frac{1}{r}$. By theorems 6 and 8 the sets $G_{p,q,r}$ are B-measurable. Now, $G = \bigcap_{r=1}^{\infty} \bigcup_{p=1}^{\infty} \bigcap_{q=p}^{\infty} G_{p,q,r}$, so that G is B-measurable.

THEOREM 10. *If (U_n') and (U_n'') are sequences of B-measurable mappings and if, for each $x \in E$, we have $\overline{\lim\limits_{n\to\infty}}\, d\big(U_n'(x), U_n''(x)\big) < \infty$, the functional $\overline{\lim\limits_{n\to\infty}}\, d\big(U_n'(x), U_n''(x)\big)$ is B-measurable.*

Proof. For each pair of natural numbers p, q and each point x, put:

$$F_{p,q}(x) = \max_{p \leqq n \leqq p+q-1} d\big(U_n'(x),\ U_n''(x)\big).$$

One clearly has, for each x:

$$\overline{\lim_{n\to\infty}}\, d\big(U_n'(x), U_n''(x)\big) = \lim_{p\to\infty} \lim_{q\to\infty} F_{p,q}(x).$$

It is therefore enough to show that each of the functionals $F_{p,q}$ is B-measurable. Now, by Theorem 8, each of the functionals $F_{p,1}(x) = d\big(U_p'(x), U_p''(x)\big)$ is B-measurable and since one has for each $q > 1$

$$2F_{p,q+1}(x) = F_{p,q}(x) + F_{p+q,1}(x) + |F_{p,q}(x) - F_{p+q,1}(x)|,$$

it follows by induction, again applying Theorem 8, that all the functionals $F_{p,q}$ are B-measurable.

THEOREM 11. *Given a sequence of non-negative continuous functionals* (F_n) *such that* $\overline{\lim_{n\to\infty}} F_n(x) < \infty$ *for each element* x *of a set* $G \subseteq E$ *of category* II, *there exists a sphere* $K \subseteq E$ *and a number* N *such that* $F_n(x) \leqq N$ *for each* $x \in K$ *and each* $n=1,2,\ldots$

Proof. The sets G_i of points x such that $F_n(x) \leqq i$ for $n=1,2,\ldots$ are clearly closed and $G \subseteq \bigcup_{i=1}^{\infty} G_i$; there therefore exists an index N such that G_N is of category II. Since it is a closed set, it contains a sphere K as required.

CHAPTER I

Groups

§1. Definition of *G-spaces*.

Let a complete metric space E be given. Suppose that to each ordered pair (x,y) of elements of the space E there corresponds a unique element z of this space called *the sum of x and y* and which we will denote by the symbol $x+y$.

Suppose further that E is a *group* under this sum operation, i.e. that

I_1. $$(x+y)+z = x+(y+z),$$

I_2. *there exists in E a zero-element Θ such that one has*
$$\Theta + x = x + \Theta = x \quad \textit{for every } x \in E,$$

I_3. *to each element x of E there corresponds an element* (which we will denote by $-x$) *which satisfies the equation*
$$x + (-x) = \Theta.$$

It follows easily from these axioms that:

a) *there exists only one zero-element Θ in E,*

b) *one has $(-x)+(x) = \Theta$ for each $x \in E$.*

c) $x+y = x+z$ *implies* $y = z$.

Suppose further that the following axioms are satisfied:

II_1. $\lim\limits_{n\to\infty} x_n = x$ *implies* $\lim\limits_{n\to\infty}(-x_n) = -x$,

II_2. $\lim\limits_{n\to\infty} x_n = x$ *and* $\lim\limits_{n\to\infty} y_n = y$ *imply* $\lim\limits_{n\to\infty}(x_n + y_n) = x + y$.

The complete metric spaces satisfying these axioms will be called *G-spaces*.

Remark. We will write $x-y$ instead of $x+(-y)$ and $-x+y$ instead of $(-x)+y$.

§2. Properties of sub-groups.

Let E be a G-space. For an element $x \in E$ and a set $H \subseteq E$, we will denote by xH and Hx respectively the set of all elements $y \in E$ such that $y = x+z$ ($z+x$, respectively) where $z \in H$.

Clearly, one always has the identities

$$x(H_1 \cup H_2) = xH_1 \cup xH_2,$$
$$x(H_1 \setminus H_2) = xH_1 \setminus xH_2,$$
$$x(H_1 \cap H_2) = xH_1 \cap xH_2,$$

and the analogous identities for H_1x and H_2x.

It is easily shown that if H has any of the properties closed, open, nowhere dense, of category I, of category II or B-measurable then the set xH also has the same properties. If z is an interior point of H, $x+z$ is an interior point of xH.

A non-empty set $H \subseteq E$ is called a *subgroup* of E, when the conditions $x \in H$ and $y \in H$ imply $x + y \in H$ and $-x \in H$. Clearly then also $\Theta \in H$.

A set is said to be *connected* when it cannot be expressed as the union of two non-empty disjoint relatively closed subsets of itself. If E is a connected set and H is a subset of E which is both open and closed, one has $H = E$, for otherwise the set $E \setminus H$ would also be non-empty and closed.

THEOREM 1. *Every subgroup $H \subseteq E$ which is of category* II *and satisfies the Baire condition is both open and closed in E.*

Proof. By theorem 1, p. 8, there exists an open sphere K in which H is everywhere of category II. One can clearly assume that the centre y_0 of K belongs to H. As H satisfies the Baire condition, the set $K \setminus H$ is of category I. Now, as y_0 is an interior point of K, the point $\Theta = -y_0 + y_0$ is an interior point of $(-y_0)K$. Hence there exists an open sphere $K_1 \subseteq (-y_0)K$ of centre Θ. We have $(-y_0)[K \setminus H] = (-y_0)K \setminus (-y_0)H$ and as $(-y_0)H = H$, since H is a subgroup, it follows that $(-y_0)[K \setminus H] = (-y_0)K \setminus H \supseteq K_1 \setminus H$, from which it follows that, $K \setminus H$ and consequently $(-y_0)[K \setminus H]$, being of category I, $K_1 \setminus H$ is also of category I.

Moreover, for each $x \in K_1$ we have $x \in xK_1$, since $\Theta \in K_1$ and $x + \Theta = x$. Consequently $K_1 \cap xK_1 \neq \emptyset$. Hence there exists an open sphere $K_2 \subseteq K_1 \cap xK_1$ of centre x. We have $K_2 \setminus H \subseteq K_1 \setminus H$ and $K_2 \setminus xH \subseteq xK_1 \setminus xH = x[K_1 \setminus H]$, from which it follows that the sets $K_2 \setminus H$ and $K_2 \setminus xH$ are also of category I.

It follows from this that $H \cap xH \neq \emptyset$; hence there exists a y such that $y \in H$ and $y \in xH$, whence $-x + y \in H$ and thus, H being a subgroup, $-x = -x + y - y \in H$, hence $x \in H$.

It is thus proved that $K_1 \subseteq H$ and, consequently, that Θ is an interior point of H. Since for each $y \in H$ we have $yH = H$ and $y = y + \Theta$, each point y of H is an interior point. H is therefore an *open* set.

To show that it is also *closed*, put $\lim_{n\to\infty} y_n = y$ where $y_n \in H$ for each $n = 1, 2, \cdots$. Now, as $\lim_{n\to\infty} (y - y_n) = \Theta \in K_1 \subseteq H$, there exists an n such that $y - y_n \in K_1 \subseteq H$, whence $y = y - y_n + y_n \in H$, q.e.d.

This theorem implies the following

THEOREM 2. *If the space E is connected, every subgroup $H \subseteq E$ which is of category* II *and satisfies the Baire condition coincides with E.*

Remark. Since every B-measurable set satisfies the Baire condition, theorems 1 and 2 are valid, in particular, when H is a B-measurable set.

§3. Additive and linear operators.

Let E and E_1 be G-spaces and U an operator defined in E whose codomain lies in E_1.

The operator U is said to be *additive* when

$$U(x + y) = U(x) + U(y) \quad \text{for all } x, y \in E.$$

We then have $U(x) = U(x + \Theta) = U(x) + U(\Theta)$, whence

$$U(\Theta) = \Theta,$$

and as $\Theta = U(\Theta) = U(x - x) = U(x) + U(-x)$, we have

$$U(-x) = -U(x).$$

A continuous additive operator is said to be *linear*.

Remark. The word *operator* is more generally used than *mapping* in this context.

THEOREM 3. *Every additive operator that is continuous at one point is a linear mapping.*

Proof. Let x_0 be a point of continuity of the additive operator U. Let $x_n \in E$, $x \in E$ and $\lim_{n\to\infty} x_n = x$. We have $\lim_{n\to\infty} (x_n - x + x_0) = x_0$, whence $\lim_{n\to\infty} U(x_n - x + x_0) = U(x_0)$ and $\lim_{n\to\infty}[U(x_n) - U(x) + U(x_0)] = U(x_0)$, so that $\lim_{n\to\infty} U(x_n) = U(x)$, from which it follows that the operator in question is also continuous at the arbitrary point x, i.e. that it is linear.

THEOREM 4. *Every B-measurable additive operator is a linear operator.*

Proof. By theorem 4, p. 10, such an operator U satisfies the Baire condition. It is therefore continuous on a certain set H where $E \setminus H$ is of category I. Let $\lim_{n\to\infty} x_n = \Theta$. As the set $x_n[E \setminus H] = E \setminus x_n H$ is of category I for each $n = 1,2,\ldots$, so also is the set

$$(E \setminus H) \cup \bigcup_{n=1}^{\infty} x_n[E \setminus H] = (E \setminus H) \cup \bigcup_{n=1}^{\infty} (E \setminus x_n H) \supseteq E \setminus \left(H \cap \bigcap_{n=1}^{\infty} x_n H\right),$$

which consequently, by theorem 2, p. 8, does not exhaust the space E.

There thus exists a point x such that

$$x \in H \text{ and } x \in x_n H \quad \text{for each } n=1,2,\ldots,$$

whence $(-x_n + x) \in H$, and as $\lim_{n\to\infty}(-x_n + x) = x$, it follows that $\lim_{n\to\infty} U(-x_n + x) = U(x)$, so that $\lim_{n\to\infty}[U(-x_n) + U(x)] = U(x)$ and finally $\lim_{n\to\infty} U(x_n) = \Theta$. The operator $U(x)$ is therefore continuous at the point Θ of E and consequently it is linear by theorem 3, just proved.

Remark. It follows from the nature of the argument employed in its proof that the theorem still holds for additive operators that satisfy the Baire condition.

THEOREM 5. *If the space E is connected and (U_n) is a sequence of linear operators, the set of points x for which the limit $\lim_{n\to\infty} U_n(x)$ exists is either of category* I *or is equal to all of E.*

The proof follows easily from theorem 2, p. 8, as the set of points x where the sequence of operators (U_n) is convergent is B-measurable by theorem 9, p. 10; by theorem 3, p. 9, this set therefore satisfies the Baire condition while, further, every set of points of convergence forms a group.

§4. A theorem on the condensation of singularities.

THEOREM 6. *Suppose that a connected space E and a double sequence of linear operators $(U_{p,q})$ are given, and that (x_p) is a sequence of points such that the limit $\lim_{q\to\infty} U_{p,q}(x_p)$ does not exist for any* $p=1,2,\cdots$. *Then the set H of points x such that the limit $\lim_{q\to\infty} U_{p,q}(x)$ does not exist for any $x \in H$, for any value of $p=1,2,\ldots$, is of category* II *and its complement $E \setminus H$ is of category* I.

Proof. For each $p=1,2,\ldots$, let H_p be the set of points of convergence of the sequence $(U_{p,q})$. We have $H_p \neq E$, since by hypothesis $x_p \in E \setminus H_p$. By theorem 5, p. 10, the set H_p is of category I. Hence the same is true of the set $\bigcup_{p=1}^{\infty} H_p$, which completes the proof, because $H = E \setminus \bigcup_{p=1}^{\infty} H_p$.

CHAPTER II

General vector spaces

§1. Definition and elementary properties of vector spaces.

Suppose that a non-empty set E is given, and that to each ordered pair (x,y) of elements of E there corresponds an element $x+y$ of E (called the *sum* of x and y) and that for each number t and $x \in E$ an element tx of E (called the *product* of the number t with the element x) is defined in such a way that these operations, namely *addition* and *scalar multiplication* satisfy the following conditions (where x,y and z denote arbitrary elements of E and a,b are numbers):

1) $x + y = y + x$,

2) $x + (y + z) = (x + y) + z$,

3) $x + y = x + z$ *implies* $y = z$,

4) $a(x + y) = ax + ay$,

5) $(a + b)x = ax + bx$,

6) $a(bx) = (ab)x$,

7) $1.x = x$.

Under these hypotheses, we say that the set E constitutes a *vector* or *linear* space. It is easy to see that there then exists exactly one element, which we denote by Θ, such that $x + \Theta = x$ for all $x \in E$ and that the equality $ax = bx$ where $x \neq \Theta$ yields $a = b$; furthermore, that the equality $ax = ay$ where $a \neq 0$ implies $x = y$.

Put, further, by definition:

$$-x = (-1)x \quad \text{and} \quad x - y = x + (-y).$$

Examples 1-10 of metric spaces, described on pp. 5 and 6, also serve as examples of vector spaces, when the usual definitions of addition and scalar multiplication are adopted.

When $x \neq y$, we understand by the *line segment* joining x and y the set of all elements of the form $tx + (1 - t)y$ where t is any number in the interval $[0,1]$.

A set $G \subseteq E$ is said to be *convex*, when it contains every line segment joining arbitrary pairs of elements of G.

If $x_1, x_2, \ldots, x_n$ are elements of a vector space, the expression

$$\alpha_1 x_1 + \alpha_2 x_2 + \ldots + \alpha_n x_n = \sum_{i=1}^{n} \alpha_i x_i,$$

where $\alpha_1, \alpha_2, \ldots, \alpha_n$ are arbitrary real numbers, is called a *linear combination* of the elements $x_1, x_2, \ldots, x_n$.

§2. Extension of additive homogeneous functionals.

Let E and E_1 be two vector spaces and f a mapping in E whose co-domain lies in E_1.

The mapping f is said to be *additive* when for each pair of elements x,y we have:

$$f(x+y) = f(x) + f(y);$$

it is called *homogeneous* when for each element x and each number t:

$$f(tx) = tf(x).$$

THEOREM 1. *Suppose that there are given*

1° *a functional* p *defined in* E *such that for all* $x,y \in E$

$$p(x+y) \leqq p(x) + p(y) \text{ and } p(tx) = tp(x) \text{ for } t \geqq 0,$$

and

2° *an additive homogeneous functional* f *defined in a vector subspace* $G \subseteq E$ (i.e. a subset of E that is itself a vector space with the same definitions of the basic operations) *such that for each* $x \in G$

$$f(x) \leqq p(x),$$

then there always exists an additive homogeneous functional F *defined in* E *such that*

$$F(x) \leqq p(x) \text{ for each } x \in E \text{ and } F(x) = f(x) \text{ for each } x \in G.$$

Proof. We can assume that $G \neq E$; let $x_0 \in E \setminus G$. By 2°, for $x', x'' \in G$ we have:

$$f(x'') - f(x') = f(x''-x') \leqq p(x''-x') = p[(x''+x_0) + (-x'-x_0)]$$
$$\leqq p(x''+x_0) + p(-x'-x_0),$$

whence

$$-p(-x'-x_0) - f(x') \leqq p(x''+x_0) - f(x'').$$

The numbers

$$m = \sup_{x \in G} [-p(-x-x_0) - f(x)] \text{ and } M = \inf_{x \in G} [p(x+x_0) - f(x)]$$

are therefore finite and $m \leqq M$. If r_0 is any number such that $m \leqq r_0 \leqq M$, we have for each $x \in G$

(1) $$-p(-x-x_0) - f(x) \leqq r_0 \leqq p(x+x_0) - f(x).$$

Consider the set G_0 of all elements y of the form

(2) $$y = x + tx_0 \text{ where } x \in G \text{ and } t \text{ is a number.}$$

Clearly G_0 is a vector space. Put

(3) $$g(y) = f(x) + tr_0,$$

where the element y is given by (2); as $x_0 \in E \setminus G$, each $y \in G_0$ has a unique representation in the form of (2) so that the functional g is well-defined on G_0. We also see that g is additive and homogeneous on G_0 and coincides with f on G. We now show that

(4) $$g(y) \leqq p(y) \text{ for each } y \in G_0 .$$

In fact, if one writes y in the form (2) it can be assumed that $t \neq 0$. Putting x/t in place of x in the inequality (1) and multiplying its right- or left-hand side, according as $t > 0$ or $t < 0$, by t, one obtains $tr_0 \leqq p(x+tx_0) - f(x)$ which by (3) implies the inequality (4).

This established, it now suffices to well-order the set $E \setminus G$, obtaining, by successive extensions of f, following the procedure described above, a functional F satisfying the conclusion of the theorem.

COROLLARY. *Given a functional p defined in E such that for* $x,y \in E$

$$p(x+y) \leq p(x) + p(y) \text{ and } p(tx) = tp(x) \text{ for } t \geq 0,$$

there exists an additive homogeneous functional F defined in E such that, for each $x \in E$

$$F(x) \leq p(x).$$

In fact, consider an $x_0 \in E$ and denote by G the set of all elements of the form tx_0 where t is an arbitrary number. G is then a vector space. Putting $f(tx_0) = tp(x_0)$ in G, we will have $f(tx_0) \leq p(tx_0)$ for any t, since $t \geq 0$ implies $tp(x_0) = p(tx_0)$ and $t < 0$ implies $0 = p(0) \leq p(x_0) + p(-x_0)$, whence $p(x_0) \geq -p(-x_0)$ and finally $tp(x_0) \leq -tp(-x_0) = p(tx_0)$; the result now follows on applying theorem 1.

§3. Applications: generalisation of the notions of integral, of measure and of limit.

We are now going to discuss several interesting applications of theorem 1 and its corollary.

1. Let E be the set of bounded real-valued functions $x(s)$ defined on a circle of unit circumference where s denotes arc-length measured from some fixed point, always in the same sense. With the usual algebraic operations, E is a vector space.

Now, for each element $x = x(s)$ of E, let us define $p(x)$ to be the *infimum* of all the numbers $M(x;\alpha_1,\alpha_2,\ldots,\alpha_n)$ of the form

$$M(x;\alpha_1,\alpha_2,\ldots,\alpha_n) = \sup_{-\infty<s<\infty} \left(\frac{1}{n}\sum_{k=1}^{n} x(s+\alpha_k)\right),$$

where $\alpha_1,\alpha_2,\ldots,\alpha_n$ is an arbitrary finite sequence of numbers. The functional p then satisfies all the hypotheses of the corollary. In fact, it is plain that, firstly, one always has $p(tx) = tp(x)$ for $t \geq 0$.

Secondly, given two elements $x = x(s)$ and $y = y(s)$ of E and a number $\varepsilon > 0$, there exist finite sequences of numbers $\alpha_1,\alpha_2,\ldots,\alpha_u$ and $\beta_1,\beta_2,\ldots,\beta_v$ such that

(5) $M(x;\alpha_1,\alpha_2,\ldots,\alpha_u) \leq p(x) + \varepsilon$ and $M(y;\beta_1,\beta_2,\ldots,\beta_v) \leq p(y) + \varepsilon$.

Arranging all the numbers $\alpha_i + \beta_j$ where $i=1,2,\ldots,u$ and $j=1,2,\ldots,v$ as a single sequence $\gamma_1,\gamma_2,\ldots,\gamma_{uv}$. in some way, one has

(6) $$p(x+y) \leq M(x+y;\gamma_1,\gamma_2,\ldots,\gamma_{uv})$$

and it is easily verified that

(7) $$M(x+y;\gamma_1,\gamma_2,\ldots,\gamma_{uv}) \leq M(x;\alpha_1,\alpha_2,\ldots,\alpha_u) + M(y;\beta_1,\beta_2,\ldots,\beta_v).$$

The relations (5) - (7) imply $p(x+y) \leq p(x) + p(y) + 2\varepsilon$, which proves the number $\varepsilon > 0$ being arbitrary, that $p(x+y) \leq p(x) + p(y)$.

This established, consider therefore the functional F which exists by the corollary.

Now, if $x(s) = 1$, we have $p(x) = 1$ and $p(-x) = -1$ and as $F(x) \leq p(x)$ and $F(x) = -F(-x) \geq -p(-x)$, one obtains $F(x) = 1$. If $x(s) \geq 0$, we have $p(-x) \leq 0$ and moreover $F(x) = -F(-x) \geq -p(-x)$, so that $F(x) \geq 0$ also.

Furthermore, the functional F has the property of satisfying, for each number s_0, the equality $F[x(s+s_0)] = F[x(s)]$. In fact, if one puts $y(s) = x(s+s_0) - x(s)$ and $\alpha_k = (k-1)s_0$ for $k=1,2,\ldots$, one has for each n:

$$p(y) \leq M(y;\alpha_1,\alpha_2,\ldots,\alpha_n) = \frac{1}{n}\sup_{-\infty<s<\infty} [x(s+ns_0) - x(s)],$$

so that $p(y) \leqq 0$; one similarly obtains $p(-y) \leqq 0$. But $F(y) \leqq p(y)$ and $F(y) = -F(-y) \geqq -p(-y)$, whence $F(y) = 0$.

Thus, using the symbol $\int x(s)ds$ to denote the functional $\frac{1}{2}\{F[x(s)] + F[x(1-s)]\}$, one has the theorem:

To every function $x(s)$ of the class E one can associate a number $\int x(s)ds$ in such a way that the following conditions (where $x(s)$ and $y(s)$ are arbitrary functions of the class E and a, b, s_0 are numbers) are satisfied:

1) $\int [ax(s) + by(s)]ds = a\int x(s)ds + b\int y(s)ds$,

2) $\int x(s)ds \geqq 0$ *when* $x(s) \geqq 0$,

3) $\int x(s + s_0)ds = \int x(s)ds$,

4) $\int x(1 - s)ds = \int x(s)ds$,

5) $\int 1ds = 1$.

It is easy to check that the functional $\int x(s)ds$, satisfying conditions 1) - 5), always lies between the lower and upper Riemann integrals of the function $x(s)$. Consequently, for every R-integrable function, this functional coincides with the integral of the function.

For L-summable functions, the functional in question does not always coincide with their L-integral. Nevertheless, starting with the vector space G of such (L-summable) functions and defining the functional $f(x)$ in G to be the L-integral of the function $x(s) \in G$, theorem 1 furnishes a functional F defined in the space E such that the functional $\int x(s)ds = \frac{1}{2}\{F[x(s)] + F[x(1-s)]\}$ clearly satisfies conditions 1) - 5) and, furthermore, coincides, for every L-summable function, with the integral of that function.

2. Consider now the class $\mathcal{K}$ of all subsets of the circumference of the circle in question and denote by A_0 the circumference itself. Putting, for each set A of this class, $\mu(A) = \int x(s)ds$, where $x(s)$ is the characteristic function of the set A and therefore a function of the space E discussed in 1, we obtain the theorem:

To each set A of the class $\mathcal{K}$ one can assign a number $\mu(A)$ in such a way that the following conditions (where A and B are arbitrary sets of the class $\mathcal{K}$) are satisfied:

1) $\mu(A \cup B) = \mu(A) + \mu(B)$ whenever $A \cap B = \emptyset$,

2) $\mu(A) \geqq 0$,

3) $\mu(A) = \mu(B)$ *if* $A \cong B$,

4) $\mu(A_0) = 1$.

The functional $\mu(A)$, which satisfies conditions 1) - 4) lies between the inner and outer Jordan measures of the set A. Consequently, for every J-measurable set, this functional coincides with the measure of the set.

For arbitrary L-measurable sets, the functional in question does not always coincide with their L-measure, but, just as before, one can arrange things in such a way that this property also holds.

3. Let E be the set of all bounded real-valued functions $x(s)$ defined in $[0,+\infty]$; with the usual definitions of algebraic operations, this is a vector space.

For each element $x = x(s)$ of E, denote by $p(x)$ the infimum of all the numbers $\overline{\lim\limits_{s\to\infty}}\ \frac{1}{n} \cdot \sum\limits_{k=1}^{n} x(s + \alpha_k)$, where $\alpha_1, \alpha_2, \ldots, \alpha_n$ is an arbitrary finite sequence of positive numbers. One easily verifies that the functional p thus defined in the space E satisfies the hypotheses of the corollary. Denoting by the symbol $\operatorname*{Lim}_{s\to\infty} x(s)$ the functional F, which exists by the corollary, one therefore has the theorem:

To every function $x(s) \in E$ one can associate a number $\operatorname*{Lim}_{s\to\infty} x(s)$ in such a way that the following conditions (where $x(s)$ and $y(s)$ are arbitrary functions of E and a, b and $s_0 \geq 0$ are numbers) are satisfied:

1) $\operatorname*{Lim}_{s\to\infty} [ax(s) + by(s)] = a \operatorname*{Lim}_{s\to\infty} x(s) + b \operatorname*{Lim}_{s\to\infty} y(s)$,
2) $\operatorname*{Lim}_{s\to\infty} x(s) \geq 0$ *whenever* $x(s) \geq 0$,
3) $\operatorname*{Lim}_{s\to\infty} x(s + s_0) = \operatorname*{Lim}_{s\to\infty} x(s)$,
4) $\operatorname*{Lim}_{s\to\infty} 1 = 1$.

The functional $\operatorname*{Lim}_{s\to\infty} x(s)$ satisfying conditions 1) - 4) always lies between $\underline{\lim}_{s\to\infty} x(s)$ and $\overline{\lim}_{s\to\infty} x(s)$. It consequently coincides with $\lim_{s\to\infty} x(s)$ whenever this limit exists in the usual sense. Note that Lim here denotes a certain generalised "limit", while lim is reserved exclusively for limit in the usual sense.

4. Let (ξ_n) be any bounded sequence. Define the function $x(s)$ in $(0,+\infty)$ by: $x(s) = \xi_n$ for $n-1 < s \leq n$, and $n=1,2,\ldots$ The function $x(s)$ thus belongs to the set E discussed in 3. Putting $\operatorname*{Lim}_{n\to\infty} \xi_n = \operatorname*{Lim}_{s\to\infty} x(s)$ in the sense of 3, one has the theorem:

To each bounded sequence (ξ_n) one can associate a number $\operatorname*{Lim}_{n\to\infty} \xi_n$ in such a way that the following conditions (where (ξ_n) and (η_n) are arbitrary bounded sequences and a and b are numbers) are satisfied:

1) $\operatorname*{Lim}_{n\to\infty} (a\xi_n + b\eta_n) = a \operatorname*{Lim}_{n\to\infty} \xi_n + b \operatorname*{Lim}_{n\to\infty} \eta_n$,
2) $\operatorname*{Lim}_{n\to\infty} \xi_n \geq 0$, *if* $\xi_n \geq 0$ *for* $n=1,2,\ldots$,
3) $\operatorname*{Lim}_{n\to\infty} \xi_{n+1} = \operatorname*{Lim}_{n\to\infty} \xi_n$,
4) $\operatorname*{Lim}_{n\to\infty} 1 = 1$.

Conditions 1) - 4) imply that the functional $\operatorname*{Lim}_{n\to\infty} \xi_n$ thus defined always lies between $\underline{\lim}_{n\to\infty} \xi_n$ and $\overline{\lim}_{n\to\infty} \xi_n$. Consequently, for every convergent sequence this functional coincides with the limit of the sequence in the usual sense.

CHAPTER III

F-spaces

§1. Definitions and preliminaries.

Let E be a vector space which is also a complete metric space such that the following conditions are satisfied, where x, x_n, y are elements of E and h, h_n are numbers:

1° $d(x,y) = d(x-y, \Theta)$,

2° $\lim_{n\to\infty} h_n = 0$ implies $\lim_{n\to\infty} h_n x = \Theta$ for each x,

3° $\lim_{n\to\infty} x_n = \Theta$ implies $\lim_{n\to\infty} hx_n = \Theta$ for each h.

The spaces E with properties 1° - 3° will be called *F-spaces*. All the examples 1 - 10 of metric spaces, described in §7 of the Introduction, pp. 6,7, are, it is easy to see, also F-spaces.

Conditions 1° - 3° immediately imply the following properties of the limit:

a) *when* $\lim_{n\to\infty} x_n = x$ *and* $\lim_{n\to\infty} y_n = y$, *one has* $\lim_{n\to\infty} (x_n + y_n) = x + y$.

It is enough, in fact, to remark that one always has

$$d(x_n + y_n, x + y) = d(x_n + y_n - x - y, \Theta) \leq d(x_n - x + y_n - y, y_n - y) + d(y_n - y, \Theta) =$$
$$d(x_n - x, \Theta) + d(y_n - y, \Theta) = d(x_n, x) + d(y_n, y).$$

b) *if* $\lim_{n\to\infty} h_n = h$, *one has* $\lim_{n\to\infty} h_n x = hx$, *for any* $x \subset E$.

It is always true that $d(h_n x, hx) = d\big((h_n - h)x, \Theta\big)$.

We thus see that *every F-space is at the same time a G-space*. It therefore follows that all the theorems of Chapter I continue to hold when E is taken to be an F-space.

Now, observe that the (*vector*) F-*spaces* are *connected*, since for every x and y of E the set of elements of the form $hx + (1-h)y$ where $0 \leq h \leq 1$ is a connected set containing the elements x and y.

Given an arbitrary sphere K (see p. 8) in the F-space E, it is easy to see that the set xK (see the definition p. 13) is also a sphere.

Let $h \neq 0$. Then the operator $U(x) = hx$ is a continuous bijective operator of the space E onto itself and one easily sees that closed, open, nowhere dense, category I, category II and B-measurable sets are transformed respectively to sets of the same type.

In particular, one has the following theorem, which follows from theorem 2 (Chapter I, §2) and from the remark, p.14, every F-space being connected:

THEOREM 1. *If E is an F-space, every linear subspace $H \subseteq E$ which is B-measurable and of category* II *is equal to* E.

§2. Homogeneous operators.

We are now going to concern ourselves with additive operators defined in an F-space E whose co-domains lie in another F-space E_1.

For all operators of this kind, theorems 3,4,5 and 6 of Chapter I continue to hold. Furthermore, if one defines a *homogeneous* operator to be an operator U which satisfies $U(hx) = hU(x)$ for all numbers h, one has the

THEOREM 2. *Every linear operator is also homogeneous.*

Proof. Suppose the operator U is linear. Then it is plain that for each $x \in E$ and each rational number r, $U(rx) = rU(x)$. Now, if (r_n) is a sequence of rational numbers tending to h, we have $\lim_{n\to\infty} r_n x = hx$. The continuity of the mapping U consequently yields, $U(hx) = \lim_{n\to\infty} U(r_n x) = \lim_{n\to\infty} r_n U(x) = h(x)$; thus the operator U is homogeneous.

§3. Series of elements. Inversioı of linear operators.

Put, for short,

$$|x| = d(x,\Theta).$$

It is easily checked that the following relations hold for all x and y of E:

1° $d(x,y) = |x - y|$,

2° $|\Theta| = 0$; $x \neq \Theta$ *implies* $|x| > 0$,

3° $|-x| = |x|$,

4° $|x| - |y| \leqq |x + y| \leqq |x| + |y|$

5° $\lim_{n\to\infty} x_n = x$ *implies* $\lim_{n\to\infty} |x_n| = |x|$.

Given a sequence (x_n) of elements of E, the series $\sum_{i=1}^{\infty} x_i$ is said to be *convergent* to an element x, or that x *is the sum of this series*, if $\lim_{n\to\infty} \sum_{i=1}^{n} x_i = x$. We write $x = \sum_{i=1}^{\infty} x_i$.

The definition of the sum of a series further implies the relations:

6° $x = \sum_{i=1}^{\infty} x_i$ *implies* $|x| \leqq \sum_{i=1}^{\infty} |x_i|$.

In fact, for each $\varepsilon > 0$ there exists an n such that $|x - \sum_{i=1}^{n} x_i| < \varepsilon$, whence $|x| \leqq \varepsilon + |\sum_{i=1}^{n} x_i| \leqq \varepsilon + \sum_{i=1}^{n} |x_i|$ and, ε being arbitrary, $|x| \leqq \sum_{i=1}^{\infty} |x_i|$.

7° *If the series* $\sum_{i=1}^{\infty} |x_i|$ *is convergent, the series* $\sum_{i=1}^{\infty} x_i$ *converges to some element, (i.e. has a sum).*

Indeed, put $s_n = \sum_{i=1}^{n} x_i$. If $p < q$, we have $|s_p - s_q| = |\sum_{i=p+1}^{q} x_i| \leqq \sum_{i=p+1}^{q} |x_i|$. One thus sees that $\lim_{p\to\infty, q\to\infty} |s_p - s_q| = 0$. Hence the series $\sum_{i=1}^{\infty} x_i$ converges to some element.

This established, we shall now prove the following theorems.

THEOREM 3. *The codomain of a linear operator is either of category* I *or is equal to the whole of* E_1.

Proof. Suppose that the codomain $H \subseteq E_1$ of the linear operator U defined in E is of the second category. We shall first show that

(1) *for each* $\varepsilon > 0$ *there exists a number* $\eta > 0$ *such that the image of the open sphere* $\{x : |x| < \varepsilon\}$ *under* U *contains the open sphere* $\{y : |y| < \eta\}$.

To this end, suppose an $\bar{\varepsilon} > 0$ is given, and for each natural number n denote by G_n the set of points of the form $x = nx'$, where $|x'| < \bar{\varepsilon}/2$, and by H_n the image of G_n under U, i.e. the set $\{y : y = U(x)$ for some $x \in G_n\}$. It follows that $E = \bigcup_{n=1}^{\infty} G_n$ and that $H = \bigcup_{n=1}^{\infty} H_n$.

Now, since H is assumed to be of category II, the same is true of some H_{n_o}. Let K_1 be an open sphere centre y_1 and radius η_1 contained in H'_{n_o}.

It is immediate that the open sphere K_2 of centre $\frac{1}{n_o}y_1$ and radius $\frac{1}{n_o}\eta_1$ is contained in H'_1. Indeed, $y \in K_2$, i.e. $|y - \frac{1}{n_o}y_1| < \frac{1}{n_o}\eta_1$, implies $n_o y \in K_1$, because $|n_o y - y_1| = |n_o(y - \frac{1}{n_o}y_1)| \leq n_o|y - \frac{1}{n_o}y_1| < \eta_1$; hence there exist points $\bar{y}_n \in H_{n_o}$ such that $\lim_{n\to\infty} y_n - n_o y$, or, in other words, $\lim_{n\to\infty} \frac{1}{n_o}\bar{y}_n = y$, and so $\frac{1}{n_o}\bar{y}_n \in H_1$, whence $y \in H'_1$.

Let K_3 be an arbitrary open sphere of centre $y_3 \in H_1$ contained in K_2. The set of points $y_3 - y$ where $y \in H_1$ then has every point of an open sphere $\{y : |y| < \bar{\eta}\}$ as an accumulation point. Now, putting $y_3 = U(x_3)$ and $y = U(x)$ where x_3 and x belong to G_1, we have $|x_3 - x| \leq |x_3| + |x| < \bar{\varepsilon}$ and $U(x_3 - x) = y_3 - y$. It is thus established that the derived set of the image of the open sphere $\{x : |x| < \bar{\varepsilon}\}$ contains an open sphere $\{y : |y| < \bar{\eta}\}$.

Now let $\varepsilon_i = \varepsilon/2^i$ for $i = 1, 2, \ldots$ By the above, there exists a sequence of numbers $\eta_i > 0$ such that the derived set of the image of the open sphere $\{x : |x| < \varepsilon_i\}$ contains an open sphere $\{y : |y| < \eta_i\}$ and one can plainly assume that $\lim_{i\to\infty} \eta_i = 0$. We are now going to define by induction two sequences of points (y_n) and (x_n) as follows. Put $|y| < \eta = \eta_1$ and let:

a) y_1 be an arbitrary point of E_1 such that $|y - y_1| < \eta_2$ and x_1 a point of E such that $U(x_1) = y_1$ and $|x_1| < \varepsilon_1$,

b) y_n be an arbitrary point of E_1 such that $|y - \sum_{k=1}^{n} y_k| < \eta_{n+1}$ and x_n a point of E such that $U(x_n) = y_n$ and $|x_n| < \varepsilon_n$.

We thus have

$$\sum_{n=1}^{\infty} y_n = y \tag{2}$$

and, as $|x_n| < \varepsilon_n = \varepsilon/2^n$,

$$\sum_{n=1}^{\infty} |x_n| < \sum_{n=1}^{\infty} \frac{\varepsilon}{2^n} \tag{3}$$

By 7° the series $\sum_{n=1}^{\infty} x_n$ is convergent. Let x be the sum of this series. By (3) and 6° we have $|x| < \varepsilon$ and by (2): $U(x) = \sum_{n=1}^{\infty} U(x_n) = \sum_{n=1}^{\infty} y_n = y$. The proposition (1) is thus proved.

Now, as, for each $y \in E_1$ we have $\lim_{n\to\infty} \frac{1}{n}y = \Theta$ and there consequently exists a natural number n such that $|\frac{1}{n}y| < \eta$, one can find an x such that $U(x) = \frac{1}{n}y$, and so $U(nx) = y$. However, it follows from this that $H = E_1$, which is the claim of the theorem.

THEOREM 4. *If the linear operator U maps E onto the whole of E_1, then there exists, for every sequence of points (y_n) of E_1 convergent to $y_0 = U(x_0)$, a sequence of points (x_n) of E convergent to x_0 and such that $U(x_n) = y_n$ for $n=1,2,\ldots$*

Proof. Let (ε_n) be a sequence of positive numbers tending to 0. Since the operator U is linear, proposition (1), established in the course of theorem 3, is valid; it follows from this that the image of the open sphere $\{x: |x| < \varepsilon_n\}$ contains an open sphere $\{y: |y| < \eta_n\}$ for each $n=1,2,\ldots$

Take a natural number m_0 such that, for each $m > m_0$, the inequality $|y_m - y_0| < \eta_n$ holds for at least one value of n and let, for a given m such that $y_m \neq y_0$, n_m be the largest of these values. Finally, let x_m be the point defined by the conditions:

a) if $m \leqq m_0$, take x_m to be an arbitrary point satisfying the equation $U(x_m) = y_m$,

b) if $m > m_0$ and $y_m \neq y_0$, take x_m to be an arbitrary point of the open sphere $\{x: |x - x_0| < \varepsilon_{n_m}\}$ satisfying the same equation,

c) if $m > m_0$ and $y_m = y_0$, put $x_m = x_0$.

The sequence (x_n) thus defined satisfies the requirements of the theorem, as is easily verified.

THEOREM 5. *If the linear operator maps E bijectively onto E_1, this operator is bicontinuous (i.e. has a continuous inverse).*

The proof follows immediately from theorem 4.

THEOREM 6. *If a vector space E is an F-space under both the metric $d(x,y)$ as well as the metric $d_1(x,y)$, and if*

$$\lim_{n\to\infty} d(x_n, x) = 0 \text{ always implies } \lim_{n\to\infty} d_1(x_n, x) = 0,$$

then, conversely,

$$\lim_{n\to\infty} d_1(x_n, x) = 0 \text{ always implies } \lim_{n\to\infty} d(x_n, x) = 0,$$

from which it follows that the notion of limit is the same for both metrics.

The proof follows from theorem 5, taking E_1 to the space E with the metric $d_1(x,y)$ and the linear operator U to be the identity operator, i.e. $U(x) = x$ for $x \in E$.

THEOREM 7. *Every additive operator U which satisfies the condition*

$$\lim_{n\to\infty} x_n = x_0 \text{ and } \lim_{n\to\infty} U(x_n) = y_0 \text{ implies } y_0 = U(x_0)$$

is a linear operator.

Proof. Introduce a new metric in E defined by

$$d_1(x', x'') = d(x', x'') + d(y', y'') \tag{4}$$

where $x' \in E, x'' \in E$, $y' = U(x')$, $y'' = U(x'')$, and d is the original metric in E or E_1 as the case may be.

It is easy to see that with the metric d_1 the space E is an F-space; in particular, to check that it is complete, let (x_n) be a sequence of points such that $\lim\limits_{p,q\to\infty} d_1(x_p, x_q) = 0$; it follows from (4) that $\lim\limits_{p,q\to\infty} d(x_p, x_q) = \lim\limits_{p,q\to\infty} d(y_p, y_q) = 0$, so that there exist an x_0 and a y_0 such that $\lim\limits_{n\to\infty} d(x_n, x_0) = \lim\limits_{n\to\infty} d(y_n, y_0) = 0$, and, since by hypothesis, $y_0 = U(x_0)$, we deduce from (4) that $\lim\limits_{n\to\infty} d_1(x_n, x_0) = 0$.

Now, for each x' and x'', we have by (4), $d_1(x', x'') \geqq d(x', x'')$; consequently $\lim\limits_{n\to\infty} d_1(x_n, x) = 0$ implies $\lim\limits_{n\to\infty} d(x_n, x) = 0$; by theorem 6

$\lim_{n\to\infty} d(x_n,x) = 0$ implies, conversely, that $\lim_{n\to\infty} d_1(x_n,x) = 0$, and therefore, by (4), $\lim_{n\to\infty} U(x_n) = U(x)$. The additive operator U is thus continuous, q.e.d.

LEMMA 1. *Let U' and U'' be two linear operators defined in the F-spaces E' and E'' respectively whose co-domains lie in the F-space E_1. If, for each x, the equation $U'(x) = U''(y)$ admits exactly one solution $y = U(x)$, the operator U is linear.*

The proof follows from theorem 7, because, as one sees immediately, $\lim_{n\to\infty} x_n = x_0$ and $\lim_{n\to\infty} U(x_n) = y_0$ imply $y_0 = U(x_0)$.

LEMMA 2. *Suppose that an additive operator U and a linear operator V such that $V(y) = \Theta$ implies $y = \Theta$ are given, and that the operator $V[U(\cdot)]$ is linear. Then the operator U is also linear.*

The proof follows from lemma 1, since the equation $V[U(x)] = V(y)$ admits exactly one solution $y = U(x)$ for each x.

DEFINITION. A class $\mathfrak{T}$ of linear operators is said to be *total* if $V(x) = 0$ for each $V \in \mathfrak{T}$ implies that $x = \Theta$.

THEOREM 8. *Let U be an additive operator from E to E_1 and let $\mathfrak{T}$ be a total class of linear operators defined in E_1. If $V[U(\cdot)]$ is a linear operator for every $V \in \mathfrak{T}$, U is also a linear operator.*

Proof. Suppose that $\lim_{n\to\infty} x_n = x_0$ and $\lim_{n\to\infty} y_n = y_0$ where $y_n = U(x_n)$ for $n=1,2,\ldots$

For each $V \in \mathfrak{T}$ we have $\lim_{n\to\infty} V[U(x_n)] = \lim_{n\to\infty} V(y_n) = V(y_0)$ and, since the operator $V[U(\cdot)]$ is linear, $\lim_{n\to\infty} V[U(x_n)] = V[U(x_0)]$, whence $V[U(x_0)] - V(y_0) = 0$, so that $V[U(x_0) - y_0] = 0$ and, as the class $\mathfrak{T}$ is total, $U(x_0) = y_0$. By theorem 7, the operator U is thus linear.

THEOREM 9. *Let (U_i) and (V_i) be two sequences of linear operators defined in E' and E'' respectively with co-domains lying in an F-space E_1. If the system of equations $U_i(x) = V_i(y)$, where $i=1,2,\ldots$, admits exactly one solution $y = U(x)$ for each x, the operator U is linear.*

Proof. Indeed, suppose that $\lim_{n\to\infty} x_n = x_0$, and, for the corresponding sequence (y_n), that $\lim_{n\to\infty} y_n = y_0$. By virtue of the continuity of the operators (U_i) and (V_i), we then have $U_i(x_0) = V_i(y_0)$ for $i=1,2,\ldots$, whence $y_0 = U(x_0)$, which implies, by theorem 7, that the operator U is continuous.

§4. Continuous non-differentiable functions.

As an application, we shall now demonstrate, by an easy deduction, from theorem 4 of Chapter I, p. 15, *the existence of a continuous function without a derivative throughout some set of positive measure*.

Let C^1 denote the set of all continuous periodic functions of period 1, and put, for each pair of functions $x_1(t)$ and $x_2(t)$ of C^1:

$$d\big(x_1(t), x_2(t)\big) = \max_{0\le t\le 1} |x_1(t) - x_2(t)|.$$

It is easy to see that C^1 is then an *F*-space.

For an arbitrary number $h \neq 0$, let

$$(5) \qquad y(t) = \frac{x(t+h) - x(t)}{h} \quad \text{for } 0\le t\le 1.$$

Let S denote the space of measurable functions (cf. 1, p. 6), which is an *F*-space (cf. §1, p. 23) and suppose that $y(t) \in S$. The expression (5) then defines a linear operator with domain C^1 and co-domain lying in S.

Suppose that $\lim_{n\to\infty} h_n = 0$ where $h_n \neq 0$ and

$$(6) \qquad U_n(x) = \frac{x(t+h_n) - x(t)}{h_n} \quad \text{for } 0 \leqq t \leqq 1.$$

Now, *if every continuous function had a derivative almost everywhere*, the limit of the expression (6), as $n \to \infty$, would exist for almost all values of t. Consequently, for every $x \in C^1$ the limit $\lim_{n\to\infty} U_n(x)$ would exist, and would be defined in the space S, i.e. it would be a limit *in measure*. Putting $U(x) = \lim_{n\to\infty} U_n(x)$, one would therefore obtain a B-measurable additive operator U which, by theorem 4, Chapter I, p. 15, would be a linear operator. U is clearly the derivative of the function $x(t)$.

It follows from the continuity of the operator U that if $\lim_{n\to\infty} x_n(t) = 0$ *uniformly*, we have $\lim_{n\to\infty} x_n'(t) = 0$ *in measure*. However, for $x_n(t) = \frac{1}{n}\sin\left(\frac{nt}{2\pi}\right)$, we have $\lim_{n\to\infty} x_n(t) = 0$ uniformly, whereas the sequence of derivatives $\left(\frac{1}{2\pi}\cos\left(\frac{nt}{2\pi}\right)\right)$ does not tend to 0 in measure. Hence there exist continuous functions with no derivative in a set of positive measure.

§5. The continuity of solutions of partial differential equations.

Let $F(x) = 0$ be a linear partial differential equation of the second order, for example:

$$(7) \qquad F(x) = a_1\frac{\partial^2 x}{\partial u^2} + a_2\frac{\partial^2 x}{\partial v^2} + a_3\frac{\partial^2 x}{\partial u \partial v} + a_4\frac{\partial x}{\partial u} + a_5\frac{\partial x}{\partial v} + a_6 x = 0,$$

where a_i $(i=1,2,\ldots,6)$ are continuous functions of the variables u and v in a closed region $\overline{G}$ having a simple closed curve C as frontier.

It can happen that, with certain boundary conditions, the equation (7) *always has a unique solution* $x(u,v)$ which is continuous in $\overline{G}$ and which possess those partial derivatives that appear in (7), i.e. those of first and second order, in the interior G of $\overline{G}$.

Within this hypothesis, the boundary conditions can be quite varied. For example, the solution may be specified on a part of the frontier (hyperbolic or parabolic type) or the normal derivative along C may be specified, etc.

Suppose now that, with t denoting the parameter which describes C, the equation (7) admits, for every function $\xi(t)$ that, along with all its derivatives up to order r, say, is continuous, a solution $x(u,v)$ that agrees with the function $\xi(t)$ on C.

Having stipulated this, we are going to show that

If the sequence $(\xi_n(t))$ satisfies the conditions (imposed on $\xi(t)$) and if $\lim_{n\to\infty} \xi_n(t) = 0$ and $\lim_{n\to\infty} \xi_n^{(i)}(t) = 0$ uniformly for $i=1,2,\ldots,r$, then, if $(x_n(u,v))$ denotes the sequence of corresponding solutions of the equation $F(x) = 0$, we have: $\lim_{n\to\infty} x_n(u,v) = 0$ uniformly in $\overline{G}$ and $\lim_{n\to\infty} (\partial^2 x_n/\partial u^2) = 0$, $\lim_{n\to\infty} (\partial^2 x_n/\partial v^2) = 0, \ldots$, etc. (i.e. for all the partial derivatives appearing in (7)) uniformly in every closed region contained in G.

For the proof, denote by E the set of all functions $x(u,v)$ satisfying (7), continuous in $\overline{G}$ and having both first-order partial derivatives (i.e. those that appear in (7)) continuous in G. Let

$(\overline{G}_n)$ be a sequence of closed regions lying in G such that $G = \bigcup_{k=1}^{\infty} G_k$.
For each pair $x(u,v) \in E$ and $y(u,v) \in E$, put:

$$d(x,y) = \max_{u,v\in\overline{G}} |x(u,v) - y(u,v)| + \sum_{k=1}^{\infty} \frac{1}{2^k} \frac{\max_{u,v\in\overline{G}_k}\left|\frac{\partial^2 x}{\partial u^2} - \frac{\partial^2 y}{\partial u^2}\right| + \ldots}{1 + \max_{u,v\in\overline{G}_k}\left|\frac{\partial^2 x}{\partial u^2} - \frac{\partial^2 y}{\partial u^2}\right| + \ldots},$$

where in each term of the series the differences of all the partial derivatives occurring in (7) appear.

Thus metrised, E constitutes an F-space and $\lim_{n\to\infty} x_n = x$, in the sense of this metric, means that x_n tends uniformly to x in G and that the partial derivatives $\frac{\partial^2 x_n}{\partial u^2}, \ldots$ (appearing in (7)) tend uniformly to the corresponding partial derivatives of the function x in every closed region $\overline{G}_k$, for $k=1,2,\ldots$

Let E_1 be the set of functions $\xi(t)$, where t describes C, which are continuous along with their first r derivatives. For each pair $\xi(t) \in E_1$ and $\eta(t) \in E_1$, put:

$$d(\xi,\eta) = \max_{t\in C} |\xi(t) - \eta(t)| + \sum_{i=1}^{r} \max_{t\in C} |\xi^{(i)}(t) - \eta^{(i)}(t)|.$$

Now denote by $\xi = U(x)$ the operator defined in E which is the operator of restriction to C, i.e. for each function $x = x(u,v) \in E$, the function $\xi = \xi(t)$ is the restriction of $x(u,v)$ to the frontier C of G. Thus defined, the operator U is plainly additive and continuous.

Now, as the co-domain of the operator U is an F-space, the inverse operator U^{-1}, which exists by hypothesis, is continuous by theorem 5, p. 26, which implies the proposition to be proved.

Remark. If we drop the hypothesis of the uniqueness of the solution of equation (7), we would merely be able to conclude (by theorem 4, p. 26) that: $(\xi_n(t))$ having the same meaning as before, *there exists* a sequence of functions $(x_n(u,v))$ satisfying (7), which restrict to the functions $\xi_n(t)$ on C and such that $\lim_{n\to\infty} x_n(u,v) = 0$ uniformly in $\overline{G}$ and $\lim_{n\to\infty} \frac{\partial^2 x_n}{\partial u^2} = 0, \ldots$ uniformly in every closed region contained in G.

§6. Systems of linear equations in infinitely many unknowns.

Let (a_{ik}), $i=1,2,\ldots$; $k=1,2,\ldots$, be an arbitrary array (double sequence) of real numbers and let E_1 be an F-space whose elements are sequences of numbers.

THEOREM 10. *If for each sequence $y = (\eta_i) \in E_1$ the system of equations*

$$(8) \qquad \sum_{k=1}^{\infty} a_{ik}\xi_k = \eta_i \quad \textit{for } i=1,2,\ldots,$$

always has a unique solution (ξ_k), there exist linear functionals $\xi_k = f_k(y)$ for $k=1,2,\ldots$, defined in E_1 such that

$$\sum_{k=1}^{\infty} a_{ik} f_k(y) = \eta_i \quad \textit{for every } y \in E_1 \textit{ and } i=1,2,\ldots$$

Proof. Denote by E the set of all sequences $x = (\xi_k)$ which satisfy the conditions

a) *the series $\sum_{k=1}^{\infty} a_{ik}\xi_k$ is convergent for each $i=1,2,\ldots$,*

b) *the sequence $(\eta_i) = \left(\sum_{k=1}^{\infty} a_{ik}\xi_k\right)$ belongs to E_1.*

For each pair $x' = (\xi'_k)$ and $x'' = (\xi''_k)$ of elements of E, put:

$$d_i(x',x'') = \sup_{n \geq 1}\left(\left|\sum_{k=1}^{n} a_{ik}(\xi'_k - \xi''_k)\right|\right),$$

$d_0(x',x'')$ = distance between the sequences $\left(\sum_{k=1}^{\infty} a_{ik}\xi'_k\right)$ and $\left(\sum_{k=1}^{\infty} a_{ik}\xi''_k\right)$ in E_1 and define the distance $d(x',x'')$ in E by means of the formula

$$d(x',x'') = \sum_{i=0}^{\infty} \frac{1}{2^i} \frac{d_i(x',x'')}{1 + d_i(x',x'')} .$$

Observe that

(9) *For every* $k=1,2,\ldots$, *if* $\lim_{n\to\infty} x_n = \Theta$ *where* $x_n = \left(\xi_k^{(n)}\right) \in E$, *we have* $\lim_{n\to\infty} \xi_k^{(n)} = 0$.

In fact, by the uniqueness of the solutions of the system of equations (8) the kth column contains at least one term $a_{ik} \neq 0$. Suppose therefore that

(10) $$a_{i_k k} \neq 0 \quad \text{for } k=1,2,\ldots .$$

Since $\lim_{n\to\infty} x_n = \Theta$, we have $\lim_{n\to\infty} d_{i_1}(x_n,\Theta) = 0$, whence $\lim_{n\to\infty} \xi_1^{(n)} = 0$, because (10) gives $a_{i_1 1} \neq 0$, and it is now easy to show by induction that one has generally $\lim_{n\to\infty} \xi_k^{(n)} = 0$ for each natural number k.

This established, the proposition (9) enables us to show that E is a complete vector space.

To this end, suppose that the sequence (x_n), where $x_n = \left(\xi_k^{(n)}\right)$ satisfies the condition $\lim_{p\to\infty,q\to\infty} (x_p,x_q) = 0$. Consequently $\lim_{p\to\infty,q\to\infty} d(x_p - x_q) = 0$, whence by (9), $\lim_{p\to\infty,q\to\infty} \left(\xi_k^{(p)} - \xi_k^{(q)}\right)$ for $k=1,2,\ldots,$ from which it follows that $\lim_{n\to\infty} \xi_k^{(n)} = \xi_k$ exists for each natural number k. Let $x = (\xi_k)$. It is easily checked that $x \in E$ and that $\lim_{n\to\infty} d_i(x_n,x) = 0$ for each $i=0,1,\ldots$, whence $\lim_{n\to\infty} d(x_n,x) = 0$; the space E is thus complete as claimed.

It follows from this that E is an F-space.

This established, put

$$y = U(x)$$

for every pair of sequences $x = (\xi_k) \in E$ and $y = (\eta_i) \in E_1$ which satisfy the system of equations (8).

One sees immediately that

(11) $$d(y,\Theta) = d_0(x,\Theta) \leq d(x,\Theta),$$

where it is understood that $d(y,\Theta)$ denotes the metric in E_1 and $d(x,\Theta)$, that in E.

By (11), $\lim_{n\to\infty} x_n = \Theta$ implies $\lim_{n\to\infty} U(x_n) = \Theta$. The operator U is therefore linear and as it maps E bijectively onto E_1, the inverse operator U^{-1} is also linear by theorem 5, p. 26. Consequently, if, for $k=1,2,\ldots$, one puts $\xi_k = f_k(y)$ where $x = U^{-1}(y) = (\xi_k)$, one sees that $\lim_{n\to\infty} y_n = \Theta$, where $x_n = U^{-1}(y_n) = \left(\xi_k^{(n)}\right)$, implies that $\lim_{n\to\infty} x_n = \Theta$, hence $\lim_{n\to\infty} \xi_k^{(n)} = 0$; thus the additive functionals f_k are linear functionals in E , q.e.d.

This theorem implies, as we shall see, the following theorem:

If the system of equations (8) *has exactly one solution for each sequence* (η_i) *belonging*

1° *to the space* c_0 *of sequences converging to* 0,

2° *to the space* s,

3° *to the space* l,

4° *to the space* l^p *where* $p > 1$,

there exists an array (b_{ki}) *such that*

$$\xi_k = \sum_{i=1}^{\infty} b_{ki}\eta_i \quad \textit{for } k=1,2,\ldots,$$

where the sequences (ξ_k) *and* (η_i) *satisfy the system of equations* (8), *and which satisfies the respective conditions:*

1° $\sum_{i=1}^{\infty} |b_{ki}| < \infty$ *for* $k=1,2,\ldots$,

2° *each row is eventually zero (i.e. there exists a sequence of natural numbers* (n_k) *such that* $b_{ki} = 0$ *for every* $i > n_k$*),*

3° $|b_{ki}| < m_k$, *for* $i=1,2,\ldots$, *for some sequence* (m_k),

4° $\sum_{i=1}^{\infty} |b_{ki}|^{p/p-1} < \infty$ *for* $k=1,2,\ldots$

Remark. If one assumes that the system of equations (8) has exactly one solution *for every convergent sequence* (η_i) (not necessarily convergent to 0), there exists, as well as the array (b_{ki}) satisfying 1°, a sequence (c_k) such that

$$\xi_k = c_k \lim_{i\to\infty} \eta_i + \sum_{i=1}^{\infty} b_{ki}\eta_i \quad \textit{for } k=1,2,\ldots$$

All these theorems follow from the general theorem established at the beginning of this section (theorem 10, p. 29) by means of a suitable representation of the linear functionals in each particular space (see the theorems below and on p. 40-42).

§7. The space s.

We are going to establish the general form of linear functionals in the space s of sequences of numbers (see the Introduction §7,2, p. 6).

THEOREM 11. *Every linear functional* f *defined in the space* s *is of the form*

$$f(x) = \sum_{i=1}^{N} a_i\xi_i, \tag{12}$$

where N *is a natural number depending on* f.

Proof. Let $x_n = \left(\xi_i^{(n)}\right)$ where $\xi_i^{(n)} = 0$ for $i \neq n$ and $\xi_n^{(n)} = 1$. Put $f(x_n) = a_n$. For every sequence $x = (\xi_n)$, we have $x = \sum_{n=1}^{\infty} \xi_n x_n$, whence $f(x) = \sum_{n=1}^{\infty} \xi_n f(x_n) = \sum_{n=1}^{\infty} a_n\xi_n$. Now, since this series is convergent for every sequence (ξ_n), there exists a natural number N such that $a_n = 0$ for each $n > N$, whence follows the form (12) of f.

M. O. Toeplitz has proved the following theorem:

THEOREM 12. *For there to exist a sequence of numbers* (ξ_k) *satisfying the system of equations* (8), *it is necessary and sufficient that, for every finite sequence of numbers* $h_1, h_2, \ldots h_r$, *the condition* $\sum_{i=1}^{r} h_i a_{ik} = 0$ *where* $k=1,2,\ldots$ *implies the equality* $\sum_{i=1}^{r} h_i\eta_i = 0$.

In particular, if the condition $\sum_{i=1}^{r} h_i a_{ik} = 0$ *where* $k=1,2,\ldots$

implies that $h_1 = h_2 = \ldots = h_r = 0$, *the system of equations* (8) *has one solution for every sequence* (η_i).

We shall now prove the

THEOREM 13. *If the system of equations* (8) *has exactly one solution for every sequence* $y = (\eta_i)$, *there exists, for every natural number* i, *a natural number* N_i *such that* $a_{ik} = 0$ *for every* $k > N_i$.

Proof. Put $\xi_k = f_k(y)$ for $\sum_{k=1}^{\infty} a_{ik}\xi_k = \eta_i$ where $i=1,2,\ldots$ By theorem 10, p. 29, f_k is a linear functional defined in the space of real sequences (cf. p. 6). There therefore exists, for every natural number k, a finite sequence of numbers $\alpha_{1k}, \alpha_{2k}, \ldots, \alpha_{N_k k}$ such that

$$f_k(y) = \sum_{i=1}^{N_k} \alpha_{ik}\eta_i = \xi_k \tag{13}$$

Moreover, the equations of the system (13) *are linearly independent.*

In fact, suppose, on the contrary, that there exists a finite sequence of numbers $h_1, h_2, \ldots, h_r$ such that $\sum_{k=1}^{r} h_k \alpha_{ik} = 0$ where $i=1,2,\ldots$ Hence by (13) one would have

$$\sum_{k=1}^{r} h_k\xi_k = \sum_{k=1}^{r} h_k f_k(y) = 0 \text{ for every sequence } y = (\eta_i). \tag{14}$$

Putting $\eta_i^\circ = a_{ij}$ for some $j \leqq r$, where r is an arbitrary fixed natural number, one immediately ascertains that, for the corresponding solution (ξ_k°) of the system of equations (8), one has: $\xi_j^\circ = 1$ and $\xi_k^\circ = 0$ for every $k \neq j$. Substituting these values in (14) yields $h_j = 0$; consequently, all the coefficients h_k vanish, which proves the linear independence of the equations (13).

It follows from theorem 12 that, for every sequence (ξ_k), there exists a sequence of numbers (η_i) satisfying the equations (13). The series $\sum_{k=1}^{\infty} a_{ik}\xi_k$ is consequently convergent for every sequence (ξ_k), from which follows the existence, for each $i=1,2,\ldots$, of an N_i meeting the requirements of the theorem.

Remark. If we drop the uniqueness hypothesis, i.e. that there exists just one solution, the theorem ceases to be true.

Indeed, for an arbitrary sequence (η_j), there exists a power series defining an entire function $\sum_{k=0}^{\infty} \xi_k z^k$ with real coefficients ξ_k such that

$$\sum_{k=0}^{\infty} \xi_k j^k = \eta_j \text{ for } j=1,2,\ldots$$

Thus, this system of equations has a solution for each sequence (η_i); this solution is clearly not unique, for there exists an entire function given by a power series not identically equal to 0 such that

$$\sum_{k=0}^{\infty} \xi_k j^k = 0 \text{ for } j=1,2,\ldots.$$

CHAPTER IV

Normed spaces

§1. Definition of normed vector spaces and of Banach spaces.

A vector space E is said to be *normed* if there is a functional on E, called a *norm* and denoted by $|x|$ or, more usually, $\|x\|$, satisfying the conditions:

1) $\|\Theta\| = 0$ and $\|x\| > 0$ if $x \neq \Theta$,

2) $\|x + y\| \leq \|x\| + \|y\|$,

3) $\|tx\| = |t| . \|x\|$ for every scalar t.

If one defines the distance between two elements x and y of E by the formula

$$d(x,y) = \|x - y\|,$$

one clearly obtains a metric space. If, further, it is complete, that is, recall, whenever (x_p) is a sequence such that $\lim_{p,q\to\infty} \|x_p - x_q\| = 0$, it is said to be a *space of type B*, i.e. a *Banach space* or *B*-space. [trans.]

It is immediate that every *B*-space is also an *F*-space, but not conversely: the examples of spaces given in the Introduction, p. 6, which are all *F*-spaces, are also *B*-spaces, with the exception of the spaces s and S.

§2. Properties of linear operators. Extension of linear functionals.

We are first going to discuss normed spaces E, which are not necessarily complete.

THEOREM 1. *For an additive operator U defined in a vector space $G \subseteq E$ to be linear, it is necessary and sufficient that there exists a number M such that*

(1) $$\|U(x)\| \leq M\|x\| \text{ for every } x \in G.$$

Proof. The condition is necessary. In fact, if no such M existed, there would exist a sequence (x_n) such that $\|U(x_n)\| > M_n\|x_n\|$ where $M_n \to +\infty$. Putting $Y_n = \frac{1}{M_n\|x_n\|} . x_n$, we would thus have $\|Y_n\| = \frac{1}{M_n}$, whence $\lim_{n\to\infty} Y_n = \Theta$ and consequently $\lim_{n\to\infty} U(Y_n) = \Theta$, which is impossible as

$$\|U(Y_n)\| = \frac{1}{M_n\|x_n\|} . \|U(x_n)\| > 1.$$

The condition is sufficient. In fact, for any sequence $(x_n) \subseteq G$ and $x \in G$, $\lim_{n\to\infty} x_n = x$ implies $\lim_{n\to\infty}\|x - x_n\| = 0$, so that $\lim_{n\to\infty}\|U(x_n) - U(x)\| = \lim_{n\to\infty}\|U(x - x_n)\| \leq \lim_{n\to\infty} M\|x - x_n\| = 0$ and finally $\lim_{n\to\infty} U(x_n) = U(x)$, q.e.d.

For a given linear operator U defined in a vector space $G \subseteq E$, the *norm* of the operator U in G, denoted by $\|U\|_G$, is the smallest number M satisfying condition (1). If $G = E$, one can write, simply, $\|U\|$ instead of $\|U\|_E$.

We therefore have $\|U(x)\| \leq \|U\|_G.\|x\|$ for every $x \in G$ and it is easy to see that

$$\|U\|_G = \sup\{\|U(x)\| : x \in G, \|x\| \leq 1\}.$$

Remark. In modern terminology, the term *linear* does not generally imply that the operator U referred to is *continuous*. If it is continuous, or, equivalently, $\|U\| < \infty$, we say U is a *continuous* or *bounded* linear operator. [Trans.]

The question arises as to whether there exists, for every normed vector space, a (continuous or bounded) [trans.] linear functional defined in this space which is not identically zero. The affirmative answer results from the following theorems, the first of which is an easy consequence of theorem 1, Chapter II, §2, p. 18.

THEOREM 2. *Suppose that f is a (bounded) linear functional defined in a vector space $G \subseteq E$. Then there exists a linear functional F defined in E which satisfies the conditions:*

$$F(x) = f(x) \text{ for } x \in G,$$

$$\text{and} \quad \|F\| = \|f\|_G.$$

For the proof, it is enough to put $p(x) = \|x\|.\|f\|_G$ in theorem 1 of Chapter II.

THEOREM 3. *For every $x_0 \in E$, $x_0 \neq \Theta$, there exists a linear functional F defined in E such that*

$$F(x_0) = \|x_0\| \text{ and } \|F\| = 1.$$

For the proof, use the preceding theorem 2, taking $G = \{hx_0 : h \text{ scalar}\}$ and putting $f(hx_0) = h.\|x_0\|$.

In particular, this result implies *the existence in every normed vector space of a continuous linear functional which is not identically zero.*

THEOREM 4. *Let f be an arbitrary functional defined in a set $G \subseteq E$. For there to exist a linear functional F defined in E and satisfying the conditions*

1° $f(x) = F(x)$ *for* $x \in G$,

2° $\|F\| \leq M$ *for some given number* $M > 0$,

it is necessary and sufficient that the inequality

$$\left|\sum_{i=1}^{r} h_i f(x_i)\right| \leq M.\left\|\sum_{i=1}^{r} h_i x_i\right\|$$

should hold, for every finite sequence $x_1, x_2, \ldots, x_r$ of elements of G and every finite sequence $h_1, h_2, \ldots, h_r$ of real numbers.

Proof. The condition is necessary. In fact, we have

$$\left|F\left(\sum_{i=1}^{r} h_i x_i\right)\right| \leq \|F\|.\left\|\sum_{i=1}^{r} h_i x_i\right\|,$$

whence, by 2°,

$$\left|\sum_{i=1}^{r} h_i F(x_i)\right| \leq M.\left\|\sum_{i=1}^{r} h_i x_i\right\|$$

and since, by 1°, $F(x_i) = f(x_i)$ for every $x_i \in G$, the desired inequality follows.

The condition is sufficient. In fact, let H be the vector space of elements of the form $z = \sum_{i=1}^{r} h_i x_i$, for r any natural number, h_i

arbitrary numbers and $x_i \in G$. Put $\phi(z) = \sum_{i=1}^{r} h_i f(x_i)$.

For $z = \sum_{i=1}^{r} h_i x_i = \sum_{i=1}^{s} h'_i x'_i$, we have, by hypothesis,

$$\left| \sum_{i=1}^{r} h_i f(x_i) - \sum_{i=1}^{s} h'_i f(x'_i) \right| \leq M . \left\| \sum_{i=1}^{r} h_i x_i - \sum_{i=1}^{s} h'_i x'_i \right\| = 0.$$

The functional ϕ is thus well-defined in H and is easily seen to be additive there. Furthermore, $|\phi(z)| = \left| \sum_{i=1}^{r} h_i f(x_i) \right| \leq M . \| \sum_{i=1}^{r} h_i x_i \|$ so that $\|\phi\|_H \leq M$. The existence in E of the functional F with properties 1° and 2° is thus obtained from theorem 2, p. 34, putting $f = \phi$ and $G = H$.

In particular, if G is a sequence (x_n) of elements of E and (c_n) denotes the corresponding sequence of values of the functional f, i.e. $c_n = f(x_n)$, $n=1,2,\ldots$, we have the

THEOREM 5. *For there to exist a linear functional F in E satisfying the conditions*

1° $F(x_n) = c_n$ *for* $n=1,2,\ldots$,

and

2° $\|F\| \leq M$,

for some given number $M > 0$, a given sequence (x_n) of elements of E and a given sequence (c_n) of real numbers, it is necessary and sufficient that the inequality

$$\left| \sum_{i=1}^{r} h_i c_i \right| \leq M . \left\| \sum_{i=1}^{r} h_i x_i \right\|$$

hold for every finite sequence $h_1, h_2, \ldots, h_r$ of real numbers.

§3. Fundamental sets and total sets.

We are now going to establish several theorems which, in the theory of normed spaces, play a role analogous to that played by Weierstrass' theorem on the approximation of continuous functions by polynomials in the theory of functions of a real variable.

LEMMA. *Given a vector space $G \subseteq E$ and an element y_0 of E, whose distance from G is $d > 0$, there exists a linear functional F defined in E and satisfying the conditions*

1) $F(y_0) = 1$,

2) $F(x) = 0$ *for* $x \in G$,

3) $\|F\| = \frac{1}{d}$.

Proof. Let H be the set of elements x of the form

(2) $x = x' + \alpha y_0$ where α is an arbitrary real number and $x' \in G$.

Thus defined, H is clearly a linear space and as $d > 0$, the representation (2) of an $x \in H$ is unique. We define the linear functional f in H by putting $f(x) = \alpha$ for x of the form (2). As $\|x\| = \|x' + \alpha y_0\| = |\alpha| . \|\frac{x'}{\alpha} + y_0\| \geq |\alpha| . d$, it follows that $|f(x)| = |\alpha| \leq \frac{\|x\|}{d}$, whence $\|f\|_H \leq \frac{1}{d}$. Moreover, if $(x_n) \subseteq G$ and $\lim_{n\to\infty} \|x_n - y_0\| = d$, then $|f(x_n - y_0)| = 1 \leq \|x_n - y_0\| . \|f\|_H$, whence $1 \leq d . \|f\|_H$, so that $\frac{1}{d} \leq \|f\|_H$. Consequently $\|f\|_H = \frac{1}{d}$.

We conclude from this, by virtue of theorem 2, p. 34, replacing G by H, that there exists a linear functional F defined in E and such that $F(x) = f(x)$ for $x \in H$ and $\|F\| = \|f\|_H = \frac{1}{d}$ (condition 3), so that, in particular, $F(x) = 0$ for $x \in G$ (condition 2) and $F(y_0) = 1$ (condition 1), q.e.d.

THEOREM 6. *For any subset $G \subseteq E$ and an arbitrary element y_0 of E, a necessary and sufficient condition for the existence of a sequence (g_n) of linear combinations of elements of G such that $\lim_{n\to\infty} g_n = y_0$, is that $f(x) = 0$ for $x \in G$ implies $f(y_0) = 0$, for any (bounded) linear functional f.*

Proof. The condition is necessary, since $f(x) = 0$, for every $x \in G$, implies that $f(g_n) = 0$ for $n=1,2,\ldots$, whence $f(\lim_{n\to\infty} g_n) = f(y_0) = 0$.

The condition is also sufficient by virtue of the above lemma, if one understands the 'G' of the lemma to denote the set of all linear combinations of elements of the set G considered here.

A set $G \subseteq E$ is called *fundamental* when the set of all linear combinations of elements of G is dense in E; it is called *total* when every (bounded) linear functional f which vanishes on G vanishes on the whole of E.

From theorem 6 easily follows the

THEOREM 7. *For a set $G \subseteq E$ to be fundamental, it is necessary and sufficient that it be total.*

A linear functional f is said to be *orthogonal to an element x_0*, when $f(x_0) = 0$; it is said to be *orthogonal to G*, when $f(x) = 0$ for every $x \in G$.

The lemma at the start of this section implies, *for every proper closed linear subspace $G \subsetneqq E$, the existence in E of a continuous linear functional, not identically zero and orthogonal to G.*

§4. The general form of bounded linear functionals in the spaces C, L^r, c, l^r, m and in the subspaces of m.

We shall now establish the general form of the bounded linear functionals in certain particular normed spaces.

1. *The space C.* Since the norm defined in the space M, of (essentially) bounded measurable functions on $[0,1]$, coincides, for continuous functions, with that of the space C, we can regard C as a vector subspace of M.

Given a bounded linear functional f defined in C, there exists, by theorem 2, p. 34, a linear functional F defined in M satisfying the conditions

(1) $$F(x) = f(x) \text{ for every } x \in C,$$

(2) $$\|F\|_M = \|f\|_C.$$

Put

$$\xi_t = \xi_t(u) = \begin{cases} 1 & \text{for } 0 \leq u \leq t, \\ 0 & \text{for } t < u \leq 1, \end{cases}$$

and

(3) $$F(\xi_t) = g(t).$$

We shall show that $g(t)$ is a function of bounded variation. Let $a = t_0 < t_1 < \ldots < t_n = b$ and $\varepsilon_i = \text{sign}[g(t_i) - g(t_{i-1})]$ for $i=1,2,\ldots,n$. We have $\sum_{i=1}^{n} |g(t_i) - g(t_{i-1})| = \sum_{i=1}^{n} \{g(t_i) - g(t_{i-1})\}\varepsilon_i = F[\sum_{i=1}^{n} \{\xi_{t_i} - \xi_{t_{i-1}}\}\varepsilon_i] \leq \|F\|_M \cdot \|\sum_{i=1}^{n} \{\xi_{t_i} - \xi_{t_{i-1}}\}\varepsilon_i\|$ and it is easily seen that the norm of this sum is equal to 1. It follows from this, together with (2), that

(4) $$\underset{0\le t\le 1}{\text{variation}}\ g(t) \le \|F\|_M = \|f\|_C.$$

This established, let $x(t) \in C$ and

(5) $$z_n = z_n(u) = \sum_{r=1}^{n} x\left(\frac{r}{n}\right)\cdot\left\{\xi_{\frac{r}{n}}(u) - \xi_{\frac{r-1}{n}}(u)\right\}.$$

The function $z_n(u)$ thus takes the values $x(\frac{r}{n})$ in the intervals $\left(\frac{(r-1)}{n},\frac{r}{n}\right]$ respectively. As the function $x = x(u)$ is continuous, we have $\lim_{n\to\infty}\|x - z_n\| = 0$, whence, in view of (1):

(6) $$\lim_{n\to\infty} F(z_n) = F(x) = f(x).$$

Moreover, (3) and (5) give

$$F(z_n) = \sum_{r=1}^{n} x\left(\frac{r}{n}\right)\cdot\left[g\left(\frac{r}{n}\right) - g\left(\frac{r-1}{n}\right)\right]$$

so that, as $x(t) \subset C$ and $g(t)$ is a function of bounded variation, $\lim_{n\to\infty} F(z_n) = \int_0^1 x(t)dg$, whence, by (6)

(7) $$f(x) = \int_0^1 x(t)dg \text{ for every } x(t) \in C.$$

Since, consequently,

$$|f(x)| = \left|\int_0^1 x(t)dg\right| \le \underset{0\le t\le 1}{\text{var}}\ g(t) \max_{0\le t\le 1}|x(t)|,$$

we have, in view of (4), putting $\|f\| = \|f\|_C$:

$$\underset{0\le t\le 1}{\text{var}}\ g(t) = \|f\|.$$

We have thus obtained the theorem:

Every (bounded) linear functional defined in the space C is of the form

$$f(x) = \int_0^1 x(t)dg,$$

where $g(t)$ is a function independent of $x(t)$ of variation $\|f\|$.

Conversely, given a function $g(t)$ of bounded variation, the formula (7) clearly defines a bounded linear functional f on C.

2. *The space L^r where $r \ge 1$.* Given a bounded linear functional f defined in the space L^r, put:

$$\xi_t = \xi_t(u) = \begin{cases}1 & \text{for } 0\le u\le t,\\ 0 & \text{for } t<u\le 1,\end{cases}$$

and

$$f(\xi_t) = g(t).$$

We shall show that $g(t)$ is an absolutely continuous function.

Indeed, let $\delta_1, \delta_2, \ldots, \delta_n$ be non-overlapping intervals with end-points t_i and t_i' where $t_i < t_i'$ and $i = 1, 2, \ldots, n$. Putting $\varepsilon_i = \text{sign}[g(t_i') - g(t_i)]$, we have

(8) $$\sum_{i=1}^{n} |g(t_i') - g(t_i)| = \sum_{i=1}^{n} \{g(t_i') - g(t_i)\}\varepsilon_i = f\left(\sum_{i=1}^{n}\{\xi_{t_i}' - \xi_{t_i}\}\varepsilon_i\right)$$

$$\le \|f\|\cdot\left\|\sum_{i=1}^{n}\{\xi_{t_i}' - \xi_{t_i}\}\varepsilon_i\right\|.$$

Since the function $\left(\xi'_{t_i} - \xi_{t_i}\right)\varepsilon_i$ takes the value $\varepsilon_i = \pm 1$ in the interval δ_i and vanishes elsewhere, it follows from the hypothesis that the intervals δ_i are non-overlapping that

$$\left\| \sum_{i=1}^{n} \left(\xi'_{t_i} - \xi_{t_i}\right)\varepsilon_i \right\| = \left(\sum_{i=1}^{n} |\delta_i|\right)^{\frac{1}{r}},$$

where $|\delta_i|$ denotes the length of δ_i. We therefore have, by (8),

$$\sum_{i=1}^{n} |g(t'_i) - g(t_i)| \leq \|f\| \left(\sum_{i=1}^{n} |\delta_i|\right)^{\frac{1}{r}},$$

which proves the absolute continuity of $g(t)$.

This established, put $g'(t) = \alpha(t)$. The function $\alpha(t)$ is integrable and, as $\xi_0 = 0$, we clearly have $f(\xi_t) = \int_0^t \alpha(u)\,du$, whence

(9) $$f(\xi_t) = \int_0^1 \xi_t(u)\alpha(u)\,du.$$

Let $c_1, c_2, \ldots, c_n$ be arbitrary numbers, $0 = t_0 < t_1 < \ldots < t_n = 1$ and $x(t) = c_i$ for $t_{i-1} \leqq t < t_i$ and $i = 1, 2, \ldots, n$. It is plain that $x(t) = \sum_{i=1}^{n} c_i\left(\xi_{t_i} - \xi_{t_{i-1}}\right)$, whence, by (9),

(10) $$f(x) = \int_0^1 x(t)\alpha(t)\,dt.$$

Thus (10) holds for every step-function $x(t)$.

If $x = x(t)$ is now an arbitrary bounded measurable function, there exists a uniformly bounded sequence of step-functions $(x_n(t))$ which converges to $x(t)$ almost everywhere. Consequently

$$\lim_{n\to\infty} \int_0^1 |x_n(t) - x(t)|^r\,dt = 0,$$

whence $\lim\limits_{n\to\infty} \|x_n - x\| = 0$ and, in view of (10),

$$f(x) = \lim_{n\to\infty} \int_0^1 x_n(t)\alpha(t)\,dt = \int_0^1 x(t)\alpha(t)\,dt.$$

The equality (10) thus holds for every bounded measurable function $x(t)$.

This established, consider now the case $r > 1$.

Put

$$x_n(t) = \begin{cases} |\alpha(t)|^{s-1}\operatorname{sign}\alpha(t), & \text{if } |\alpha(t)|^{s-1} \leqq n, \\ n \operatorname{sign}\alpha(t), & \text{if } |\alpha(t)|^{s-1} > n, \end{cases}$$

where $\frac{1}{r} + \frac{1}{s} = 1$. We have $|f(x_n)| = \left|\int_0^1 x_n(t)\alpha(t)\,dt\right| \leqq \|f\|\left(\int_0^1 |x_n(t)|^r dt\right)^{\frac{1}{r}}$ and as $x_n(t)\alpha(t) = |x_n(t)|.|\alpha(t)| \geqq |x_n(t)|.|x_n(t)|^{1/(s-1)}$, we have

$$\int_0^1 |x_n(t)|^{s/(s-1)}\,dt \leqq \|f\|\left(\int_0^1 |x_n(t)|^r dt\right)^{1/r}$$

from which, since $s/(s-1) = r$, we conclude that

$$\left(\int_0^1 |x_n(t)|^r dt\right)^{1-\frac{1}{r}} \leqq \|f\|.$$

As this inequality holds for every natural number n, and as $|x_n(t)|^r \leqq |\alpha(t)|^{rs-r} = |\alpha(t)|^s$ and, almost everywhere, $\lim\limits_{n\to\infty} |x_n(t)|^r = |\alpha(t)|^s$, we obtain

(11) $$\left(\int_0^1 |\alpha(t)|^s dt\right)^{\frac{1}{s}} \leq \|f\|,$$

from which it follows that $\alpha(t)$ is an sth-power-summable function. Hence, if $x(t)$ is any measurable rth-power-summable function, the product $x(t)\alpha(t)$ is an integrable function.

Now define the sequence $(x_n(t))$ as follows:

(12) $$x_n = x_n(t) = \begin{cases} x(t) & \text{for } |x(t)| \leq n, \\ n \operatorname{sign} x(t) & \text{for } |x(t)| > n. \end{cases}$$

We then have

(13) $$\|x - x_n\| = \left(\int_0^1 |x(t) - x_n(t)|^r dt\right)^{\frac{1}{r}} \text{ and } \lim_{n\to\infty}\|x - x_n\| = 0,$$

from which it follows that

$$\left|\int_0^1 x(t)\alpha(t)dt - f(x_n)\right| = \left|\int_0^1 [x(t) - x_n(t)]\alpha(t)dt\right|$$
$$\leq \left(\int_0^1 |x(t) - x_n(t)|^r dt\right)^{\frac{1}{r}} \cdot \left(\int_0^1 |\alpha(t)|^s dt\right)^{\frac{1}{s}},$$

whence, using (13), $f(x) = \lim_{n\to\infty} f(x_n) = \int_0^1 x(t)\alpha(t)dt$ and as

$$|f(x)| = \left|\int_0^1 x(t)\alpha(t)dt\right| \leq \left(\int_0^1 |\alpha(t)|^s dt\right)^{\frac{1}{s}} \cdot \|x\|,$$

it follows, in view of (11), that we have

$$\|f\| = \left(\int_0^1 |\alpha(t)|^s dt\right)^{\frac{1}{s}}.$$

We have thus proved the following theorem:

Every bounded linear functional f defined in the space L^r where $r > 1$ is of the form

$$f(x) = \int_0^1 x(t)\alpha(t)dt$$

where $\alpha(t) \in L^s$ and $\|f\| = \left(\int_0^1 |\alpha(t)|^s dt\right)^{\frac{1}{s}}$.

We now pass to the case $r = 1$. Suppose $0 \leq u < u + h \leq 1$ and set

$$x(t) = \begin{cases} 1/h & \text{for } u \leq t \leq u + h, \\ 0 & \text{for } 0 \leq t < u \text{ and } u + h < t \leq 1. \end{cases}$$

We have, by (10), $|f(x)| = \left|\int_0^1 x(t)\alpha(t)dt\right| = \frac{1}{h}\left|\int_u^{u+h}\alpha(t)dt\right|$ and as $|f(x)| \leq \|f\| \cdot \|x\| = \|f\| \cdot 1$, it follows that $\left|\int_u^{u+h}\alpha(t)dt\right| \leq \|f\| \cdot h$. The function $g(u) = \int_0^u \alpha(t)dt$ thus satisfies a Lipschitz condition and since $g'(t) = \alpha(t)$ almost everywhere, we conclude from this that

(14) $$|\alpha(t)| \leq \|f\| \text{ almost everywhere.}$$

If now $x = x(t)$ is an arbitrary integrable function and the sequence $(x_n(t))$ is defined as in (12), we have

$$\|x - x_n\| = \int_0^1 |x(t) - x_n(t)|dt \to 0,$$

whence

$$f(x) = \lim_{n\to\infty} f(x_n) = \lim_{n\to\infty} \int_0^1 x_n(t)\alpha(t)dt = \int_0^1 x(t)\alpha(t)dt,$$

since $|x_n(t)\alpha(t)| \leq |x(t)\alpha(t)|$. Now, as

$$\left|\int_0^1 x(t)\alpha(t)\,dt\right| \leq \int_0^1 |x(t)|\,dt . \operatorname*{ess.sup}_{0\leqq t\leqq 1}|\alpha(t)|,$$

we obtain, in view of (14), the equality

$$\|f\| = \operatorname*{ess.sup}_{0\leqq t\leqq 1}|\alpha(t)|.$$

We have thus proved the following theorem:

Every bounded linear functional f defined in the space L^1 is of the form

$$f(x) = \int_0^1 x(t)\alpha(t)\,dt,$$

where $\alpha(t)$ is an essentially bounded function and $\|f\| = \operatorname*{ess.sup}_{0\leqq t\leqq 1}|\alpha(t)|$.

3. *The space c.* Let

$$(15)\qquad \xi_i^n = \begin{cases} 1 \text{ for } n = i, \\ 0 \text{ for } n \neq i. \end{cases}$$

$$x_n = (\xi_i^n) \text{ and } x' = (\xi_i^i).$$

For f a given (bounded) linear functional on c, put

$$(16)\qquad f(x_n) = C_n \text{ and } f(x') = C'.$$

Putting $\alpha = \lim_{n\to\infty} \xi_n$, where $x = (\xi_n) \in c$, we thus have

$$\left\| x - \alpha x' - \sum_{n=1}^{r} (\xi_n - \alpha)x_n \right\| = \sup_{n>r}|\xi_n - \alpha|,$$

whence

$$x = \alpha x' + \lim_{r\to\infty}\sum_{n=1}^{r} (\xi_n - \alpha)x_n$$

and consequently

$$x = \alpha x' + \sum_{n=1}^{\infty} (\xi_n - \alpha)x_n.$$

Hence

$$f(x) = \alpha f(x') + \sum_{n=1}^{\infty} (\xi_n - \alpha) f(x_n),$$

whence, using (16),

$$(17)\qquad f(x) = \alpha C' + \sum_{n=1}^{\infty} (\xi_n - \alpha)C_n.$$

If $x = (\xi_n)$ now stands for the sequence given by

$$\xi_n = \begin{cases} \text{sign } C_n & \text{for } n \leqq r, \\ 0 & \text{for } n > r, \end{cases}$$

we have $\|x\| = 1$, $\alpha = \lim_{n\to\infty} \xi_n = 0$ and $f(x) = \sum_{n=1}^{r} |C_n|$, and as $|f(x)| \leqq \|f\| . \|x\|$, it follows that $\sum_{n=1}^{r} |C_n| \leqq \|f\|$. Since the natural number r is arbitrary, the series $\sum_{n=1}^{r} |C_n|$ is convergent. Putting

$$C' - \sum_{n=1}^{\infty} C_n = C,$$

we have, generally, in view of (17),

$$(18)\qquad f(x) = \alpha C + \sum_{n=1}^{\infty} C_n \xi_n \text{ where } \alpha = \lim_{n\to\infty} \xi_n.$$

Now let

$$\xi_n = \begin{cases} \text{sign } C_n & \text{for } n \leqq r, \\ \text{sign } C & \text{for } n > r. \end{cases}$$

Then $\|x\| = 1$, $\alpha = \lim_{n\to\infty} \xi_n = \text{sign } C$ and

$$f(x) = |C| + \sum_{n=1}^{r} |C_n| + \sum_{n=r+1}^{\infty} C_n \cdot \text{sign } C \leqq \|f\|$$

and since this inequality holds for every natural number r, we obtain

$$|C| + \sum_{n=1}^{\infty} C_n \leqq \|f\|.$$

Since, moreover,

$$f(x) \leqq \left[|C| + \sum_{n=1}^{\infty} |C_n|\right]\|x\|$$

it follows that we have the equality

(19) $$|C| + \sum_{n=1}^{\infty} |C_n| = \|f\|.$$

Because of (18) and (19), the following theorem has thus been proved:

Every bounded linear functional f defined in c is of the form

$$f(x) = C\lim_{n\to\infty} \xi_n + \sum_{n=1}^{\infty} C_n\xi_n, \text{ for } x = (\xi_n) \in c,$$

and we have

$$|C| + \sum_{n=1}^{\infty} |C_n| = \|f\|.$$

The space l^r where $r \geq 1$. As before, let $x_n = (\xi_i^n)$ where the ξ_i^n are defined as in (15). We thus have, for any $x = (\xi_i) \in l^r$,

$$\|x - \sum_{i=1}^{n} \xi_i x_i\| = \left(\sum_{i=n+1}^{\infty} |\xi_i|^r\right)^{\frac{1}{r}} \to 0,$$

whence

(20) $$x = \sum_{i=1}^{\infty} \xi_i x_i.$$

For a given bounded linear functional f defined in l^r, put $f(x_i) = C_i$, so that, by (20)

(21) $$f(x) = \sum_{i=1}^{\infty} \xi_i C_i.$$

Consider now the case $r = 1$.

Let $\xi_n = \text{sign } C_n$ and $\xi_i = 0$ for $i \neq n$. We then have $f(x) = |C_n| \leqq \|f\|$.

Moreover, for every sequence $x = (\xi_i) \in l^1$, we have the inequality $|f(x)| \leq \left(\sum_{i=1}^{\infty} |\xi_i|\right) \cdot \sup_{1\leq i<\infty} |C_i|$ and consequently $\|f\| = \sup_{1\leqq i<\infty} |C_i|$.

We have therefore proved the following theorem:

Every bounded linear functional f defined in l^1 is of the form

$$f(x) = \sum_{i=1}^{\infty} C_i\xi_i, \text{ for } x = (\xi_i) \in l^1,$$

where $\|f\| = \sup_{1\leq i<\infty} |C_i|$.

Let us now consider the case $r > 1$. Let $x^\circ = (\xi_i^\circ)$ where

$$\xi_i^\circ = \begin{cases} |C_i|^{s-1}\operatorname{sign} C_i & \text{for } i \leq n, \\ 0 & \text{for } i > n \end{cases}$$

and $\frac{1}{r} + \frac{1}{s} = 1$. We then have

$$\|x^\circ\| = \left(\sum_{i=1}^{n} |C_i|^{rs-r}\right)^{\frac{1}{r}} = \left(\sum_{i=1}^{n} |C_i|^{s}\right)^{\frac{1}{r}}$$

whence, because of (21),

$$f(x^\circ) = \sum_{i=1}^{n} |C_i|^s \leq \|f\|.\left(\sum_{i=1}^{n} |C_i|^{s}\right)^{\frac{1}{r}},$$

so that

$$\left(\sum_{i=1}^{n} |C_i|^{s}\right)^{\frac{1}{s}} \leq \|f\|,$$

and, since n is arbitrary,

$$\left(\sum_{i=1}^{\infty} |C_i|^{s}\right)^{\frac{1}{s}} \leq \|f\|.$$

Furthermore, for every sequence $x = (\xi_i) \in l^r$, we have

$$|f(x)| = \left|\sum_{i=1}^{\infty} \xi_i C_i\right| \leq \left(\sum_{i=1}^{\infty} |\xi_i|^{r}\right)^{\frac{1}{r}}.\left(\sum_{i=1}^{\infty} |C_i|^{s}\right)^{\frac{1}{s}},$$

whence we finally obtain the equality

$$\|f\| = \left(\sum_{i=1}^{\infty} |C_i|^{s}\right)^{\frac{1}{s}}$$

We have thus proved the following theorem:

Every bounded linear functional f defined in the space l^r, where $r > 1$, is of the form

$$f(x) = \sum_{i=1}^{\infty} C_i\xi_i, \quad \textit{where } x = (\xi_i) \in l^r,$$

and we have

$$\|f\| = \left(\sum_{i=1}^{\infty} |C_i|^{s}\right)^{\frac{1}{s}} \quad \textit{where } \frac{1}{r} + \frac{1}{s} = 1.$$

5. *The space m and its separable linear subspaces.* Let E be a separable linear subspace of m, so that its elements are bounded sequences of numbers. Endow E with the norm inherited from m, in other words, set

$$\|x\| = \sup_{1\leq i<\infty} |\xi_i| \quad \text{where } x = (\xi_i) \in E.$$

Let (x_n), where $x_n = (\xi_i^n)$, be a dense sequence in E. Consider x_1 and x_2. We are going to establish, for every $\varepsilon_2 > 0$, the existence of a natural number k_2 with the property that, for every pair λ_1 and λ_2 of real numbers,

$$|\lambda_1 x_1 + \lambda_2 x_2| \leq \max_{1\leq i\leq k_2} |\lambda_1\xi_i^1 + \lambda_2\xi_i^2|.(1 + \varepsilon_2). \tag{22}$$

Putting aside the trivial case in which x_1 is a multiple of x_2, or vice versa, suppose, on the contrary, that there exists, for every natural number k, a pair λ_1^k and λ_2^k such that

$$|\lambda_1^k x_1 + \lambda_2^k x_2| > \max_{1\leq i\leq k} |\lambda_1^k\xi_i^1 + \lambda_2^k\xi_i^2|.(1 + \varepsilon_2).$$

Denoting the larger of the two numbers $|\lambda_1^k|$ and $|\lambda_2^k|$ by m_k and putting $l_1^k = \lambda_1^k/m_k$ and $l_2^k = \lambda_2^k/m_k$, we would thus have

(23) $$|l_1^k x_1 + l_2^k x_2| > \max_{1\le i\le k}|l_1^k \xi_i^1 + l_2^k \xi_i^2|.(1+\varepsilon_2).$$

Since $1 \le |l_1^k| + |l_2^k| \le 2$ for every natural number k, the sequence (l_1^k) and (l_2^k) have convergent subsequences. In fact, there exists an increasing sequence (k_j) of natural numbers such that the sequences $(l_1^{k_j})$ and $(l_2^{k_j})$ converge to, say, l_1 and l_2 respectively, where $1 \le |l_1| + |l_2| \le 2$. As $k_j \to +\infty$ as $j \to \infty$ and, besides,

$$\lim_{j\to\infty}|(l_1 x_1 + l_2 x_2) - (l_1^{k_j} x_1 + l_2^{k_j} x_2)| = 0,$$

we would have, in view of (23),

$$|l_1 x_1 + l_2 x_2| \ge \max_{i\ge 1}|l_1 \xi_i^1 + l_2 \xi_i^2|.(1+\varepsilon_2),$$

which is impossible because, by definition,

$$|l_1 x_1 + l_2 x_2| \ge \sup_{i\ge 1}|l_1 \xi_i^1 + l_2 \xi_i^2|.$$

Having proved the existence of a natural number k_2 satisfying (22) for every pair of numbers λ_1 and λ_2, it is easy to deduce by induction that, for any given sequence (ε_n) of positive numbers, there exists, for every $n > 1$, a natural number k_n such that, for every finite sequence $\lambda_1, \lambda_2, \ldots, \lambda_n$ of real numbers, one has

(24) $$|\lambda_1 x_1 + \lambda_2 x_2 + \ldots + \lambda_n x_n| \le \max_{1\le i\le k_n}|\lambda_1 \xi_i^1 + \lambda_2 \xi_i^2 + \ldots + \lambda_n \xi_i^n|.(1+\varepsilon_n).$$

This established, for every natural number n, denote by x_i', for $i=1,2,\ldots,n$, the sequence

(25) $$\xi_1^i, \xi_2^i, \ldots, \xi_{k_n}^i, 0, 0, 0, \ldots;$$

so that, by (24), we have, for arbitrary numbers $\lambda_1, \lambda_2, \ldots, \lambda_n$,

(26) $$|\lambda_1 x_1 + \lambda_2 x_2 + \ldots + \lambda_n x_n| \le |\lambda_1 x_1' + \lambda_2 x_2' + \ldots + \lambda_n x_n'|.(1+\varepsilon_n).$$

Now let f be a bounded linear functional defined in E. Then $|f(\lambda_1 x_1 + \lambda_2 x_2 + \ldots + \lambda_n x_n)| \le \|f\|.|\lambda_1 x_1 + \lambda_2 x_2 + \ldots + \lambda_n x_n|$, and consequently, by (26), $|\lambda_1 f(x_1) + \lambda_2 f(x_2) + \ldots + \lambda_n f(x_n)| \le \|f\|.(1+\varepsilon_n).|\lambda_1 x_1' + \lambda_2 x_2' + \ldots + \lambda_n x_n'|$.

Since by the definition (25) of x_i' we have $x_i' \in c$, by theorem 5, p. 35, there exists a linear functional f_n defined in c and satisfying the conditions:

$$f_n(x_i') = f(x_i) \text{ for } i=1,2,\ldots,n \text{ and } \|f_n\| \le \|f\|.(1+\varepsilon_n).$$

In view of the general form of bounded linear functionals in the space c, established on p. 41, and because all the terms beyond the k_nth of the sequences x_i', for $i=1,2,\ldots,n$, are zero, we conclude that there exists a finite sequence of numbers $a_{n_1}, a_{n_2}, \ldots, a_{n_{k_n}}$ satisfying the conditions:

$$\sum_{j=1}^{k_n} a_{n_j}\xi_j^i = f_n(x_i') = f(x_i) \text{ for } i=1,2,\ldots,n,$$

and

$$\sum_{j=1}^{k_n} |a_{n_j}| = \|f_n\| \le \|f\|.(1+\varepsilon_n),$$

whence, putting

$$(27)\qquad \alpha_{n_j} = \begin{cases} \dfrac{a_{n_j}}{1+\varepsilon_n} & \text{for } j \leqq k_n, \\ 0 & \text{for } j > k_n, \end{cases}$$

we obtain

$$(28)\qquad \sum_{j=1}^{\infty} \alpha_{n_j} \xi_j^i = \frac{1}{1+\varepsilon_n} f(x_i) \text{ for } i=1,2,\dots,n,$$

and

$$(29)\qquad \sum_{j=1}^{\infty} |\alpha_{n_j}| \leqq \|f\|.$$

Assume that the sequence (ε_n) has been chosen so that $\lim_{n\to\infty} \varepsilon_n = 0$. It then follows from (28) that $\lim_{n\to\infty} \sum_{j=1}^{\infty} \alpha_{n_j} \xi_j^i = f(x_i)$ for $i=1,2,\dots,$ and we shall show that this still holds, with the same doubly-infinite array (α_{n_j}), for every $x \in E$.

To this end put $x = (\xi_i)$. Since the sequence (x_n), where $x_n = (\xi_i^n)$, is dense in E by definition, there exists, for every $\varepsilon > 0$, an $x_i = (\xi_j^i)$ such that $\|x - x_i\| < \varepsilon$, whence

$$\left| \sum_{j=1}^{\infty} \alpha_{n_j} \xi_j - f(x) \right| \leqq \left| \sum_{j=1}^{\infty} \alpha_{n_j} (\xi_j - \xi_j^i) \right| + \left| \sum_{j=1}^{\infty} \alpha_{n_j} \xi_j^i - f(x_i) \right| + \left| f(x_i) - f(x) \right|$$

and as

$$\left| \sum_{j=1}^{\infty} \alpha_{n_j} (\xi_j - \xi_j^i) \right| \leqq \left(\sum_{j=1}^{\infty} |\alpha_{n_j}| \right) . \|x - x_i\| \leqq \|f\| . \varepsilon,$$

we have, for sufficiently large n,

$$\left| \sum_{j=1}^{\infty} \alpha_{n_j} \xi_j - f(x) \right| \leqq \|f\| . \varepsilon + \varepsilon + \|f\| . \varepsilon = (2\|f\| + 1)\varepsilon$$

and consequently we have

$$(30)\qquad \lim_{n\to\infty} \sum_{j=1}^{\infty} \alpha_{n_j} \xi_j = f(x) \text{ for every } x = (\xi_j) \in E.$$

We finally show that

$$(31)\qquad \lim_{n\to\infty} \sum_{j=1}^{\infty} |\alpha_{n_j}| = \|f\|.$$

Indeed, if we put

$$(32)\qquad M = \lim_{n\to\infty} \sum_{j=1}^{\infty} |\alpha_{n_j}|,$$

we have, by (30), $|f(x)| \leqq M \sup_{j \geqq 1} |\xi_j| = M . \|x\|$ for every $x \in E$, whence, as $\|f\|$ is the smallest number such that $|f(x)| \leqq \|f\| . \|x\|$ for every $x \in E$, we conclude that $\|f\| \leqq M$, which, by virtue of (32) and (29), yields the equality (31).

Gathering together formulae (27), (29), (30) and (31), we see that the following theorem has been established:

Every bounded linear functional f defined in a separable vector subspace E of m is of the form

$$f(x) = \lim_{n\to\infty} \sum_{j=1}^{\infty} \alpha_{n_j} \xi_j$$

where $x = (\xi_j)$ and (α_{n_j}) is an array of real numbers satisfying the conditions:

1° $\alpha_{n_j} = 0$ *for* $j > k_n$ *where* (k_n) *is an increasing sequence of natural numbers,*

2° $\sum_{j=1}^{\infty} |\alpha_{n_j}| \leq \|f\|$ *for* $n=1,2,\ldots,$

3° $\lim_{n\to\infty} \sum_{j=1}^{\infty} |\alpha_{n_j}| = \|f\|$.

§5. Closed and complete sequences in the spaces C, L^r, c and l^r.

We are here going to apply some earlier results to various ideas and problems associated with properties of the particular spaces that we have just discussed.

A sequence of functions $(x_n(t))$ where $x_n(t) \in C$, $0 \leq t \leq 1$, is said to be *closed in* C when, for every function $x(t) \in C$, there exists a sequence of linear combinations $\left(\sum_{i=1}^{k_n} \alpha_i^{(n)} x_i(t)\right)$ which converges uniformly to $x(t)$.

The sequence $(x_n(t))$ is called *complete in* C when, for any function $g(t)$ of bounded variation, the conditions $\int_0^1 x_n(t)\,dg = 0$ for $n=1,2,\ldots$ imply that $g(0) = g(t) = g(1)$ for all but at most countably many values of t.

A sequence of functions $(x_n(t))$ where $x_n(t) \in L^r$ and $0 \leq t \leq 1$ is said to be *closed in* L^r when, for every function $x(t) \in L^r$, there exists a sequence (g_n) of functions of the form

$$g_n(t) = \sum_{i=1}^{k_n} \alpha_i^{(n)} x_i(t)$$

which converges in mean with power r to $x(t)$, i.e. such that

$$\lim_{n\to\infty} \int_0^1 |x(t) - g_n(t)|^r dt = 0.$$

The sequence $(x_n(t))$ is called *complete in* L^r when, for any function $g(t)$ which is bounded and measurable or which belongs to L^s where $\frac{1}{r} + \frac{1}{s} = 1$, according as $r = 1$ or $r > 1$, the conditions

$$\int_0^1 x_n(t)g(t)\,dt = 0 \text{ for } n=1,2,\ldots$$

imply that $g(t) = 0$ almost everywhere.

Both notions occur in the theory of orthogonal series.

It is easy to see that a necessary and sufficient condition for a sequence of functions to be closed in C, or in L^r, is that it be fundamental, in the sense defined in §3, p. 35, of this chapter. Similarly, for it to be complete, it is necessary and sufficient that it be total, in the sense defined in the same place. To see this, it is enough to recall the general form of bounded linear functionals in C and L^r established on pp. 37-39.

Finally, theorem 7, p. 36, immediately implies that *for a sequence of functions to be complete in* C, *or* L^r *respectively, it is necessary and sufficient that it be closed.*

In a similar way, one can define the notions of closed and complete sequences for the spaces c and l^r.

§6. Approximation of functions belonging to C and L^r by linear combinations of functions.

Theorem 6, p. 36, can also be interpreted in various particular normed spaces. Here are two examples of this:

1. *The space C. For there to exist polynomials in the terms of the sequence $(x_n(t))$ where $x_n(t) \in C$, $0 \leq t \leq 1$, which uniformly approximate a given function $x(t) \in C$ arbitrarily closely, it is necessary and sufficient that, for any function $g(t)$ of bounded variation, the conditions $\int_0^1 x_n(t)\,dg = 0$ for $n=1,2,\ldots$ imply that $\int_0^1 x(t)\,dg = 0$.*

2. *The spaces L^r. For there to exist linear combinations of terms of the sequence $(x_n(t))$ where $x_n(t) \in L^r$, $0 \leq t \leq 1$, which approximate in the mean with the rth power a given function $x(t) \in L^r$ arbitrarily closely, it is necessary and sufficient that, for any function $g(t)$ which is bounded and measurable when $r = 1$ and which belongs to L^s where $\frac{1}{r} + \frac{1}{s} = 1$ when $r > 1$, the conditions $\int_0^1 g(t)x_n(t)\,dt = 0$ for $n=1,2,\ldots$ imply that $\int_0^1 g(t)x(t)\,dt = 0$.*

§7. The problem of moments.

We now discuss applications of theorem 5, p. 35.

The *problem of moments* is the name given to the problem of finding conditions for the existence of a function f satisfying the infinite set of equations

$$\int_a^b f\phi_i\,dt = c_i \text{ where } i=1,2,\ldots, \tag{33}$$

for a given sequence of functions (ϕ_i) and a given sequence of numbers (c_i).

We give here the solution of this problem in two special cases of normed spaces: it is obtained by means of the appropriate interpretation of theorem 5, p. 35, in these spaces.

I. *The space C.* Let $x_i = x_i(t)$ where $0 \leq t \leq 1$ be a continuous function. Every bounded linear functional f in C being (cf. p. 37) of the form $f(x) = \int_0^1 x(t)\,dg$ where $\|f\| = \operatorname{var}_{0\leq t\leq 1} g(t)$, theorem 5, p. 35, yields the following theorem:

For there to exist a function $g(t)$ with

$$\operatorname{var}_{0\leq t\leq 1} g(t) \leq M$$

and satisfying the equations

$$\int_0^1 x_i(t)\,dg = c_i \text{ for } i=1,2,\ldots,$$

it is necessary and sufficient that, for every finite sequence of real numbers $h_1, h_2, \ldots, h_r$, one has

$$\left|\sum_{i=1}^{r} h_i c_i\right| \leq M.\max_{0\leq t\leq 1}\left|\sum_{i=1}^{r} h_i x_i(t)\right|.$$

II. *The space L^r.* For $r > 1$, proceeding in a similar way, one arrives at the following theorem:

For there to exist a function $\alpha(t)$ where $0 \leq t \leq 1$ such that

$$\int_0^1 |\alpha(t)|^s\,dt \leq M^s \text{ where } \frac{1}{r} + \frac{1}{s} = 1$$

and which satisfies the equations

$$\int_0^1 x_i(t)\alpha(t)\,dt = c_i \text{ where } x_i(t) \in L^r \text{ and } i=1,2,\ldots,$$

it is necessary and sufficient that, for every finite sequence of real numbers $h_1, h_2, \ldots, h_k$ one has

$$\left|\sum_{i=1}^{k} h_i c_i\right| \leq M\left(\int_0^1 \left|\sum_{i=1}^{k} h_i x_i(t)\right|^r dt\right)^{\frac{1}{r}}$$

For $r=1$, the functions $x_i(t)$ must be integrable and the function $\alpha(t)$ which is sought is (essentially) bounded and such that

$$\operatorname*{ess\,sup}_{0\leq t\leq 1}|\alpha(t)| \leq M.$$

The necessary and sufficient condition is then the following:

$$\left|\sum_{i=1}^{k} h_i c_i\right| \leq M.\int_0^1 \left|\sum_{i=1}^{k} h_i x_i(t)\right| dt.$$

§8. Condition for the existence of solutions of certain systems of equations in infinitely many unknowns.

We consider another problem, namely, given an array (α_{ik}) and a sequence of numbers (c_i), we seek to establish conditions for the existence of a sequence of numbers (z_i) satisfying the infinite set of equations

$$\sum_{k=1}^{\infty} \alpha_{ik} z_k = c_i \text{ where } i=1,2,\ldots \tag{34}$$

We here give, again with the help of theorem 5, p. 35, the solution of this problem in two particular spaces:

III. *The space c.* Let $x_i = (\alpha_{ik})$ and

$$\lim_{k\to\infty} \alpha_{ik} = 0 \text{ for } i=1,2,\ldots \tag{35}$$

Since every bounded linear functional in the space c is of the form

$$f(x) = C \lim_{i\to\infty} \xi_i + \sum_{i=1}^{\infty} C_i \xi_i$$

where

$$x = (\xi_i) \text{ and } \|f\| = |C| + \sum_{i=1}^{\infty} |C_i|$$

(cf. p. 40), theorem 5, p. 35, yields, in view of (35), the following result:

For there to exist a sequence of numbers (z_k) which satisfy the equations (34) *as well as the condition $\sum_{k=1}^{\infty}|z_k| \leq M$, it is necessary and sufficient that, for every finite sequence of numbers $h_1, h_2, \ldots, h_r$, one has the inequality*

$$\left|\sum_{i=1}^{r} h_i c_i\right| \leq M \sup_{k\geq 1}\left|\sum_{i=1}^{r} h_i \alpha_{ik}\right|.$$

IV. *The space l^1.* Let $x_i = (\alpha_{ik})$ and assume that $\sum_{k=1}^{\infty}|\alpha_{ik}| < \infty$ for $i=1,2,\ldots$ Since every bounded linear functional in l^1 is of the form $f(x) = \sum_{i=1}^{\infty} z_i \xi_i$ where $x = (\xi_i)$ and $\|f\| = \sup_{i\geq 1}|z_i|$ (cf. p.41), theorem 5, p. 35 immediately yields the following result:

For there to exist a bounded sequence (z_k) which satisfies the equations (34) *as well as the condition $\sup_{k\geq 1}|z_k| \leq M$, it is necessary and sufficient that, for every finite sequence of real numbers $h_1, h_2, \ldots, h_r$, one has the inequality*

$$\left|\sum_{i=1}^{r} h_i c_i\right| \leq M.\sum_{k=1}^{\infty}\left|\sum_{i=1}^{r} h_i \alpha_{ik}\right|.$$

CHAPTER V

Banach spaces

§1. Linear operators in Banach spaces

We are here going to establish several general theorems concerning Banach or B-spaces E, already defined at the beginning of Chapter IV, in which their property of being *complete* as well as just normed plays an essential role.

THEOREM 1. *Let F be a B-measurable mapping and U an additive operator, both defined in the Banach space E, such that $\|F(x)\| \geq \|U(x)\|$ for every $x \in E$. Then U is a bounded linear operator.*

Proof. By theorem 4 of the introduction, p. 10, there exists a set $H \subseteq E$ of category I such that F is continuous in $E \setminus H$. Consequently, for $x_0 \in E \setminus H$ there exists an $r > 0$ and an $M > 0$ such that

(1) $\|x - x_0\|$ implies $\|U(x)\| \leq \|F(x)\| \leq M$ for every $x \in E \setminus H$.

Since the set $\{x : x \in E \setminus H$ and $\|x - x_0\| \leq \frac{r}{2}\}$ is of category II, so even more so, is the set $G = \{x' + x : x \in E \setminus H,\ \|x - x_0\| \leq \frac{r}{2}$ and $\|x'\| < \frac{r}{2}\}$. Hence, in particular, G is non-empty and contains an element of the form $x' + x_1 \in E \setminus H$ where $x_1 \in E \setminus H$. As $\|x_1 - x_0\| \leq \frac{r}{2}$, we have $\|x' + x_1 - x_0\| \leq r$, whence, by (1), $\|U(x')\| \leq \|U(x' + x_1)\| + \|U(x_1)\| \leq 2M$. Thus U is (norm)-bounded in the sphere $\{x : \|x\| \leq \frac{r}{2}\}$, and, therefore, of course, in every sphere. It then follows, by theorem 1 (Chapter IV, §2), p. 33, that the mapping U is continuous and thus U is a bounded linear operator.

THEOREM 2. *Let U be an additive operator such that, for every $x \in E$, whenever (x_n) is a sequence in E with $\lim_{n\to\infty} x_n = x$, $\underline{\lim}_{n\to\infty}\|U(x_n)\| \geq \|U(x)\|$. Then U is a bounded linear operator.*

Proof. The set $G_n = \{x \in E:\ \|U(x)\| \leq n\}$ is closed for $n = 1,2,\ldots$ and since $E = \bigcup_{n=1}^{\infty} G_n$, at least one of the sets G_n contains a sphere in which U is bounded, whence, as in the previous theorem, it follows that U is continuous.

THEOREM 3. *Let (U_n) be a sequence of bounded linear operators defined in E. Suppose that $\big(U_n(x)\big)$ is convergent for all x in a set G which is dense in a sphere K and that the sequence $\big(\|U_n\|\big)$ is bounded. Then the sequence $\big(U_n(x)\big)$ is convergent at every point $x \in E$.*

Proof. For a given $x_0 \in K$, there exists, by hypothesis, a sequence (x_n) such that $x_n \in G$ for $n=1,2,\ldots$ and $\lim_{n\to\infty} x_n = x_0$.

Now, for any three natural numbers n, p and q, we have:

$$\|Up(x_0) - Uq(x_0)\| \leq \|Up(x_0 - xn)\| + \|Uq(xn - x_0)\| + \|Up(xn) - Uq(xn)\|$$

and so

$$\overline{\lim_{p,q\to\infty}}\, \|U_p(x_0) - U_q(x_0)\| \leq 2M\|x_0 - x_n\| \text{ where } M = \overline{\lim_{n\to\infty}}\|U_p\|,$$

whence, as $\lim_{n\to\infty} \|x_0 - x_n\| = 0$, we have $\lim_{p,q\to\infty} \|U_p(x_0) - U_q(x_0)\| = 0$, which implies the convergence of the sequence $(U_n(x_0))$.

Now let x be an arbitrary element of E. Letting x_0' denote the centre of the sphere K, there exists an $\varepsilon > 0$ such that $x_0' + \varepsilon x \in K$, and so the sequence $(U_n(x_0' + \varepsilon x))$ is convergent. The convergence of $(U_n(x))$ now follows from that of the sequences $(U_n(x_0'))$ and $(U_n(x_0' + \varepsilon x))$.

THEOREM 4. *Let* (U_n) *be a sequence of bounded linear operators defined in* E. *Then the set* $H = \{x \in E : \overline{\lim_{n\to\infty}}\|U_n(x)\| < \infty\}$ *is either of category* I *or is the whole of* E.

The proof follows from theorem 1 (Chapter III, §1) p. 23, seeing that H is a B- measurable linear subspace of E.

THEOREM 5. *If the sequence* (U_n) *of bounded linear operators in* E *is such that* $\overline{\lim_{n\to\infty}}\|U_n(x)\| < \infty$ *for every* $x \in E$, *the sequence of norms* $(\|U_n\|)$ *is bounded.*

Proof. By theorem 11 of the introduction, p. 11, there exists a sphere $K \subseteq E$ and a number N such that $\|U_n(x)\| \leq N$ for every $x \in K$ and $n=1,2,\ldots$ Letting r denote the radius of the sphere K, one easily deduces that $\|U_n\| \leq 2\frac{N}{r}$ for every $n=1,2,\ldots$

THEOREM 6. *If the sequence* $(x_n) \subseteq E$ *is such that* $\overline{\lim_{n\to\infty}}|f(x_n)| < \infty$ *for every bounded linear functional* f *defined in* E, *the sequence of norms* $(\|x_n\|)$ *is bounded.*

Proof. The set E^* of all bounded linear functionals on E, with the norm previously defined for such functionals, constitutes a Banach space. We define a sequence (F_n) of functionals on E^*, by putting $F_n(f) = f(x_n)$ for each $f \in E^*$. The hypothesis that $\overline{\lim_{n\to\infty}}\, |f(x_n)| < \infty$ consequently implies that $\overline{\lim_{n\to\infty}}\, |F_n(f)| < \infty$ for every $f \in E^*$. By virtue of the preceding theorem 5, there thus exists a number N such that $|F_n(f)| \leq N.\|f\|$ for every $n=1,2,\ldots$ Moreover, as $x_n \in E$, for every $n=1,2,\ldots$ there exists, by theorem 3 (Chapter IV, §2) p. 33, a bounded linear functional f_n defined in E such that $|f_n(x_n)| = \|x_n\|$ and $\|f_n\| = 1$. We therefore have $\|x_n\| = |f_n(x_n)| = |F_n(f_n)| \leq N.\|f_n\| = N$, for every n, q.e.d.

§2. The principle of condensation of singularities.

THEOREM 7. *Let* (U_{pq}) *be a double sequence of bounded linear operators in* E *such that*

$$(2) \qquad \overline{\lim_{q\to\infty}}\, \|U_{pq}\| = \infty \text{ for every } p=1,2,\ldots$$

Then there exists a set $G \subseteq E$, *independent of* p, *of category* II *in* E, *such that, for every* $x \in G$:

$$(3) \qquad \overline{\lim_{q\to\infty}}\, \|U_{pq}(x)\| = \infty \text{ for every } p=1,2,\ldots$$

Proof. The set $H_p = \{x \in E : \overline{\lim_{q\to\infty}}\|U_{pq}(x)\| < \infty\}$ cannot be the whole of E, because, by theorem 5 above, the hypothesis (2) would then be contradicted. It follows from this, by theorem 4 above, that H_p and consequently the set $H = \bigcup_{p=1}^{\infty} H_p$ of all the elements $x \in E$ for which the condition (3) *fails to hold*, is of category I in E. All that remains, therefore, is to take $G = E \setminus H$.

Remark. The codomains E_p of the operators U_{pq} can vary with $p=1,2,\ldots$, whereas, for a given p, they must plainly be assumed to be the same for all values of q, since, in the statement of theorem 7, we make use of the notion of convergence of the sequences $(U_{pq}(x))$ as q tends to ∞.

The above theorem 7, together with theorem 6 (Chapter I, §4) p. 15, constitute, in the setting of functional analysis, what is known as the *principle of condensation of singularities*. We are going to elaborate on this with the help of some examples.

Let $(g_k(t))$ be an orthonormal sequence of square-integrable functions in [0,1]. For any given integrable function $x(t)$ in [0,1], the series

$$\sum_{k=1}^{\infty} g_k(t)\int_0^1 g_k(s)x(s)\,ds$$

is called the *development of the function* $x(t)$ *with respect to the sequence* $(g_k(t))$, provided, of course, that the integrals $\int_0^1 g_k(s)x(s)\,ds$ exist for each $k=1,2,\ldots$

The following theorems are available, for example in the spaces C and L^1:

In C. Let (t_p) *be a given sequence of points in* [0,1]. *Then the existence, for each* $p=1,2,\ldots$, *of a continuous function* $x_p(t)$ *whose development is divergent or unbounded, respectively, at the point* t_p *implies the existence of a continuous function* $x(t)$ *whose development is divergent or unbounded, respectively, at each point* t_p, *for* $p=1,2,\ldots$

The proof follows from theorem 7, p. 50, together with theorem 6 (Chapter 1, §4), p. 15, if one puts

$$U_{pq}(x) = \sum_{k=1}^{q} g_k(t_p)\int_0^1 g_k(s)x(s)\,ds,$$

considering the U_{pq} as linear functionals on C.

In L^1. *Let* $([\alpha_p,\beta_p])$ *be a given sequence of sub-intervals of* [0,1]. *Then the existence, for each* $p=1,2,\ldots$, *of an integrable function* $x_p(t)$ *whose development has the property that*

$$\overline{\lim_{n\to\infty}} \int_{\alpha_p}^{\beta_p} |s_n(t)|\,dt = \infty \quad \textit{where } s_n(t) = \sum_{k=1}^{n} g_k(t)\int_0^1 g_k(t)x_p(t)\,dt$$

implies the existence of an integrable function $x(t)$ *whose development has the same property in all the intervals* $[\alpha_p,\beta_p]$ *simultaneously.*

The proof follows from theorem 7, p. 50, if one puts

$$U_{pq}(x) = \sum_{k=1}^{q} g_k(t)\int_0^1 g_k(t)x(t)\,dt \quad \text{for } \alpha_p \leq t \leq \beta_p$$

and regards the U_{pq} as linear operators defined in the space of integrable functions on [0,1] whose codomains lie, respectively, in the spaces of integrable functions in $[\alpha_p,\beta_p]$.

Remark. In particular, if the set of points $\{(\alpha_p,\beta_p):p=1,2,\ldots\}$ form a dense subset of the unit square $[0,1]\times[0,1]$, the property in question holds for $x(t)$ in every subinterval $[\alpha,\beta]$ of [0,1].

Based on this remark, one can prove, for the case of Fourier series, *the existence of an integrable function* $x(t)$ *such that*

$\overline{\lim_{n\to\infty}}\left|\int_\alpha^\beta s_n(t)\,dt\right| = \infty$ *for every subinterval* $[\alpha,\beta]$ *of* $[0,2\pi]$.

§3. Compactness in Banach spaces.

LEMMA. *Given a closed linear subspace G which is a proper subset of a linear subspace $D \subseteq E$, there exists, for every number $\varepsilon > 0$, an $x_0 \in D$ such that*

$$\|x_0\| = 1 \text{ and } \|x_0 - x\| \geqq 1 - \varepsilon \text{ for every } x \in G.$$

Proof. Let $x' \in D \setminus G$, let d be the distance of x' from G and let η be an arbitrary positive number. Then there exists a $y' \in G$ such that $d \leqq \|x' - y'\| \leqq d + \eta$. Put $x_0 = \frac{x'-y'}{\|x'-y'\|}$. For every $x \in G$ one then has $\|x_0 - x\| = \frac{1}{\|x'-y'\|}\cdot\big\|x' - y' - \|x'-y'\|.x\big\|$ and since $x \in G$ and $y' \in G$ imply that $y' + \|x'-y'\|x \in G$, this yields $\|x_0 - x\| \geqq \frac{1}{\|x'-y'\|}.d \geqq \frac{d}{d+\eta}$, whence, putting $\eta = d.\frac{\varepsilon}{1-\varepsilon}$, we have $\|x_0 - x\| \geqq 1 - \varepsilon$.

Plainly, we also have $\|x_0\| = \|x'-y'\|/\|x'-y'\| = 1$ and $x_0 \in D$, because $x' \in D$ and $y' \in G \subseteq D$.

THEOREM 8. *If every norm-bounded subset of E is relatively compact, then there is a finite set $x_1, x_2, \ldots, x_r$ of elements of E such that every $x \in E$ may be written in the form*

$$x = \alpha_1 x_1 + \alpha_2 x_2 + \ldots + \alpha_r x_r, \tag{4}$$

where $\alpha_1, \alpha_2, \ldots, \alpha_r$ are numbers (which depend on x).

Proof. Let x_1 be an arbitrary element of E such that $\|x_1\| = 1$ and, for $r > 1$, let x_{r+1} be any element of E such that

$$\|x_{r+1}\| = 1 \text{ and } \|x_{r+1} - x_i\| \geqq \tfrac{1}{2} \text{ for } i=1,2,\ldots,r. \tag{5}$$

For each $r \geqq 1$, let G_r denote the set of all elements $x \in E$ which are of the form (4) and put $D = E$. If the theorem is false, we should have $G_r \neq D$ for every r, whence, according to the above lemma, with $G = G_r$, $\varepsilon = \frac{1}{2}$ and $x_0 = x_{r+1}$, we deduce the existence, for each natural number r, of an $x_{r+1} \in E$ satisfying (5), i.e. of an *infinite* sequence (x_r) such that $\|x_r\| = 1$ and $\|x_p - x_q\| \geqq \frac{1}{2}$ for $p \neq q$.

This sequence could not have any convergent subsequences, and would therefore constitute a non-relatively compact set, although it is norm-bounded. One would thus have a contradiction with the hypothesis of the theorem.

§4. A property of the spaces L^r, c and l^r.

By applying theorem 4 (Chapter I, §3) p. 15, to the general form of the bounded linear functionals defined in these spaces, one obtains the following theorems:

For L^r, where $r \geqq 1$. If $\alpha(t)$, $0 \leqq t \leqq 1$, is a measurable function and if the integral $\int_0^1 x(t)\alpha(t)dt$ exists for every function $x(t) \in L^r$, then $\int_0^1 |\alpha(t)|^{r/(r-1)}dt < \infty$, for $r > 1$, and $\alpha(t)$ is an (essentially) bounded function for $r = 1$.

Proof. Put, for n a natural number

$$\alpha_n(t) = \begin{cases} \alpha(t) & \text{for } |\alpha(t)| \leqq n, \\ n \operatorname{sign} \alpha(t) & \text{for } |\alpha(t)| > n. \end{cases}$$

We then have $|x(t)\alpha(t)| \geqq |x(t)\alpha_n(t)|$, so that, as $\lim_{n\to\infty}\alpha_n(t) = \alpha(t)$,

$$\lim_{n\to\infty} \int_0^1 x(t)\alpha_n(t)dt = \int_0^1 x(t)\alpha(t)dt.$$

Since, for $n=1,2,\ldots$, $\alpha_n(t)$ is a bounded function, the expression

$\int_0^1 x(t)\alpha_n(t)dt$ defines a bounded linear functional in L^r, and so, by theorem 4, p. 15, $\int_0^1 x(t)\alpha(t)dt$ is also a bounded linear functional. Consequently, by the theorem on the general form of bounded linear functionals in the spaces L^r where $r > 1$ (see p. 39), there exists a function $\bar{\alpha}(t) \in L^{r/(r-1)}$ such that $\int_0^1 x(t)\bar{\alpha}(t)dt = \int_0^1 x(t)\alpha(t)dt$ for every $x(t) \in L^r$. Thus, putting

$$x(t) = \begin{cases} 1 \text{ for } 0 \leq t \leq t_0, \\ 0 \text{ for } t_0 < t \leq 1, \end{cases}$$

we have $\int_0^{t_0} \bar{\alpha}(t)dt = \int_0^{t_0} \alpha(t)dt$ for every $0 \leq t_0 \leq 1$, and so $\bar{\alpha}(t) = \alpha(t)$ *almost everywhere.*

One argues in an analogous way for $r = 1$.

For c. If the series $\sum_{i=1}^{\infty} \alpha_i \xi_i$ *is convergent for every convergent sequence* $x = (\xi_i)$, *then* $\sum_{i=1}^{\infty} |\alpha_i| < \infty$.

Proof. Since, for each $n=1,2,\ldots$, $\sum_{i=1}^{n} \alpha_i \xi_i$ is a bounded linear functional in the space c and $\lim_{n\to\infty} \sum_{i=1}^{n} \alpha_i \xi_i = \sum_{i=1}^{\infty} \alpha_i \xi_i$, one deduces from theorem 4, p. 15, that $\sum_{i=1}^{\infty} \alpha_i \xi_i$ is also a bounded linear functional on c. Consequently there exists an $M > 0$ such that

$$\left| \sum_{i=1}^{\infty} \alpha_i \xi_i \right| \leq M . \|x\| = M . \sup_{i \geq 1} |\xi_i| .$$

Therefore, putting

$$\xi_i = \begin{cases} \text{sign } \alpha_i \text{ for } i \leq n \text{ and } \alpha_i \neq 0, \\ 0 \quad\quad \text{for } i > n \text{ or } \alpha_i = 0, \end{cases}$$

we obtain $\sum_{i=1}^{n} |\alpha_i| \leq M$ for each $n=1,2,\ldots$, whence $\sum_{i=1}^{\infty} |\alpha_i| \leq M$.

For l^r *where* $r \geq 1$. *If the series* $\sum_{i=1}^{\infty} \alpha_i \xi_i$ *is convergent for every sequence* $x = (\xi_i) \in l^r$, *then* $\sum_{i=1}^{\infty} |\alpha_i|^{r/(r-1)} < \infty$, *for* $r > 1$, *and the sequence* (α_i) *is bounded for* $r = 1$.

The proof is analogous to that of the preceding theorem.

§5. Banach spaces of measurable functions.

We pause here to discuss several properties of B-spaces satisfying further special conditions. To this end, let E be a space of measurable functions defined in the closed interval $[0,1]$ and such that for any sequence $(x_n) = (x_n(t)) \subseteq E$, $0 \leq t \leq 1$:

1. $\lim_{n\to\infty} \|x_n\| = 0$ *implies* $\lim_{n\to\infty} x_n(t) = 0$ *almost everywhere;*
2. $\lim_{n\to\infty} \|x_n\| = 0$ *implies the existence in* E *of a subsequence* $(x_{n_i}(t))$ *and an* x *such that* $|x_{n_i}(t)| \leq |x(t)|$ *for every* $i=1,2,\ldots$, *and almost every* t, $0 \leq t \leq 1$;
3. $\lim_{n\to\infty} x_n(t) = x(t)$ *almost everywhere implies* $\lim_{n\to\infty} \|x_n\| \geq \|x\|$.

Particular examples of such spaces are M, C and L^r, which have been discussed repeatedly (cf. p. 6-7, 36-39, and 45-47); as far as condition 2 is concerned, in the case where $\sum_{i=1}^{\infty} \|x_{n_i}\| < \infty$, one merely has to define the function $x(t)$ by the equality

$$x(t) = \sum_{i=1}^{\infty} |x_{n_i}(t)|.$$

THEOREM 9. *Let E and E_1 be two B-spaces satisfying conditions* 1,2 *and* 3 *and let $K(s,t)$ be a function defined in the square* $[0,1] \times [0,1]$. *If the integral*

$$(6) \qquad u(s) = \int_0^1 K(s,t)x(t)dt$$

exists for every $x \in E$ and for almost every value of s and if $u(s) \in E_1$, then u is a bounded linear operator.

Proof. Suppose that (x_n) and x are such that

$$(7) \qquad \lim_{n\to\infty} \|x_n - x\| = 0$$

and let $(\bar{x}_n)$ be any subsequence of (x_n). By (7) as well as condition (2), there exists a subsequence $(\bar{x}_{n_i} - x)$ of $(\bar{x}_n - x)$ and a $z \in E$ such that, for every $i=1,2,\ldots$, one has $|\bar{x}_{n_i}(t) - x(t)| \leq z(t)$ almost everywhere. Clearly $\lim_{i\to\infty} K(s,t)[\bar{x}_{n_i}(t) - x(t)] = 0$ almost everywhere, and $|K(s,t).(\bar{x}_{n_i} - x)| \leq |K(s,t)| \times z(t)$. Furthermore the integral $\int_0^1 K(s,t)z(t)dt$ exists in a set H of measure 1, as $z \in E$. We therefore have $\lim_{i\to\infty} \int_0^1 K(s,t)[\bar{x}_{n_i}(t) - x(t)]dt = 0$ and consequently $\lim_{i\to\infty} \int_0^1 K(s,t)\bar{x}_{n_i}(t)dt = \int_0^1 K(s,t)x(t)dt$ for $s \in H$.

Now, since every subsequence of (u_n), where $u_n(s) = \int_0^1 K(s,t)x_n(t)dt$, has a subsequence which converges to $u(s)$ almost everywhere, we have $\lim_{n\to\infty} u_n(s) = u(s)$ almost everywhere, which, in view of 3 and theorem 2, p. 49, implies that the operator u defined by (6) is a bounded linear operator.

§6. Examples of bounded linear operators in some special Banach spaces.

We here give several applications of theorem 9, proved above, to the spaces M, C and L^r.

The space M. If $K(s,t)$, for $0 \leq s \leq 1$ and $0 \leq t \leq 1$, is a measurable function and if, for every s, $\int_0^1 |K(s,t)|dt < N < \infty$, the expression

$$(8) \qquad U(x) = \int_0^1 K(s,t)x(t)dt$$

defines a bounded linear operator from M into M.

The space C. If the function $K(s,t)$ is continuous for $0 \leq s \leq 1$ and $0 \leq t \leq 1$, expression (8) defines a bounded linear operator from C into C.

The space L^1. If $K(s,t)$ is a function which is measurable in the square $0 \leq s \leq 1$ and $0 \leq t \leq 1$ and is such that $\int_0^1 C(s)ds < N < \infty$ where $C(s) = \operatorname{ess.sup}_{0\leq t\leq 1} |K(s,t)|$, then (8) defines a bounded linear operator from L^1 into L^1.

The spaces L^p. If $K(s,t)$ is a function, measurable in the square $0 \leq s \leq 1$ and $0 \leq t \leq 1$, and such that, for every pair of functions $x(t) \in L^p$ and $y(s) \in L^{q/(q-1)}$ where $p \geq 2$ and $q \geq 2$ we have

$$(9) \qquad \int_0^1\int_0^1 |K(s,t)x(t)y(s)|dsdt < \infty,$$

then (8) defines a bounded linear operator from L^p into L^q.

Indeed, for any $x \in L^p$, we have, for every $y \in L^{q/(q-1)}$:

$$\int_0^1\int_0^1 K(s,t)x(t)y(s)\,ds\,dt = \int_0^1 y(s)\left[\int_0^1 K(s,t)x(t)\,dt\right]ds,$$

whence (cf. p. 52) $\int_0^1 K(s,t)x(t)\,dt \in L^q$ and consequently, by (8), U is a bounded linear operator.

For condition (9) to be satisfied, it is sufficient that $\int_0^1\int_0^1 |K(s,t)|^{r/(r-1)}\,ds\,dt < \infty$, where r is the smaller of the numbers p and $q/(q-1)$; in particular, that the function $K(s,t)$ be bounded, when $r = 1$, and integrable when $p = q = +\infty$.

In fact, one has, by Riesz' inequality:

$$\left|\int_0^1\int_0^1 K(s,t)x(t)y(s)\,ds\,dt\right| \leqq \left\{\int_0^1\int_0^1 |K(s,t)|^{\frac{r}{r-1}}\,ds\,dt\right\}^{\frac{r-1}{r}} \cdot \left\{\int_0^1 |x(s)|^r ds\right\}^{\frac{1}{r}} \cdot \left\{\int_0^1 |y(t)|^r dt\right\}^{\frac{1}{r}}.$$

In particular, for $p = q = 2$, condition (9) can therefore be replaced by $\int_0^1\int_0^1 |K(s,t)|^2\,ds\,dt < \infty$, which implies that the operator given by (8) is a bounded linear operator from L^2 into itself.

The same remark applies to the cases $p = q = 1$ and $p = q = \infty$.

§7. Some theorems on summation methods.

Given an infinite array of numbers

$$\text{(A)}\qquad \begin{array}{cccccc} a_{11}, & a_{12}, & \ldots, & a_{1k}, & \ldots \\ a_{21}, & a_{22}, & \ldots, & a_{2k}, & \ldots \\ \cdot & \cdot & \cdot & \cdot & \cdot \\ a_{i1}, & a_{i2}, & \ldots, & a_{ik}, & \ldots \\ \cdot & \cdot & \cdot & \cdot & \cdot \end{array},$$

we shall say that a sequence of numbers $x = (\xi_k)$ is *summable* (to $A(x)$) *by the method* A (which corresponds to the array A), when each series $A_i(x) = \sum_{k=1}^{\infty} a_{ik}\xi_k$ is convergent and the sequence $(A_i(x))$ also converges (to $A(x)$).

The method A is said to be *permanent*, when every convergent sequence is summable to its limit by this method. It is called *reversible* when, to every convergent sequence (η_i), there corresponds exactly one sequence x, not necessarily convergent, such that $A_i(x) = \eta_i$ for $i=1,2,\ldots$ We shall say that a method B, corresponding to the array B = (b_{ik}) *is not weaker than* A, when every sequence summable by method A is also summable by method B.

Finally, a method A is known as a *perfect* method, when it is both permanent, reversible and such that the conditions

$$\text{(10)}\qquad \sum_{i=1}^{\infty} |\alpha_i| < \infty \text{ and } \sum_{i=1}^{\infty} \alpha_i a_{ik} = 0 \text{ for } k = 1,2,\ldots$$

imply that

$$\text{(11)}\qquad \alpha_i = 0 \text{ for every } i = 1,2,\ldots$$

THEOREM 10. *For a method* A *to be permanent, it is necessary and sufficient that the following conditions be simultaneously satisfied:*

1° $\sum_{k=1}^{\infty} |a_{ik}| \leqq M$ *for every* $i=1,2,\ldots,$

2° $\lim_{i\to\infty} a_{ik} = 0$ *for every* $k=1,2,\ldots,$

3° $\lim_{i\to\infty} \sum_{k=1}^{\infty} a_{ik} = 1.$

Proof. Necessity. The convergence of the series $\sum_{k=1}^{\infty} a_{ik}\xi_k$ for every convergent sequence $x = (\xi_k)$ and for every $i=1,2,\ldots$ implies (cf. p. 53, for c) the absolute convergence of the series $\sum_{k=1}^{\infty} a_{ik}$. The A_i, defined in the space c, are consequently bounded linear functionals and since, for $x \in c$, $(A_i(x))$ is a convergent sequence, we conclude, by theorem 5, p. 50, that condition 1° is satisfied.

Now, let $\xi_i^0 = 1$ for $i=1,2,\ldots,\xi_i^n = 0$ for $i \neq n$ and $\xi_n^n = 1$ for every $n=1,2,\ldots$ Put $x_n = (\xi_i^n)$ for $n=0,1,2,\ldots$ We have $A_i(x_0) = \sum_{k=1}^{\infty} a_{ik}$ and $A_i(x_n) = a_{in}$ for i and n natural numbers, so that $A(x_0) = 1$ and $A(x_n) = 0$ for $n > 0$, from which it follows that conditions 2° and 3° are also satisfied.

The sufficiency of the three conditions follows from theorem 3, p. 49, together with the fact that the sequence (x_n) just defined is fundamental in the space c.

LEMMA 1. *Let* A *be a permanent method and let* $y_0 = (\eta_i^0)$ *be a convergent sequence. If, for every sequence of numbers* (α_i) *the conditions* (10) *imply that* $\sum_{i=1}^{\infty} \alpha_i \eta_i^0 = 0$, *there exists, for every* $\varepsilon > 0$, *a convergent sequence* x *such that*

$$|A_i(x) - \eta_i^0| < \varepsilon \textit{ for every } i=1,2,\ldots \tag{12}$$

Proof. Let G denote the set of all convergent sequences (η_i) to which there correspond convergent sequences x such that $\eta_i = A_i(x)$ for $i=1,2,\ldots$ Regarded as a subset of the space c, the set G thus defined is clearly a vector subspace. If y_0 is not an accumulation point of G, there exists, by the lemma of Chapter IV, §3, p. 35, a bounded linear functional F defined in c such that $F(y_0) = 1$ and $F(y) = 0$ for every $y \in G$. Recalling the general form of bounded linear functionals in the space c (cf. Chapter IV, §4, p. 36), there thus exists a sequence of numbers (α_i) such that the series $\sum_{i=1}^{\infty} \alpha_i$ is absolutely convergent and such that:

$$\sum_{i=1}^{\infty} \alpha_i \eta_i + \alpha \lim_{i\to\infty} \eta_i = 0 \text{ for } (\eta_i) \in G, \tag{13}$$

$$\sum_{i=1}^{\infty} \alpha_i \eta_i^0 + \alpha \lim_{i\to\infty} \eta_i^0 = 1. \tag{14}$$

Since the method A is permanent, (13) implies

$$\sum_{i=1}^{\infty} \alpha_i A_i(x) + \alpha \lim_{k\to\infty} \xi_k = 0 \text{ for every } x = (\xi_k) \in c \tag{15}$$

and the preceding theorem 10 implies the existence of an M satisfying condition 1°. We consequently have

$$\sum_{i=1}^{\infty} \sum_{k=1}^{\infty} |\alpha_i| \cdot |a_{ik}| \cdot |\xi_k| \leqq M \cdot \Big(\sum_{i=1}^{\infty} |\alpha_i|\Big) \|x\|,$$

whence

$$\sum_{i=1}^{\infty} \alpha_i A_i(x) = \sum_{k=1}^{\infty} \xi_k \cdot \sum_{i=1}^{\infty} \alpha_i a_{ik}. \tag{16}$$

Putting, for a fixed natural number k, $\xi_k = 1$ and $\xi_n = 0$ for $n \neq k$, we conclude from (15) that

(17) $$\sum_{i=1}^{\infty} \alpha_i a_{ik} = 0 \text{ for } k = 1,2,\ldots$$

Then putting $\xi_k = 1$ for $k=1,2,\ldots$, one deduces from (15), (16) and (17) that $\alpha = 0$, whence, by (14), that $\sum_{i=1}^{\infty}\alpha_i\eta_i^\circ = 1$, which, in view of (17), contradicts the hypothesis.

LEMMA 2. *If the method* A *is permanent and the conditions* (10) *imply condition* (11), *there exists, for every convergent sequence* (η_i°) *and every number* $\varepsilon > 0$, *a convergent sequence* x *satisfying condition* (12).

The proof is immediate by virtue of the preceding lemma 1.

LEMMA 3. *If* $x_0 = (\overset{\circ}{\xi}_k)$ *is a bounded sequence which is summable by a permanent method* A, *there exists, for every* $\varepsilon > 0$, *a convergent sequence* x *such that*

(18) $$|A_i(x) - A_i(x_0)| < \varepsilon \text{ for every } i=1,2,\ldots$$

Proof. Put

(19) $$\eta_i^\circ = A_i(x_0) \text{ for } i = 1,2,\ldots,$$

and denote by (α_i) any sequence satisfying the conditions (10), p. 55. By (19), we have

(20) $$\sum_{i=1}^{\infty} \alpha_i\eta_i^\circ = \sum_{i=1}^{\infty} \alpha_i A_i(x_0)$$

and, as A is a permanent method, there exists by theorem 10, p. 55-6, a number M satisfying 1°, whence

$$\sum_{i=1}^{\infty}\sum_{k=1}^{\infty} |\alpha_i|\cdot|a_{ik}|\cdot|\overset{\circ}{\xi}_k| \leq M\Big(\sum_{i=1}^{\infty}|\alpha_i|\Big)\sup_{k\geq 1}|\overset{\circ}{\xi}_k|$$

and by (20)

$$\sum_{i=1}^{\infty}\alpha_i\eta_i^\circ = \Big(\sum_{k=1}^{\infty}\overset{\circ}{\xi}_k\Big)\Big(\sum_{i=1}^{\infty}\alpha_i a_{ik}\Big),$$

so that, by (10), $\sum_{i=1}^{\infty}\alpha_i\eta_i^\circ = 0$. This established, the assertion of the lemma follows immediately from lemma 1.

LEMMA 4. *Let* x_0 *be a sequence which is summable by a method* A *which is both permanent and reversible. If, for any* $\varepsilon > 0$, *there exists a convergent sequence* x *satisfying condition* (18), *the sequence* x_0 *is summable to the same number by every permanent method* B *which is not weaker than* A.

Proof. The reversibility of A implies (see Chapter III, §6, theorem 10, p. 29, and the remark, p. 31) the existence of a sequence (α_i) and an array (β_{ik}) having the following properties:

(21) $$\sum_{i=1}^{\infty} |\beta_{ik}| < \infty \text{ for } k = 1,2,\ldots,$$

(22) *if, for a convergent sequence* $y = (\eta_i)$, *one puts:*

$$\xi_k = f_k(y) = \sum_{i=1}^{\infty} \beta_{ik}\eta_i + \alpha_k \lim_{i\to\infty} \eta_i \text{ where } k = 1,2,\ldots,$$

one has

$$\sum_{k=1}^{\infty} a_{ik}\xi_k = \eta_i \text{ for } i = 1,2,\ldots$$

Thus defined, the f_k are bounded linear functionals in the space c. Since, for every convergent sequence y the corresponding sequence $x = (\xi_k)$ is, by hypothesis, summable by the permanent method B, each of the series $\sum_{k=1}^{\infty} b_{ik}\xi_k$ is convergent and so is the sequence of their sums $(B_i(x))$.

For every $y \in c$ put:

$$F_i(y) = \sum_{k=1}^{\infty} b_{ik} f_k(y) \text{ for } i = 1,2,\ldots \text{ and } F(y) = \lim_{i\to\infty} F_i(y).$$

Thus defined, the F_i are bounded linear functionals; by theorem 4 (Chapter I, §3), p. 15, F is also a bounded linear functional.

This established, following the hypothesis, let x_0 be the given sequence and x a convergent sequence satisfying (18). Putting $y = (A_i(x_0))$ and $y = (A_i(x))$, we have $y_0 \in c$, $y \in c$ and $\|y - y_0\| \leq \varepsilon$, so that

$$(23) \qquad |B(x) - B(x_0)| = |F(y) - F(y_0)| \leq \|F\|\varepsilon$$

and, as $A(x) = B(x)$, it follows that

$$|A(x_0) - B(x_0)| \leq |A(x_0) - A(x)| + |B(x) - B(x_0)|,$$

whence, by (18) and (23),

$$|A(x_0) - B(x_0)| \leq \|F\|\varepsilon + \varepsilon,$$

which implies that $A(x_0) = B(x_0)$, q.e.d.

Lemmas 3 and 4 yield the following

THEOREM 11. *If the permanent method* B *is not weaker than the permanent reversible method* A, *every bounded sequence which is summable by* A *is also summable by* B *to the same number.*

Furthermore, lemmas 2 and 4 yield

THEOREM 12. *If* A *is a perfect method and* B *a permanent method, not weaker than* A, *every sequence summable by* A *is also summable by* B *to the same number.*

CHAPTER VI

Compact operators

§1. Compact operators.

A bounded linear operator (between Banach spaces) U is said to be *compact* if it maps (norm) bounded sets to relatively compact sets.

EXAMPLE. Let, for $i=1,2,\dots,n$, X_i be a bounded linear functional and x_i an element of a space. Then U, given by $U(x) = \sum_{i=1}^{n} X_i(x)x_i$ is compact.

THEOREM 1. *The codomain of any compact operator is separable.*

Proof. The set $G_n = \{U(x) : \|x\| \leq n\}$ is relatively compact and therefore separable, and hence so is the set $\bigcup_{n=1}^{\infty} G_n$ which is equal to the codomain of U.

THEOREM 2. *If* (U_n) *is a sequence of compact operators and the bounded linear operator* U *is such that* $\lim_{n\to\infty} \|U_n - U\| = 0$, *then* U *is also compact.*

Proof. Let (x_i) be a bounded sequence and $(\bar{x}_i)$ a subsequence of (x_i) obtained by a diagonal argument such that $\lim_{i\to\infty} U_n(\bar{x}_i)$ exists for each natural number n. Consequently, for $n=1,2,\dots$, we have:

$$\|U(\bar{x}_p) - U(\bar{x}_q)\| \leq \|U(\bar{x}_p) - U_n(\bar{x}_p)\| + \|U_n(\bar{x}_p) - U_n(\bar{x}_q)\| + \|U_n(\bar{x}_q) - U(\bar{x}_q)\|$$

so that

$$\|U(\bar{x}_p) - U(\bar{x}_q)\| \leq \|U - U_n\|(\|\bar{x}_p\| + \|\bar{x}_q\|) + \|U_n(\bar{x}_p) - U_n(\bar{x}_q)\|,$$

whence, clearly $\overline{\lim}_{p,q\to\infty} \|U(\bar{x}_p) - U(\bar{x}_q)\| = 0$. Hence $(U(\bar{x}_i))$ is a Cauchy sequence from which the compactness of the operator U follows.

§2. Examples of compact operators in some special spaces.

If $K(s,t)$ is a continuous function for $0 \leq s \leq 1$ and $0 \leq t \leq 1$, the function (of the variable s) given by

$$(1) \qquad u(s) = \int_0^1 K(s,t)x(t)\,dt$$

is continuous, for any integrable function $x(t)$. Regarded as an operator defined in one of the spaces

$$(2) \qquad M,\ C,\ L^1 \text{ and } L^r \text{ where } r > 1,$$

with codomain also in any of these spaces, the *operator given by* (1) *is compact.*

The proof rests on the following theorem of Arzelà:

For a bounded sequence of continuous functions $(u_n(s))$ *to have a uniformly convergent subsequence, it is sufficient that, for every number* $\varepsilon > 0$, *there exists a number* $\eta > 0$ *such that the inequality* $|s_1 - s_2| < \eta$ *implies the inequality* $|u_n(s_1) - u_n(s_2)| \leq \varepsilon$ *for every* $n=1,2,\ldots$

Indeed, assume that $\|x_n\| \leq 1$ and that, for $0 \leq s \leq 1$ and $n=1,2,\ldots$, $u_n(s) = \int_0^1 K(s,t)x_n(t)dt$. The continuity of $K(s,t)$ implies the existence, for every $\varepsilon > 0$, of an $\eta > 0$ such that the inequality $|s_1 - s_2| < \eta$ implies that $|K(s_1,t) - K(s_2,t)| \leq \varepsilon$ for $0 \leq t \leq 1$. Consequently

$$|u_n(s_1) - u_n(s_2)| \leq \left|\int_0^1 [K(s_1,t) - K(s_2,t)]x_n(t)dt\right| \leq \varepsilon\int_0^1 |x_n(t)|dt,$$

which yields, by virtue of the inequality $\int_0^1 |x_n(t)|dt \leq \|x_n\|$, easily seen to hold in the spaces (2), $|u_n(s_1) - u_n(s_2)| \leq \varepsilon$, so that, by Arzelà's theorem, one can extract a uniformly convergent subsequence from the sequence $(u_n(s))$. Now, since any sequence of functions in any of the spaces (2) which is uniformly convergent is also convergent in the norm of the space, we have shown that the operator (1) is compact in such spaces.

In particular, we have the following theorems:

The space C. *For the operator* (1) *to be compact in* C, *it is sufficient that, for every* s_0, *one has:*

$$\lim_{s\to s_0} \int_0^1 |K(s_0,t) - K(s,t)|dt = 0. \tag{3}$$

In fact, for every $\varepsilon > 0$ one easily deduces from (3) the existence of an $\eta > 0$ such that $|s_1 - s_2| \leq \eta$ implies $\int_0^1 |K(s_1,t) - K(s_2,t)|dt \leq \varepsilon$, which implies, as before, the compactness of the operator (1) in C.

Condition (3) will be satisfied, for example, when the function $K(s,t)$ is bounded and such that $\lim_{s\to s_0} K(s,t) = K(s_0,t)$ for every s_0 and almost every t.

Note finally that the operator given by

$$y(s) = \int_0^s K(s,t)x(t)dt$$

is also compact in C, if condition (3) is satisfied.

The spaces L^p. *Let* $K(s,t)$ *be a measurable function for* $0 \leq s \leq 1$ *and* $0 \leq t \leq 1$ *and let* r *denote the smaller of the numbers* p *and* $q/(q-1)$ *where* $p > 1$ *and* $q > 1$. *If*

$$\int_0^1\int_0^1 |K(s,t)|^{\frac{r}{r-1}} dsdt < \infty, \tag{4}$$

then (1) *defines a compact operator from* L^p *into* L^q.

Indeed let $(K_n(s,t))$ be a sequence of continuous functions such that

$$\lim_{n\to\infty} \int_0^1\int_0^1 |K_n(s,t) - K(s,t)|^{\frac{r}{r-1}} dsdt = 0. \tag{5}$$

Since, for $n=1,2,\ldots$, the operator U_n given by $y = U_n(x) = \int_0^1 K_n(s,t)x(t)dt$ is compact from L^p into L^q, it follows that

$$\|U_n(x) - U(x)\|^q \leq \int_0^1 \left|\int_0^1 (K_n - K)x(t)dt\right|^q ds$$

$$\leq \left\{\int_0^1 ds\left[\int_0^1 |K_n - K|^{r/(r-1)} dt\right]^{q(r-1)/r}\right\}\cdot\left(\int_0^1 |x(t)|^r dt\right)^{q/r}.$$

Now, as $r \leq p$, we have

$$\left(\int_0^1 |x(t)|^r dt\right)^{\frac{1}{r}} \leq \left(\int_0^1 |x(t)|^p dt\right)^{\frac{1}{p}}$$

and since $r \leq \frac{q}{q-1}$, i.e. $\frac{q(r-1)}{r} \leq 1$, we have

$$\|U_n(x) - U(x)\| \leq \left\{\int_0^1\int_0^1 |K_n - K|^{\frac{r}{r-1}} dtds\right\}^{\frac{r-1}{r}} \|x\|,$$

and so

$$\|U_n - U\| \leq \left[\int_0^1\int_0^1 |K_n - K|^{\frac{r}{r-1}} dtds\right]^{\frac{r-1}{r}},$$

whence, by (5), $\lim_{n\to\infty} \|U_n - U\| = 0$, which implies, by theorem 2, p. 59, the compactness of the operator U, i.e. of the operator given by (1), where $u(s) = U(x) \in L^q$.

Remark. In particular, for $p = q = 2$ the condition

$$\int_0^1\int_0^1 K^2(s,t)\,dsdt < +\infty$$

therefore implies the compactness of the operator (1) as an operator from L^2 into itself.

§3. Adjoint (conjugate) operators

As usual, let E and E_1 be two Banach spaces and U a bounded linear operator from E to E_1.

We shall denote by X and Y bounded linear functionals on E and E_1 respectively.

Consider the expression $Y[U(x)]$ where Y is an arbitrary bounded linear functional on E_1. This expression can clearly be regarded as a functional defined on E. Indeed, let us put

(6) $$X(x) = Y[U(x)].$$

Thus defined, the functional X is additive and continuous, because we have $|X(x)| = |Y[U(x)]| \leq \|Y\|.\|U\|.\|x\|$, whence

(7) $$\|X\| \leq \|Y\|.\|U\|.$$

Now, the relation (6) serves to define a new operator U^*, given by

$$X = U^*(Y),$$

from the space E_1^* of bounded linear functionals on E to the space E^* of those defined on E.

The operator U^* is called the *adjoint* or *conjugate* of U. By (7), it is additive and continuous.

THEOREM 3. *For the adjoint U^* of the bounded linear operator U, we have $\|U^*\| = \|U\|$.*

Proof. Firstly, for every $x \in E$, we have $|Y[U(x)]| \leq \|Y\|.\|U\|.\|x\|$, whence $\|U^*(Y)\| = \|Y(U)\| \leq \|Y\|.\|U\|$ and consequently

(8) $$\|U^*\| \leq \|U\|.$$

Moreover, given an arbitrary $x_0 \in E$, there exists, by theorem 3, Chapter IV, §2, p. 34, a bounded linear functional Y_0 on E_1 such that $\|Y_0\| = 1$ and $|Y_0[U(x_0)]| = \|U(x_0)\|$, and so $\|U(x_0)\| = |Y_0[U(x_0)]| \leq \|U^*\|.\|Y_0\|.\|x_0\| = \|U^*\|.\|x_0\|$, whence $\|U(x_0)\| \leq \|U^*\|.\|x_0\|$ and consequently

(9) $$\|U\| \leq \|U^*\|.$$

The result now follows from the inequalities (8) and (9).

THEOREM 4. *If U is a compact operator, so is its adjoint U^*: in other words, if (Y_n) is a bounded sequence in E_1^*, i.e. $\|Y_n\| < M$, say, then there exists a subsequence (Y_{n_i}) and an element X of E^* such that*

$$(10) \qquad \lim_{i\to\infty} \|U^*(Y_{n_i}) - X\| = 0.$$

Proof. By theorem 1, p. 59, the codomain $G \subseteq E_1$ of the operator U contains a countable dense set, and so, by theorem 3, Chapter V, §1, p. 49, we can extract, from the sequence $(Y_n) \subseteq E_1^*$, where $\|Y_n\| < M < \infty$ for every n, a subsequence (Y_{n_i}) which is convergent for every $y \in G$, i.e. $\big(Y_{n_i}(y)\big)$ is convergent for every $y \in G$. Now put $\lim_{i\to\infty} Y_{n_i}[U(x)] = \lim_{i\to\infty} U^*(Y_{n_i})[x] = \lim_{i\to\infty} X_{n_i}(x) = X(x)$ and let x_i be an element of E such that

$$(11) \qquad \|x_i\| = 1 \text{ and } |X(x_i) - X_{n_i}(x_i)| \geqq \tfrac{1}{2}\|X - X_{n_i}\|.$$

Now, if the theorem were false, i.e. there existed $\eta > 0$ such that $\|X - X_{\eta_i}\| > \eta$ for each $i=1,2....$, we would have by (11), putting $Y_i' = Y_{n_i}$:

$$(12) \qquad \left|Y_i'[U(x_i)] - \lim_{j\to\infty} Y_j'[U(x_i)]\right| \geqq \frac{\eta}{2}$$

and, as $\|x_i\| = 1$ for every i, there would exist a sequence of indices (k_i) such that $\lim_{i\to\infty} U(x_{k_i}) = y_0$. We would therefore be able to find, for any $\varepsilon > 0$, a natural number N such that, for every $i > N$, we had $\|y_0 - U(x_{k_i})\| < \varepsilon$ and

$$\left|Y_{k_i}'(y_0) - \lim_{j\to\infty} Y_{k_j}'(y_0)\right| < \varepsilon,$$

whence

$$\begin{aligned}\left|Y_{k_i}'[U(x_{k_i})] - \lim_{j\to\infty} Y_{k_i}'[U(x_{k_i})]\right| &\leqq \left|Y_{k_i}'[U(x_{k_i}) - y_0]\right| + \left|Y_{k_i}'(y\ ) - \lim_{j\to\infty} Y_{k_j}'(y_0)\right| \\ &\quad + \left|\lim_{j\to\infty} Y_{k_j}'[U(x_{k_i}) - y_0]\right| \\ &\leqq M.\varepsilon + \varepsilon + M.\varepsilon,\end{aligned}$$

which is impossible by (12), as the number ε can be arbitrarily small.

§4. Applications. Examples of adjoint operators in some special spaces.

The space C. If $K(s,t)$ is continuous for $0 \leqq s \leqq 1$ and $0 \leqq t \leqq 1$, the expression

$$U(x) = \int_0^1 K(s,t)x(t)\,dt$$

defines a bounded linear operator U on C.

Let Y be any bounded linear functional on C, which is therefore (cf. Chapter IV, §4, p. 36) of the form $Y(y) = \int_0^1 y(t)\,dY(t)$ where $Y(t)$ is a function of bounded variation. The functional $U^*(Y) = X$ given by $X(x) = Y[U(x)]$ is also a bounded linear functional on C, and is therefore also of the form

$$(13) \qquad X(x) = \int_0^1 x(t)\,dX(t),$$

where $X(t)$ is also a function of bounded variation, and we can assume that $X(0) = 0$. Consequently, if we put

$$(14) \qquad y(s) = U(x)(s) = \int_0^1 K(s,t)x(t)\,dt,$$

we have, for every function $x(t) \in C$:

(15) $$\int_0^1 x(s)\,dX(s) = \int_0^1 y(s)\,dY(s).$$

Consider the function defined by

$$x_{v,n}(s) = \begin{cases} 1 \text{ for } 0 \leq s \leq v \\ 0 \text{ for } v + \frac{1}{n} \leq s \leq 1, \end{cases}$$

and which is linear for $v \leq s \leq v + \frac{1}{n}$. Putting $x_{v,n}$ for x in (14) and (15), we obtain by changing the order of integration (Fubini),

$$\int_0^1 x_{v,n}(s)\,ds = \int_0^1\left[\int_0^1 K(s,t)x_{v,n}(t)\,dt\right]dY(s) = \int_0^1 x_{v,n}(t)\left[\int_0^1 K(s,t)\,dY(s)\right]dt,$$

whence, letting $n \to \infty$, we have for $s=0,1$ and all points of continuity of the function $X(s)$, thus for all except at most countably many points s,

(16) $$X(s) = \int_0^s\left[\int_0^1 K(s,t)\,dY(s)\right]dt;$$

now, as the value of the Stieltjes integral (13) remains the same when the value of the function $X(t)$ is altered at countably many points (except 0 and 1), we can assume that the function $X(s)$ is defined by the formula (16) throughout $[0,1]$, so that it is continuous for $0 \leq s \leq 1$.

The expression (16) can thus be regarded as a representation of the adjoint operator U^*, by which it is to be understood that if $Y(s)$ is a function of bounded variation which represents the bounded linear functional given by $\int_0^1 y(s)\,dY(s)$, the corresponding function of bounded variation $X(s)$ represents the bounded linear functional given by $\int_0^1 x(t)\,dX(t)$.

For the bounded linear operator U given by

$$U(x)(s) = x(s) - \int_0^1 K(s,t)x(t)\,dt,$$

with the same function $K(s,t)$, we have

$$U^*(Y)(t) = Y(t) - \int_0^t dt\int_0^1 K(s,t)\,dY(s) - X(t).$$

The L^p spaces. If the function $K(s,t)$ is measurable for $0 \leq s \leq 1$ and $0 \leq t \leq 1$ and if

(17) $$\int_0^1\int_0^1 |K(s,t)x(t)Y(s)|\,ds\,dt < \infty$$

for every pair of functions $x(t) \in L^p$ and $Y(s) \in L^{\frac{q}{q-1}}$ where $p > 1$ and $q > 1$, the operator U given by

$$U(x)(s) = y(s) = \int_0^1 K(s,t)x(t)\,dt$$

is a bounded linear operator from L^p into L^q.

The general bounded linear functional Y on the space L^q is of the form

$$Y(y) = \int_0^1 Y(s)y(s)\,ds,$$

where $Y(s)$ is a function belonging to $L^{\frac{q}{q-1}}$ and we have

$$Y(y) = \int_0^1 Y(s)ds\int_0^1 K(s,t)x(t)dt = \int_0^1 x(t)dt\int_0^1 K(s,t)Y(s)ds.$$

Putting

$$X(t) = \int_0^1 K(s,t)Y(s)ds, \tag{18}$$

we have

$$\int_0^1 Y(s)y(s)ds = \int_0^1 X(t)x(t)dt.$$

The expression (18) can be regarded as a representation of the adjoint operator U^*.

In the special case where $p = q > 1$, the adjoint of the bounded linear operator U given by

$$U(x)(s) = x(s) - \int_0^1 K(s,t)x(t)dt$$

is of the form

$$X(t) = U^*(Y)(t) = Y(t) - \int_0^1 K(s,t)Y(s)ds.$$

The space L^1. The above considerations apply equally to the space L^1. If (17) holds for $x \in L^1$ and $Y \in M$, the expression

$$y(s) = U(x)(s) = \int_0^1 K(s,t)x(t)dt$$

defines a bounded linear operator from L^1 into itself.

The adjoint operator is of the form

$$X(t) = U^*(Y)(t) = \int_0^1 K(s,t)Y(s)ds$$

where $Y(s) \in M$ represents the bounded linear functional $\int_0^1 Y(s)y(s)ds$ for $y(s) \in L^1$, whilst $X(t) \in M$ represents the bounded linear functional $\int_0^1 X(t)x(t)dt$ for $x(t) \in L^1$. For corresponding pairs X,Y and x,y, i.e. $X = U^*Y$ and $y = Ux$, we have

$$\int_0^1 X(t)x(t)dt = \int_0^1 Y(s)y(s)ds.$$

CHAPTER VII

Biorthogonal sequences

§1. Definition and general properties.

A sequence (x_i) of elements and (f_i) of bounded linear functionals is said to be *biorthogonal* when

(1) $$f_i(x_j) = \begin{cases} 1 \text{ for } i = j, \\ 0 \text{ for } i \neq j. \end{cases}$$

Given an arbitrary $x \in E$, the series

(2) $$\sum_{i=1}^{\infty} x_i \cdot f_i(x)$$

is called the *development of* x *with respect to the biorthogonal sequence* (x_i), (f_i).

In the case where the sequence (f_i) constitutes a total set of functionals (cf. Chapter III, §3, p. 27) and the series (2) is convergent for some x, then x is the sum of this series; in fact, we then have for every $j=1,2,\ldots,$

$$f_j[x - \sum_{i=1}^{\infty} x_i \cdot f_i(x)] = f_j(x) - f_j(x) = 0.$$

THEOREM 1. *If the series* (2) *is convergent for every* $x \in E$, *the series*

$$\sum_{i=1}^{\infty} f_i(x) \,.\, F(x_i)$$

is also convergent for every $x \in E$ *for any bounded linear functional* F.

Proof. Putting

$$S_n = \sum_{i=1}^{n} f_i \cdot F(x_i),$$

we have $S_n(x) = \sum_{i=1}^{n} f_i(x) \,.\, F(x_i) = F[\sum_{i=1}^{n} x_i \cdot f_i(x)]$, from which the convergence of the sequence $(S_n(x))$ for every x plainly follows.

THEOREM 2. *If the partial sums* (3) *of the series*

(4) $$\sum_{i=1}^{\infty} f_i \cdot F(x_i)$$

form a norm-bounded set for any bounded linear functional F, *the series* (2) *is convergent for every* $x \in E$ *which is the limit of some sequence of linear combinations of terms of the sequence* (x_i), (*i.e. for all* x *in the closed linear span of the* x_i).

Proof. Putting

(5) $$s_n(x) = \sum_{i=1}^{n} x_i \cdot f_i(x),$$

we have $F[s_n(x)] = \sum_{i=1}^{n} F(x_i) \cdot f_i(x) = S_n(x)$ (see (3)) and, as by hypothesis $\|S_n\| \leq M$ where M is a number independent of n, we have, for every $x \in E$, by theorem 6 (Chapter V, §1), p. 50, $\overline{\lim_{n\to\infty}} \|s_n(x)\| < \infty$.

There therefore exists by theorem 5 (Chapter V, §1) p.50, a number N, independent of n and of x, such that $\|s_n(x)\| \leq N.\|x\|$.

Now, since we have $\lim_{n\to\infty} s_n(x_i) = x_i$ for every $i=1,2,\ldots$, a simple argument establishes the existence of $\lim_{n\to\infty} s_n(x)$ for every element $x \in E$ which satisfies the condition of the theorem.

THEOREM 3. *If the partial sums* (5) *of the series* (2) *form a norm-bounded set for any $x \in E$, the series* (4) *is convergent for every functional F which is the limit of a sequence of linear combinations of terms of the sequence (f_i), (i.e. for all F in the closed linear span of the f_i).*

The proof is similar to that of theorem 2 above.

THEOREM 4. *Under the same hypotheses, if, further, (x_i) is a fundamental sequence, the series* (2) *is convergent for every $x \in E$.*

Proof. We have, by (5), $\overline{\lim_{n\to\infty}} \|s_n(x)\| < \infty$ for every $x \in E$, and, further, $\lim_{n\to\infty} s_n(x_i) = x_i$ for every $i=1,2,\ldots$ The convergence of the series (2) for every $x \in E$ now follows by virtue of theorems 5 and 3 (Chapter V, §1) pp. 49-50.

§2. Biorthogonal sequences in some special spaces.

Let us now consider how biorthogonal sequences behave in the spaces which are of particular interest to us.

Put

$$\text{(6)} \qquad \int_0^1 x_i(t)\,y_j(t)\,dt = \begin{cases} 1 & \text{for } i = j, \\ 0 & \text{for } i \neq j. \end{cases}$$

Assume further that $(x_i(t))$ is a sequence of functions in L^p where $p > 1$ and that $(y_i(t))$ is a sequence in $L^{p/(p-1)}$; assume finally that these sequences are complete (or, equivalently, closed).

THEOREM 5. *Under these hypotheses, if the series*

$$\sum_{i=1}^{\infty} x_i(t) \int_0^1 y_i(t)\,x(t)\,dt$$

is p^{th} power mean convergent for any function $x(t) \in L^p$, the series

$$\text{(7)} \qquad \sum_{i=1}^{\infty} y_i(t) \int_0^1 x_i(t)\,y(t)\,dt$$

is $\frac{p}{p-1}^{th}$ power mean convergent for every function $y(t) \in L^{p/(p-1)}$.

Proof. Let the bounded linear functional f_i on L^p be defined by

$$f_i(x) = \int_0^1 y_i(t)\,x(t)\,dt \quad \text{for } x(t) \in L^p.$$

The series $\sum_{i=1}^{\infty} x_i \cdot f_i(x)$ is then p^{th}power mean (i.e. norm-) convergent for every $x \in L^p$ by hypothesis. By theorem 3 above, the series $\sum_{i=1}^{\infty} f_i \cdot F(x_i) = \sum_{i=1}^{\infty} y_i(t) \int_0^1 x_i(t)\,y(t)\,dt$ where $y(t) \in L^{p/(p-1)}$ is consequently $\frac{p}{p-1}^{\text{th}}$ power mean, or norm-, convergent for every bounded linear functional F defined on the space L^p; hence the same is true of the series (7) for every function $y(t) \in L^{p/(p-1)}$, q.e.d.

In particular, when $x_i(t) = y_i(t) \in L^r$ where r is the larger of the two numbers p and $\frac{p}{p-1}$, the theorem just proved implies the following corollary.

If the series

$$(8)\qquad \sum_{i=1}^{\infty} x_i(t)\int_0^1 x_i(t)x(t)\,dt$$

is p^{th} power mean convergent for every $x \in L^p$, then it is $\frac{p}{p-1}^{th}$ power mean convergent for every $x \in L^{p/(p-1)}$.

For example, one could take the x_i, for $i=1,2,\ldots$, to be (essentially) bounded functions.

Now consider the case where, in the hypothesis (6), $(x_i(t))$ is a sequence of integrable functions and $(y_i(t))$ is a sequence of functions which are bounded for $0 \leq t \leq 1$. Suppose, finally, that the sequence $(x_i(t))$ is complete in L^1.

THEOREM 6. *Under these hypotheses, if the series $\sum_{i=1}^{\infty} x_i(t)\int_0^1 y_i(t)x(t)\,dt$ is mean convergent for every $x(t) \in L^1$, then the series $\sum_{i=1}^{\infty} y_i(t)\int_0^1 x_i(t)y(t)\,dt$ is bounded almost everywhere for every $y(t) \subset M$ and conversely.*

The proof is similar to that of the preceding theorem 5: one regards the x_i as elements of the domain L^1 and the y_i as representing bounded linear functionals, and then employs theorems 3 and 4, p. 67.

In particular, when $x_i(t) = y_i(t)$, we have the corollaries:

1° *If the series (8), where $x_i(t) = y_i(t) \in M$, is mean convergent for every $x(t) \in L^1$, then it is bounded for every $x(t) \in M$ and conversely.*

2° *If the series (8), where $x_i(t) = y_i(t) \in C$ and (x_i) is a complete sequence in C, is uniformly convergent for every $x(t) \in C$, then it is mean convergent for every $x(t) \in L^1$ and conversely.*

The proof is obtained, in one direction, by regarding the x_i as elements of the domain C and the $y_i = x_i$ as representing bounded linear functionals, and, in the opposite direction, by regarding the x_i as elements of the domain L^1 and the $y_i = x_i$ as representing bounded linear functionals on L^1.

§3. Bases in Banach spaces.

A sequence (x_i) of elements of E is called a (*Schauder*) *base* when, for every $x \in E$, there exists a unique sequence of numbers (η_i) such that

$$x = \sum_{i=1}^{\infty} \eta_i x_i .$$

Given a base (x_i), let E_1 be the set of sequences $y = (\eta_i)$ for which the series $\sum_{i=1}^{\infty}\eta_i x_i$ is convergent. Putting

$$\|y\| = \sup_{1 \leq n} \left\| \sum_{i=1}^{n} \eta_i x_i \right\|,$$

it is easily shown that E_1, thus normed, is a Banach space.

Now put

$$x = U(y) = \sum_{i=1}^{\infty} \eta_i x_i \text{ for every sequence } y = (\eta_i) \in E_1 .$$

Thus defined, U is a bounded linear operator, because $\|U(y)\| \leq \|y\|$, and, as it maps E_1 bijectively onto E, its inverse U^{-1} is also a bounded linear operator.

Finally, the map f_i defined by

$$f_i(x) = \eta_i \text{ where } x = \sum_{i=1}^{\infty} \eta_i x_i$$

is a bounded linear functional, because $\|\eta_i x_i\| \leqq 2\|y\|$ and

$$|f_i(x)| = |\eta_i| \leqq \frac{2}{\|x_i\|}\|y\| \leqq \frac{2}{\|x_i\|}\|U^{-1}\|.\|x\|.$$

We thus have

$$x = \sum_{i=1}^{\infty} x_i f_i(x) \text{ for every } x \in E$$

and, this development being unique, the relations (1) (p. 65) follow, so that the *sequence* $(x_i),(f_i)$ *is biorthogonal.*

Observe that for every bounded linear functional F defined on E the series $\sum_{i=1}^{\infty} f_i(x)F(x_i)$ converges to $F(x)$, because, for every $x \in E$, one has:

$$\sum_{i=1}^{\infty} f_i(x)F(x_i) = \lim_{n\to\infty} F\left[\sum_{i=1}^{n} x_i f_i(x)\right] = F(x).$$

It is not known if every separable Banach space has a base.

The question is only settled for certain special spaces. Thus, for example, in L^p, where $p \geqq 1$, a base is given by the orthogonal Haar system. In C a base has been constructed by Schauder. In l^p for $p \geqq 1$ a base is furnished by the sequence (x_i) where

$$x_i = \left(\xi_n^i\right) \text{ and } \xi_n^i = \begin{cases} 1 \text{ for } i = n, \\ 0 \text{ for } i \neq n; \end{cases}$$

we then have $f_i(x) = \xi_i$ for $x = (\xi_i)$. Finally in c a base is given by the same sequence together with the element $x_0 = (\xi_n^\circ)$ where $\xi_n^\circ = 1$ for $n=1,2,...$ We then have $f_0(x) = \lim_{i\to\infty} \xi_i$ for $x = (\xi_i) \in c$.

§4. Some applications to the theory of orthogonal expansions.

THEOREM 7. *Suppose that the sequences* $(x_i),(f_i)$ *and* $(y_i),(\phi_i)$ *are biorthogonal and that the equations* $f_i(x) = \phi_i(y)$, *for* $i=1,2,...$, *have exactly one solution* $y = U(x)$ *for every* x. *Then the convergence of the series* $\sum_{i=1}^{\infty} h_i x_i$ *implies that of the series* $\sum_{i=1}^{\infty} h_i y_i$ *for every sequence of numbers* (h_i).

Proof. It is easily seen that if $\lim_{n\to\infty} x_n = x_0$ and $\lim_{n\to\infty} y_n = y_0$, where $y_n = U(x_n)$, then $y_0 = U(x_0)$. It follows from theorem 7 (Chapter III, §3) p. 26, that the operator U is bounded and linear. Therefore putting $\|U\| = M$, we have $\|U(x)\| \leqq M\|x\|$ and since, by definition, $U(x_i) = y_i$ for $i=1,2,...$, it follows that

$$U\left(\sum_{i=1}^{n} h_i x_i\right) = \sum_{i=1}^{n} h_i y_i$$

for any real numbers h_i, from which the result immediately follows.

COROLLARY. *Suppose that* $\{x_i(t)\}$ *and* $\{y_i(t)\}$ *are orthonormal sequences of continuous functions, and that for every continuous function* $x(t)$ *there exists a unique continuous function* $y(t)$ *such that*

$$\int_0^1 x_i(t)x(t)dt = \int_0^1 y_i(t)y(t)dt.$$

Then if the series $\sum_{i=1}^{\infty} h_i x_i(t)$ *is uniformly convergent so is the series* $\sum_{i=1}^{\infty} h_i y_i(t)$.

Analogous corollaries hold for other function spaces.

THEOREM 8. *Let* $(x_i),(f_i)$ *be a biorthogonal sequence, where* (f_i) *is a total sequence, and let* (h_i) *be a sequence of numbers such that whenever* (α_i) *is the sequence of coefficients of an element* x *(i.e.* $\alpha_i = f_i(x)$ *for* $i=1,2,\ldots$*),* $(h_i\alpha_i)$ *is the coefficient sequence of an element* y.

If under these conditions, (β_i) *is the coefficient sequence of a bounded linear functional* F *(i.e.* $\beta_i = F(x_i)$ *for* $i=1,2,\ldots$*), the sequence* $(h_i\beta_i)$ *is also the coefficient sequence of some bounded linear functional* Φ.

Proof. By hypothesis, the system of equations $h_if_i(x) = f_i(y)$ for $i=1,2,\ldots$ has, for every x, exactly one solution, which we denote by $y = U(x)$.

The equalities $\lim_{n\to\infty} x_n = x_0$ and $\lim_{n\to\infty} y_n = y_0$ where $y_n = U(x_n)$ clearly imply that $y_0 = U(x_0)$. Consequently, by theorem 7 (Chapter III, §3) p. 26, U is a bounded linear operator. In particular, it is easily checked that

(9) $$U(x_i) = h_ix_i \text{ for every } i = 1,2,\ldots$$

Now, given a bounded linear functional F such that $\beta_i = F(x_i)$ for $i=1,2,\ldots$, we have, in view of (9), $F[U(x_i)] = h_iF(x_i) = h_i\beta_i$, i.e. the numbers $h_i\beta_i$ are the coefficients of the functional $\Phi = U^*(F)$, q.e.d.

Note that if $x = \lim_{i\to\infty} x_i$, $U(x)$ is, by (9), the limit of a linear combination of terms of the sequence (x_i).

As an easy application of this remark we obtain the following

THEOREM 9. *Let* $(x_i(t))$ *be an orthonormal sequence of continuous functions which is also a closed sequence in the space* C.

If the sequence of scalars (h_i) *transforms every sequence* (α_i) *of coefficients of a bounded function into the coefficient sequence* $(h_i\alpha_i)$ *of another bounded function, then it transforms every coefficient sequence* (β_i) *of any continuous function into the coefficient sequence* $(h_i\beta_i)$ *of another continuous function.*

The converse theorem is also true.

Lastly, we have

THEOREM 10. *Let* $(x_i(t))$ *be a complete orthonormal sequence of bounded functions in* $L^{p/(p-1)}$*, where* $p > 1$.

If the sequence of scalars (h_i) *transforms the coefficient sequence* (α_i) *of an arbitrary function* $x(t) \in L^p$ *into the coefficient sequence* $(h_i\alpha_i)$ *of another function* $y(t) \in L^p$*, then it also transforms every coefficient sequence* (β_i) *of an arbitrary function* $\bar{x}(t) \in L^{p/(p-1)}$ *into the coefficient sequence* $(h_i\beta_i)$ *of a function* $\bar{y}(t) \in L^{p/(p-1)}$*. If* $p = \infty$*, then* $L^p = M$.

CHAPTER VIII

Linear functionals

§1. Preliminaries.

Given a closed linear subspace $G \subseteq E$ and the dual space E^* of bounded linear functionals on E, we have already seen (cf. Chapter IV, §3, p. 35, lemma) that, for any *element* $x_0 \in E \setminus G$, there exists a *functional* $f \in E^*$ such that

$$f(x_0) = 1 \text{ and } f(x) = 0 \text{ for every } x \in G.$$

This naturally leads one to inquire whether, conversely, the analogous relationship between subspaces $\Gamma \subseteq E^*$ and elements of E holds. To be precise, we require to know if, given a closed linear subspace $\Gamma \subseteq E^*$, there exists, for an arbitrary *functional* $f_0 \in E^* \setminus \Gamma$, an *element* $x \in E$ such that

(1) $$f_0(x) = 1 \text{ and } f(x) = 0 \text{ for every } f \in \Gamma.$$

The answer is , however, negative in general.

Indeed, take $E = c$, the space of convergent sequences of real numbers, so that E^* is the dual space of c, and let Γ be the space of all elements f of E^* of the form:

(2) $$f(x) = \sum_{i=1}^{\infty} \alpha_i \xi_i \text{ where } x = (\xi_i) \in c \text{ and } \sum_{i=1}^{\infty} |\alpha_i| < \infty.$$

Thus defined, Γ is a closed linear subspace of E^*. In fact, suppose

(3) $$\lim_{n\to\infty} \|f_n - f\| = 0,$$

where

(4) $$f_n \in \Gamma \text{ and } f_n(x) = \sum_{i=1}^{\infty} \alpha_i^{(n)} \xi_i \text{ for } n = 1,2,\ldots$$

We need to show that $f \in \Gamma$. Now (3) implies that $\lim_{p,q\to\infty} \|f_p - f_q\| = 0$, whence, since by definition

$$f_p(x) - f_q(x) = \sum_{i=1}^{\infty} \left(\alpha_i^{(p)} - \alpha_i^{(q)}\right)\xi_i,$$

we conclude from (3), in view of the theorem of Chapter IV, §4, p. 41, that

$$\lim_{p,q\to\infty} \sum_{i=1}^{\infty} |\alpha_i^{(p)} - \alpha_i^{(q)}| = 0$$

and, consequently, that there exists a sequence (α_i) such that

$$\lim_{n\to\infty} \sum_{i=1}^{\infty} |\alpha_i^{(n)} - \alpha_i| = 0 \text{ and } \sum_{i=1}^{\infty} |\alpha_i| < \infty.$$

We therefore have, for every $x = (\xi_i) \in c$, the equality:

$$\lim_{n\to\infty} \sum_{i=1}^{\infty} \alpha_i^{(n)} \xi_i = \sum_{i=1}^{\infty} \alpha_i \xi_i,$$

whence, by (4),

$$\lim_{n\to\infty} f_n(x) = \sum_{i=1}^{\infty} \alpha_i \xi_i$$

and since by (3)

$$\lim_{n\to\infty} |f_n(x) - f(x)| \leq \lim_{n\to\infty} \|f_n - f\| . \|x\| = 0,$$

it follows that $f(x) = \sum_{i=1}^{\infty} \alpha_i \xi_i$ for every $x \in c$. The functional f is thus of the form (2), so that, finally, $f \in \Gamma$.

This established, let

(5) $$f_0(x) = \lim_{i\to\infty} \xi_i \text{ for } x = (\xi_i) \in c.$$

The functional f_0 thus defined plainly does not belong to Γ. However there exists no $x_0 = \left(\xi_i^{(0)}\right) \in c$ satisfying the conditions (1), because as (1) and (5) imply that $\lim_{i\to\infty} \xi_i^{(0)} = 1$, it is impossible to have $\sum_{i=1}^{\infty} \alpha_i \xi_i^{(0)} = 0$, as required by (1), for every sequence of numbers (α_i) satisfying the conditions (2).

§2. Regularly closed linear spaces of linear functionals.

A linear subspace Γ of the dual E^* of a Banach space E is said to be *regularly closed* when there exists, for every element of $E^* \setminus \Gamma$, an element $x_0 \in E$ which satisfies the conditions (1).

The preceding example shows that a closed linear subspace of E^* is not always regularly closed. However, the converse is true: every linear subspace Γ of E^* which is regularly closed is also norm-closed.

In fact, put

(6) $$f_n \in \Gamma \text{ for } n=1,2,\ldots$$

and

(7) $$\lim_{n\to\infty} \|f_n - f\| = 0.$$

If $f_0 \notin \Gamma$, a regularly closed linear subspace, there would exist an $x_0 \in E$ satisfying (1), in particular, by (6) we would then have $f_n(x_0) = 0$ for $n=1,2,\ldots$, whence by (7), $f_0(x_0) = \lim_{n\to\infty} f_n(x_0) = 0$, contradicting (1). Hence we must have $f_0 \in \Gamma$, i.e. the space Γ is (norm)-closed.

It is easy to give examples of regularly closed spaces. Indeed, let E be a Banach space and $G \subseteq E$ any linear subspace of E. The set Γ of bounded linear functionals f defined in E and such that

$$f(x) = 0 \text{ for every } x \in G$$

is easily seen to be regularly closed.

Remark. If the set Γ in question is not only a regularly closed linear subspace but is also total, it then coincides with the whole of E^*.

In fact, the definition of total subset (see Chapter III, §3, p. 36) implies that the only element of E at which all the elements $f \in \Gamma$ vanish is the (zero) element Θ.

We shall now discuss the properties of regularly closed spaces of bounded linear functionals.

§3. Transfinitely closed sets of bounded linear functionals.

Given any ordinal number θ which is a *limit ordinal*, i.e. has no immediate predecessor, and a bounded sequence of real numbers (C_ξ) *of type* θ, i.e. where $1 \leq \xi < \theta$, the *transfinite limit superior of*

(c_ξ), denoted by $\overline{\lim\limits_{\xi\to\theta}}\, c_\xi$, is defined to be the infimum of the set of real numbers $\{t$: there exists an ordinal η $(<\theta)$, depending on t, such that $c_\xi \leq t$ for all $\xi \geq \eta\}$. The *transfinite limit inferior of* (c_ξ) is then defined by the formula

$$\underline{\lim}_{\xi\to\theta}\, c_\xi = -\overline{\lim}_{\xi\to\theta}\,(-c_\xi).$$

LEMMA 1. *If, for a sequence* (f_ξ) *of type* θ *of bounded linear functionals on* E,

$$\|f_\xi\| \leq M \text{ for } 1 \leq \xi < \theta,$$

there exists a bounded linear functional f *satisfying the conditions:*

(8) $\|f\| \leq M$ *and* $\underline{\lim}_{\xi\to\theta} f_\xi(x) \leq f(x) \leq \overline{\lim}_{\xi\to\theta} f_\xi(x)$ *for every* $x \in E$.

The proof follows from theorem 1 (Chapter II, §2, p. 14), on putting $p(x) = \overline{\lim}_{\xi\to\theta} f_\xi(x)$. The functional p also satisfies $p(x) \leq M\|x\|$ for $x \in E$.

Having established this lemma, we shall call a bounded linear functional f which satisfies the conditions (8) a *transfinite limit of the sequence* (f_ξ).

In particular, when $\lim\limits_{n\to\infty}\|f_n - f\| = 0$ the functional f is clearly a transfinite limit of the sequence (f_n) because then $\underline{\lim}_{n\to\infty} f_n(x) = f(x) = \overline{\lim}_{n\to\infty} f_n(x)$ for every $x \in E$.

A linear space Γ of bounded linear functionals is said to be *transfinitely closed*, when every norm-bounded transfinite sequence (f_ξ) of elements of Γ has a transfinite limit f in Γ.

Every transfinitely closed space Γ is also norm-closed.

In fact, (6) and (7) then yield $\lim\limits_{n\to\infty} f_n(x) = f_0(x)$ for every $x \in E$ and as every functional f which here satisfies the condition $\underline{\lim}_{n\to\infty} f_n(x) \leq f(x) \leq \overline{\lim}_{n\to\infty} f_n(x)$ coincides with f_0, it follows that f_0 is the only transfinite limit of the sequence (f_n), so that $f_0 \in \Gamma$ which is consequently (norm)-closed.

LEMMA 2. *Given a transfinitely closed linear space* Γ *of bounded linear functionals on* E *and a bounded linear functional* f_0 *on* E *not belonging to* Γ, *there exists, for each number* M *satisfying the condition*

$$0 < M < \|f - f_0\| \text{ for every } f \in \Gamma,$$

an element $x_0 \in E$ *such that*

$$f_0(x_0) = 1,\ f(x_0) = 0 \text{ for every } f \in \Gamma \text{ and } \|x_0\| < \frac{1}{M}.$$

Proof. For any increasing sequence of numbers (M_i) with $M_1 = M$, which tends to infinity, let m denote the largest cardinal number satisfying the following condition: given any set $G \subseteq E$ of power $< m$, there exists an element $f \in \Gamma$ such that

(10) $\|f\| \leq M_2$ and $|f(x) - f_0(x)| \leq M_1\|x\|$ for every $x \in G$.

Note straightaway that the number m thus defined does not exceed the cardinality of E because, if there existed an $f \in \Gamma$ such that $|f(x) - f_0(x)| \leq M_1 \cdot \|x\|$ for every $x \in E$, one would have $\|f - f_0\| \leq M_1 = M$, contrary to the hypothesis (9).

This said, we shall now show that m is a *finite* number.

Indeed, suppose that m is not finite and consider any set $G \subseteq E$ of power m. Arrange the elements of G as a transfinite sequence (x_ξ) where $1 \leq \xi < \theta$ and θ is the least ordinal of power m; θ is plainly a limit ordinal. Consequently, for every ordinal number $\eta < \theta$ the power of the set of terms of the sequence (x_ξ) where $1 \leq \xi < \eta$ is $< m$.

By definition of m, there therefore exists, for every $\eta < \theta$, a linear functional $f_\eta \in \Gamma$ such that

(11) $\|f_\eta\| \leq M_2$ and $|f_\eta(x_\xi) - f_0(x_\xi)| \leq M_1 . \|x_\xi\|$ for every $\xi < \eta$

and, as Γ is transfinitely closed by hypothesis, there exists an element $f \in \Gamma$ which is a transfinite limit of the sequence (f_η), $1 \leq \eta < \theta$, and therefore, by (11), satisfies the conditions $\|f\| \leq M_2$ and $|f(x_\xi) - f_0(x_\xi)| \leq M_1 \|x_\xi\|$ for $1 \leq \xi < \theta$, i.e. the conditions (10). Thus assuming m was infinite, there would exist for every set $G \subseteq E$ of power m an $f \in \Gamma$ satisfying (10), contradicting the definition of m.

Now as m is finite, there exists a finite set $G_1 \subseteq E$ such that no functional f satisfying the conditions

$$\|f\| \leq M_2 \text{ and } |f(x) - f_0(x)| \leq M_1 \|x\| \text{ for every } x \in G_1$$

belongs to Γ.

By induction, one easily establishes the existence of a sequence (G_i) of finite subsets of E such that no functional f which, for some k, satisfies the conditions

$$\|f\| \leq M_k \text{ and } |f(x) - f_0(x)| \leq M_i \|x\| \text{ for } x \in G_i \text{ and } i < k$$

belongs to Γ. Consequently, if for some f we have:

(12) $$|f(x) - f_0(x)| \leq M_i \|x\| \text{ for } x \in G_i \text{ and } i = 1,2,\ldots,$$

the functional f does not belong to Γ.

We can assume that the elements of the sets G_i, where $i=1,2,\ldots$, have norms equal to M_1/M_i: one merely has to multiply these elements by appropriate scalars. If the elements of these sets are then arranged as a sequence (x_n), first writing down the elements of G_1, followed by those of G_2 and so on, we obtain

(13) $$\lim_{n\to\infty} x_n = \Theta \text{ and } \|x_n\| \leq 1 \text{ for } n=1,2,\ldots,$$

and if

(14) $$|f(x_n) - f_0(x_n)| \leq M_1 \text{ for } n=1,2,\ldots,$$

the functional f does not belong to Γ.

Let G_0 denote the set of all sequences of the form $(f(x_n))$ for $f \in \Gamma$. Clearly $G_0 \subseteq c$ and $(f_0(x_n)) \in c$. It follows from (14) that the distance of $(f_0(x_n))$ from the linear space G_0 is $\geq M_1$. In view of the general form of bounded linear functionals on the space c (cf. Chapter 4, §4, p. 40), there thus exists by the lemma of Chapter IV, §3, p. 35, putting $G = G_0$, a sequence of numbers (C_n) and a number C such that

(15) $$C \lim_{n\to\infty} f_0(x_n) + \sum_{n=1}^{\infty} C_n f_0(x_n) = 1,$$

(16) $$C \lim_{n\to\infty} f(x_n) + \sum_{n=1}^{\infty} C_n f(x_n) = 0 \text{ for every } f \in \Gamma$$

and

(17) $$|C| + \sum_{n=1}^{\infty} |C_n| \leq \frac{1}{M_1}.$$

Therefore, putting $x_0 = \sum_{n=1}^{\infty} C_n x_n$ it finally follows from (15)-(17), using (13), that $f_0(x_0) = 1$, $f(x_0) = 1$ for every $f \in \Gamma$ and $\|x_0\| \leq \sum_{n=1}^{\infty} |C_n| . \|x_n\| \leq \frac{1}{M_1} = \frac{1}{M}$, q.e.d.

Lemma 2 just proved implies the following

LEMMA 3. *The notions of regularly closed and transfinitely closed are the same for linear spaces of bounded linear functionals.*

Proof. If a linear space Γ of bounded linear functionals is transfinitely closed, it is (norm)-closed, which by lemma 2 immediately implies that Γ is regularly closed.

Conversely, let Γ be a regularly closed linear space of bounded linear functionals, (f_ξ) an arbitrary norm-bounded sequence of type θ of elements of Γ and f_0 any functional which is a transfinite limit of the sequence (f_ξ). We thus have

(18) $$\underline{\lim}_{\xi\to\theta} f_\xi(x) \leq f_0(x) \leq \overline{\lim}_{\xi\to\theta} f_\xi(x) \text{ for every } x \in E.$$

If, then, f_0 did not belong to Γ, there would exist, by definition of Γ, an element $x \in E$ satisfying the conditions (1), p. 71, whence, in particular, $f_\xi(x) = 0$, contradicting (18). Hence $f_0 \in \Gamma$, from which it follows that Γ is transfinitely closed.

Lemmas 2 and 3 yield the following

THEOREM 1. *Given a regularly closed linear space Γ of bounded linear functionals on E and a bounded linear functional f_0 not belonging to Γ, there exists for each number M satisfying the condition*

$$0 < M < \|f - f_0\| \text{ for every } f \in \Gamma$$

an element $x_0 \in E$ such that

$$f_0(x_0) = 1,\ f(x_0) = 0 \text{ for every } f \in \Gamma \text{ and } \|x_0\| < \frac{1}{M}$$

§4. Weak convergence of bounded linear functionals.

A sequence (f_n) of bounded linear functionals is said to converge *weakly* to the functional f when

$$\lim_{n\to\infty} f_n(x) = f(x) \text{ for every } f \in E.$$

The functional f is called a *weak limit* of the sequence (f_n).

It follows that the functional f is additive and B-measurable; according to theorem 4 (Chapter I, §3, p. 15) it is therefore a bounded linear functional. Moreover, by theorem 5 (Chapter V, §1, p. 50), the sequence of norms $(\|f_n\|)$ is bounded. Finally, we have

(19) $$\|f\| \leq \underline{\lim}_{n\to\infty} \|f_n\|,$$

because the weak convergence of the sequence (f_n) to f implies that $\lim_{n\to\infty} |f_n(x)| = |f(x)|$ for every x, and as $|f_n(x)| \leq \|f_n\|\cdot\|x\|$ for $n=1,2,\ldots$, we have $|f(x)| \leq \|x\|\cdot\underline{\lim}_{n\to\infty} \|f_n\|$, from which (19) follows.

One easily deduces the following

THEOREM 2. *For a sequence (f_n) of bounded linear functionals to converge weakly to the functional f, it is necessary and sufficient that both*

(20) *the sequence $(\|f_n\|)$ is bounded*

and

(21) $\lim_{n\to\infty} f_n(x) = f(x)$ *for every x belonging to a dense (or fundamental) set, hold.*

THEOREM 3. *If the space E is separable, every norm-bounded sequence of bounded linear functionals (f_n) has a weakly convergent subsequence.*

Proof. It suffices to extract from the sequence (f_n) a subsequence which converges at every point of a countable dense subset of E and this is easily accomplished by a diagonal procedure.

§5. Weakly closed sets of bounded linear functionals in separable Banach spaces.

Given two sets of bounded linear functionals Δ and Γ where $\Delta \subseteq \Gamma$, the set Δ is said to be *weakly dense* in Γ when, for every $f \in \Gamma$, there exists a sequence $(f_n) \subseteq \Delta$ which converges weakly to f.

The set Γ of bounded linear functionals is said to be *weakly closed* when Γ contains every functional f which is the weak limit of a sequence of functionals of Γ.

THEOREM 4. *If the space E is separable, every set Γ of bounded linear functionals on E contains a countable subset Δ which is weakly dense in Γ.*

Proof. We need only consider the case where the set Γ is norm-bounded, because every set of bounded linear functionals is the union of at most countably many sets having this property.

Let (x_n) be a dense sequence in E and, for $n=1,2,\ldots,$ let Z_n be the subset of n-dimensional space defined by

$$(22) \qquad Z_n = \{\left(f(x_1), f(x_2), \ldots, f(x_n)\right) : f \in \Gamma\}.$$

There clearly exists for every n a countable subset Δ_n of Γ such that the subset of points of Z_n for which $f \in \Delta_n$ is dense in Z_n. The set $\Delta = \bigcup_{n=1}^{\infty} \Delta_n$ is obviously countable and for every $f \in \Gamma$ there exists a sequence $(f_n) \subseteq \Delta_n \subseteq \Delta$ such that $|f_n(x_i) - f(x_i)| < \frac{1}{n}$ for every $i=1,2,\ldots,n$, and which therefore converges weakly to f, since (f_n) is norm-bounded by hypothesis as $(f_n) \subseteq \Delta_n \subseteq \Gamma$.

THEOREM 5. *For separable Banach spaces E, the notions of regularly closed and weakly closed are the same for linear spaces of bounded linear functionals on E.*

Proof. Firstly, let (f_n) be a sequence of bounded linear functionals belonging to Γ which converges weakly to a functional f_0. We thus have

$$(23) \qquad \lim_{n\to\infty} f_n(x) = f_0(x) \text{ for every } x \in E.$$

If f_0 did not belong to the set Γ, assumed to be regularly closed, there would exist, by the definition of this notion, an element $x_0 \in E$ satisfying the conditions

$$(24) \qquad f_0(x_0) = 1 \text{ and } f(x_0) = 0 \text{ for every } f \in \Gamma.$$

As $(f_n) \subseteq \Gamma$, we would consequently have $f_n(x_0) = 0$ for every $n=1,2,\ldots,$ whence, by (23), $f_0(x_0) = 0$, contradicting (24). It follows from this that $f_0 \in \Gamma$; the set Γ is therefore weakly closed.

Remark. Note that the separability of E does not play any part in this part of the proof.

Conversely, in view of lemma 3, p. 75, it is enough to show that the set Γ, assumed to be weakly closed, is transfinitely closed.

Let (f_ξ) be a sequence of type θ such that

$$(25) \qquad f_\xi \in \Gamma \text{ and } \|f_\xi\| \leq M \text{ for } 1 \leq \xi < \theta$$

and let (x_i) be a dense sequence in E. By hypothesis, there exists for every natural number n an ordinal ξ_n such that

$$(26) \qquad \underline{\lim}_{\xi\to\theta} f_\xi(x_i) - \frac{1}{n} \leq f_{\xi_n}(x_i) \leq \overline{\lim}_{\xi\to\theta} f_\xi(x_i) + \frac{1}{n} \text{ for } 1 \leq i \leq n.$$

Now, as the space E is separable, by theorem 3, p. 75, one can extract a weakly convergent subsequence from the sequence (f_{ξ_n}).

Letting f denote the weak limit of the sequence (f_{ξ_n}), it follows from (25) that $f \in \Gamma$ and, further, (26) implies that f is a transfinite limit of the sequence (f_ξ).

Theorems 1, p. 75 and 5, p. 76, which have just been established, immediately imply the following

THEOREM 6. *If the Banach space E is separable, then, given a weakly closed linear space Γ of bounded linear functionals on E and any bounded linear functionals f_0 not belonging to Γ, there exists, for each number M satisfying the condition*

$$0 < M < \|f - f_0\| \text{ for every } f \in \Gamma$$

an element $x_0 \in E$ such that

$$f_0(x_0) = 1, \quad f(x_0) = 0 \text{ for every } f \in \Gamma \text{ and } \|x_0\| < \frac{1}{M}.$$

In view of lemma 3, p. 75, theorem 5 implies that *the notions of regularly, transfinitely and weakly closed*, as applied to linear spaces of bounded linear functionals, *are all equivalent for separable Banach spaces E.*

A consequence of this, recalling the remark on p. 76, is the following

THEOREM 7. *If the Banach space E is separable and Γ is a set of bounded linear functionals on E which is not only a weakly closed linear space but is also total, then Γ contains every bounded linear functional on E (i.e. $\Gamma = E^*$).*

§6. Conditions for the weak convergence of bounded linear functionals on the spaces C, L^p, c and l^p.

We go on to study weak convergence of bounded linear functionals in several particular *separable* Banach spaces, namely, the spaces C, L^p for $p \geq 1$, c and l^p for $p \geq 1$.

For a countable dense set we take: in C and L^p, the polynomials with rational coefficients, in c and l^p respectively, the sequences of rational numbers which are eventually constant and eventually zero, respectively.

The spaces L^p for $p > 1$. Since every bounded linear functional f on L^p is of the form (cf. Chapter IV, §4, p.39)

$$(27) \qquad \int_0^1 x(t)\alpha(t)\,dt \text{ where } \alpha(t) \in L^{p/(p-1)}$$

the sequence of functionals

$$(28) \qquad (f_n) \text{ where } f_n(x) = \int_0^1 x(t)\alpha_n(t)\,dt \text{ and } \alpha_n(t) \in L^{p/(p-1)}$$

converges weakly to the functional given by (27), when

$$(29) \qquad \lim_{n\to\infty} \int_0^1 x(t)\alpha_n(t)\,dt = \int_0^1 x(t)\alpha(t)\,dt$$

for every function $x(t) \in L^p$.

Now, one can easily show that *for the sequence* (28) *to converge weakly to the functional* (27) *it is necessary and sufficient that the conditions*:

$$(30) \qquad \text{the sequence } \left(\int_0^1 |\alpha_n(t)|^{p/(p-1)}dt\right) \text{ is bounded}$$

and

$$(31) \qquad \lim_{n\to\infty} \int_0^u \alpha_n(t)\,dt = \int_0^u \alpha(t)\,dt \text{ for } 0 \leq u \leq 1,$$

both hold.

The proof follows from theorem 2, p. 75, given that

$$\|f_n\| = \left[\int_0^1 |\alpha_n(t)|^{p/(p-1)}dt\right]^{(p-1)/p},$$

further, that the functions $x_u(t)$ defined for $0 \leq u \leq 1$ by the conditions

$$x_u(t) = \begin{cases} 1 \text{ for } 0 \leq t \leq u, \\ 0 \text{ for } u < t \leq 1, \end{cases}$$

constitute a total subset of L^p and finally that

$$\int_0^1 x_u(t)\alpha_n(t)dt = \int_0^u \alpha_n(t)dt$$

for every $n=1,2,\ldots$

The space L^1. Since every bounded linear functional f on L^1 is of the form

(32) $$f(x) = \int_0^1 x(t)\alpha(t)dt \text{ where } \alpha(t) \in M$$

(cf. Chapter IV, §4, p. 40) the sequence of functionals

(33) $$(f_n) \text{ where } f_n(x) = \int_0^1 x(t)\alpha_n(t)dt \text{ and } \alpha_n(t) \in M$$

converges weakly to the functional given by (32) when

(34) $$\lim_{n\to\infty} \int_0^1 x(t)\alpha_n(t)dt = \int_0^1 x(t)\alpha(t)dt \text{ for every } x(t) \in L^1.$$

One shows as in the previous case that the sequence of linear functionals (33) converges weakly to the functional (32) if and only if we have both

(35) *the sequence $(\alpha_n(t))$ is a norm-bounded subset of M*

and

(36) $$\lim_{n\to\infty} \int_0^u \alpha_n(t)dt = \int_0^u \alpha(t)dt \text{ for every } u \text{ with } 0 \leq u \leq 1.$$

Remark. The conditions (30) and (31) are clearly necessary and sufficient for property (29) to hold. The same is true of the conditions (35) and (36) for property (34).

The spaces l^p for $p \geq 1$. Since every bounded linear functional f on l^p is of the form

(37) $$f(x) = \sum_{i=1}^{\infty} \alpha_i\xi_i \text{ where } x = (\xi_i) \in l^p \text{ and } (\alpha_i) \in \begin{cases} l^{p/(p-1)} & \text{for } p > 1, \\ M & \text{for } p = 1, \end{cases}$$

(cf. Chapter IV, §4, p. 42), the sequence of functionals

(38) $$(f_n) \text{ where } f_n(x) = \sum_{i=1}^{\infty} \alpha_{in}\xi_i \text{ and } (\alpha_{in}) \in \begin{cases} l^{p/(p-1)} & \text{for } p > 1, \\ M & \text{for } p = 1, \end{cases}$$

converges weakly to the functional (37) when

(39) $$\lim_{n\to\infty} \sum_{i=1}^{\infty} \alpha_{in}\xi_i = \sum_{i=1}^{\infty} \alpha_i\xi_i \text{ for every } x = (\xi_i) \in l^p.$$

For the sequence of linear functionals (38) *to converge weakly to the functional* (37), *it is necessary and sufficient that*

(40) *the sequence* $\begin{cases} \left(\sum_{i=1}^{\infty} |\alpha_{in}|^{p/(p-1)} \right) & \text{for } p > 1, \\ \left(\sup_{i \geq 1} |\alpha_{in}| \right) & \text{for } p = 1 \end{cases}$ *be bounded*

and

(41) $$\lim_{n\to\infty} \alpha_{in} = \alpha_i \text{ for } i = 1,2,\ldots$$

The proof follows from theorem 2, §4, p. 75, in view of the fact that the elements

$$x_j = (\xi_{ij}) \text{ where } \xi_{ij} = \begin{cases} 1 & \text{for } i = j, \\ 0 & \text{for } i \neq j, \end{cases}$$

form a total sequence in l^p and, further, that $f_n(x_j) = \alpha_{jn}$ for all natural numbers j and n; finally, one uses the expressions for the norms of bounded linear functionals on l^p, given on p. 42.

Remark. The conditions (40) and (41) are also necessary and sufficient for (39) to hold.

The space c. In view of the general form of bounded linear functionals f on c, given by

(42) $$f(x) = A \lim_{i\to\infty} \xi_i + \sum_{i=1}^{\infty} \alpha_i \xi_i \text{ where } x = (\xi_i) \in c \text{ and } (\alpha_i) \subset l^1$$

(cf. Chapter IV, §4, p. 40), the sequence of bounded linear functionals

(43) $$(f_n) \text{ where } f_n(x) = A_n \lim_{i\to\infty} \xi_i + \sum_{i=1}^{\infty} \alpha_{in}\xi_i \text{ and } (\alpha_{in}) \in l^1 \text{ for } n = 1,2,\ldots$$

converges weakly to the functional (42) when

(44) $$\lim_{n\to\infty} \left[A_n \lim_{i\to\infty} \xi_i + \sum_{i=1}^{\infty} \alpha_{in}\xi_i \right] = A \lim_{i\to\infty} \xi_i + \sum_{i=1}^{\infty} \alpha_i \xi_i \text{ for every } x = (\xi_i) \in c.$$

It is easily shown that *for the sequence* (43) *to converge weakly to the functional* (42), *it is necessary and sufficient that*

(45) *the sequence* $\left(\sum_{i=1}^{\infty} |\alpha_{in}| + |A_n| \right)$ *be bounded*

and

(46) $$\lim_{n\to\infty} \left(A_n + \sum_{i=1}^{\infty} \alpha_{in} \right) = A + \sum_{i=1}^{\infty} \alpha_i \text{ and } \lim_{n\to\infty} \alpha_{in} = \alpha_i \text{ for } i=1,2,\ldots$$

§7. Weak compactness of bounded sets in certain spaces.

The preceding results allow one to deduce, by virtue of theorem 3, p. 75, the following theorems.

For L^p where $p > 1$. Every sequence of functions $(\alpha_n(t))$, where $\alpha_n(t) \in L^p$, satisfying the condition

$$\int_0^1 |\alpha_n(t)|^p dt < M,$$

where M is a number independent of n, contains a subsequence $(\alpha_{n_i}(t))$ such that, for some function $\alpha_0(t) \in L^p$:

$$\lim_{i\to\infty} \int_0^1 \alpha_{n_i}(t)x(t)dt = \int_0^1 \alpha_0(t)x(t)dt \text{ for every } x(t) \in L^{p/(p-1)}.$$

Indeed, the expressions $\int_0^1 \alpha_n(t)x(t)dt$ for $n=1,2,\ldots$, can be regarded as bounded linear functionals $L^{p/(p-1)}$. Since they form a norm-bounded set and the space $L^{p/(p-1)}$ is separable, one can, by theorem 3, p. 75, extract a weakly convergent subsequence from the sequence $(\alpha_n(t))$; one has then merely to apply the result established for the L^p spaces in §6, p. 77.

For M. Every norm-bounded sequence of functions $(\alpha_n(t)) \subseteq M$ *contains a subsequence* $(\alpha_{n_i}(t))$ *such that, for some function* $\alpha_0(t) \in M$:

$$\lim_{i\to\infty} \int_0^1 \alpha_{n_i}(t)x(t)dt = \int_0^1 \alpha_0(t)x(t)dt \text{ for every } x(t) \in L^1.$$

The proof is analogous to the preceding one.

For l^p, $p > 1$, *and* m, analogous theorems are available.

§8. Weakly continuous linear functionals defined on the space of bounded linear functionals.

Let F be a linear functional defined on the space E^* of all bounded linear functionals on the Banach space E. Then F is said to be *weakly continuous* when $\lim_{n\to\infty} F(f_n) = F(f)$ whenever the sequence (f_n) and the element f in E^* are such that (f_n) converges weakly to f.

THEOREM 8. *If the Banach space E is separable and the bounded linear functional F on E^* is weakly continuous, there exists an element $x_0 \in E$ such that*

(47) $$F(f) = f(x_0) \text{ for every } f \in E^*.$$

Proof. If Γ denotes the set $\{f \in E^*: F(f) = 0\}$, it easily follows from the weak continuity of F that Γ is a weakly closed linear space. We can evidently assume that $\Gamma \neq E^*$ (for otherwise we need merely take $x_0 = \Theta$). Let, therefore, f_0 be a bounded linear functional satisfying the equation

(48) $$F(f_0) = 1.$$

It follows from theorem 6, p. 77, that there exists an $x_0 \in E$ such that

(49) $$f_0(x_0) = 1 \text{ and } f(x_0) = 0 \text{ for every } f \in \Gamma.$$

Now, the identity

(50) $$f = f_0.F(f) + \phi \text{ for every } f \in E^*, \text{ where } \phi = f - f_0.F(f),$$

yields, by (48), $F(\phi) = 0$, whence $\phi \in \Gamma$ and consequently, by (49), $\phi(x_0) = 0$, which by (50) implies the property (47), q.e.d.

Remark. If the space E is not separable, theorem 8 still holds provided that F is a bounded linear functional and the set Γ is regularly closed, which enables one to appeal to theorem 1, p. 75, instead of theorem 6, p. 77, in the argument.

CHAPTER IX

Weakly convergent sequences

§1. Definition. Conditions for the weak convergence of sequences of elements.

A sequence (x_n) of elements of E is said to be *weakly convergent* to the element $x \in E$ when

$$\lim_{n\to\infty} f(x_n) = f(x) \text{ for every } f \in E^*,$$

i.e. for every bounded linear functional f defined on the given space E.

THEOREM 1. *For the sequence* (x_n) *to converge weakly to* x, *it is necessary and sufficient that*

(1) *the sequence* $(\|x_n\|)$ *be bounded*

and

(2) $\lim_{n\to\infty} \phi(x_n) = \phi(x)$ *for every* $\phi \in \Delta$ *where* Δ *is a dense subset of* E^*.

Proof. The necessity of (1) follows from theorem 6 (Chapter V, §1) p. 50, while that of (2) is obvious.

To prove the sufficiency, consider an arbitrary functional $f \in E^*$. By (2) there then exists for every number $\varepsilon > 0$ a functional $\phi \in \Delta$ such that $\|\phi - f\| < \frac{\varepsilon}{2M}$ where $M = \sup(\{\|x_n\| : n=1,2,\ldots\} \cup \{\|x\|\})$, which is finite by (1). Consequently,

$$|f(x - x_n)| \leq |\phi(x - x_n)| + \frac{\varepsilon}{2M} \cdot \|x - x_n\| \leq |\phi(x - x_n)| + \varepsilon;$$

as $\lim_{n\to\infty} \phi(x_n) = \phi(x)$ and ε is arbitrary, we conclude from this that $\lim_{n\to\infty} f(x_n) = f(x)$, i.e. that the sequence (x_n) converges weakly to x.

Remark. It is enough to require of the set Δ that the set of all linear combinations of functionals of Δ be a dense subset of E^*.

THEOREM 2. *If the sequence* (x_n) *converges weakly to* x, *there exists a sequence* (g_n) *of linear combinations of terms of the sequence* (x_n) *such that* $\lim_{n\to\infty} g_n = x$.

The proof follows from theorem 6 (Chapter IV, §3) p. 36, and the definition of weak convergence.

§2. Weak convergence of sequences in the spaces C, L^p, c and l^p.

We here discuss the weak convergence of sequences in some of the more important special spaces.

The space C. In view of the general form of bounded linear functionals in C (see p. 37), *for a sequence of continuous functions* $(x_n(t))$ *to converge weakly to the continuous function* $x(t)$, *it is necessary and sufficient that*

$$\lim_{n\to\infty} \int_0^1 x_n(t)\,dg(t) = \int_0^1 x(t)\,dg(t) \tag{3}$$

for every function $g(t)$ *of bounded variation.*

It follows from this that *for a sequence of functions* $(x_n(t)) \subseteq C$ *to converge weakly to the function* $x(t) \in C$, *it is necessary and sufficient that both*

(4) *the set* $\{x_n(t): n=1,2,...\}$ *of functions is (norm)-bounded,*

and

(5) $\lim_{n\to\infty} x_n(t) = x(t)$ *for every* $t \in [0,1]$.

In fact, the necessity of (4) follows from theorem 1, p. 81, and that of (5) is a consequence of the fact that, if t_0 denotes an arbitrary point in $[0,1]$, the linear functional $f(x) = x(t_0)$ is bounded, whence $\lim_{n\to\infty} f(x_n) = f(x)$ and consequently $\lim_{n\to\infty} x_n(t_0) = x(t_0)$.

The sufficiency follows from the fact that conditions (4) and (5) imply the equality (3) for every function $g(t)$ of bounded variation (cf. Introduction §5, p. 4).

This established, one obtains, from theorem 2, the following theorem:

If a sequence of continuous functions $(x_n(t))$, $0 \leq t \leq 1$, *is (norm)-bounded and converges everywhere to a continuous function* $x(t)$, *then there exists a sequence of polynomials (linear combinations) of terms of the sequence* $(x_n(t))$ *which converges uniformly to* $x(t)$.

This is a remarkable property of the space of continuous functions which already fails, for example, for functions of the first Baire class.

The L^p *spaces for* $p > 1$. *The sequence* $(x_n(t)) \subseteq L^p$ *converges weakly to* $x(t) \in L^p$ *when*

$$\lim_{n\to\infty} \int_0^1 x_n(t)\alpha(t)\,dt = \int_0^1 x(t)\alpha(t)\,dt$$

for every function $\alpha(t) \in L^{p/(p-1)}$.

By the remark on p. 78, we have the following theorem:

For the sequence of functions $(x_n(t)) \subseteq L^p$ *to converge weakly to the function* $x(t) \in L^p$, *it is necessary and sufficient that both*

(6) *the sequence* $\left(\int_0^1 |x_n(t)|^p dt\right)$ *be bounded*

and

(7) $$\lim_{n\to\infty} \int_0^u x_n(t)\,dt = \int_0^u x(t)\,dt \text{ for } 0 \leq u \leq 1.$$

The space L^1. *The sequence* $(x_n(t)) \subseteq L^1$ *converges weakly to* $x_0 \in L^1$ *when*

(8) $$\lim_{n\to\infty} \int_0^1 x_n(t)\alpha(t)\,dt = \int_0^1 x_0(t)\alpha(t)\,dt$$

for every (essentially) bounded function $\alpha(t)$.

As a result of this, we have the following theorem:

For the sequence of functions $(x_n(t)) \subseteq L^1$ *to converge weakly to the function* $x_0(t) \in L^1$, *it is necessary and sufficient that the following conditions all be satisfied*

(9) *the sequence* $\left(\int_0^1 |x_n(t)|\,dt\right)$ *is bounded*

(10) *for every number* $\varepsilon > 0$ *there exists a number* $\eta > 0$ *such that*

$$\left|\int_H x_n(t)\,dt\right| \leq \varepsilon \text{ for } n = 1,2,...,$$

for every set $H \subseteq [0,1]$ *of measure* $< \eta$,

(11) $$\lim_{n\to\infty} \int_0^u x_n(t)\,dt = \int_0^u x_0(t)\,dt \text{ for } 0 \leq u \leq 1.$$

In fact, (8) is equivalent to the statement that $\lim_{n\to\infty} \int_0^1 [x_n(t) - x_0(t)]\alpha(t)\,dt = 0$ for $\alpha(t) \in M$; the theorem in question easily follows from this with the help of the theorem of Lebesgue given on p. 5 (see Introduction, §6).

The space c. For a sequence (x_n), *where* $x_n = \left(\xi_i^n\right) \in c$, *to converge weakly to the element* $x = (\xi_i) \in c$, *it is necessary and sufficient that*

(12) *the sequence* $(\|x_n\|)$ *be bounded,*

and

(13) $$\lim_{n\to\infty} \xi_i^n = \xi_i \text{ and } \lim_{n\to\infty}\left(\lim_{i\to\infty} \xi_i^n\right) = \lim_{i\to\infty} \xi_i.$$

The proof is immediate, given that every bounded linear functional f on c is of the form $f(x) = C \lim_{i\to\infty} \xi_i + \sum_{i=1}^{\infty} C_i \xi_i$ where $x = (\xi_i)$ and $\|f\| = |C| + \sum_{i=1}^{\infty} |C_i|$ (see p. 41) and remembering that if one puts

$$f_i(x) = \begin{cases} \lim_{i\to\infty} \xi_i & \text{for } i = 0, \\ \xi_i & \text{for } i \geq 1, \end{cases}$$

the set of linear combinations of the f_i's constitute a dense subset of the space of all bounded linear functionals on c.

The l^p *spaces,* $p > 1$. *For a sequence* (x_n) *where* $x_n = \left(\xi_i^{(n)}\right) \in l^p$ *to converge weakly to* $x = (\xi_i) \in l^p$, *it is necessary and sufficient that*

(14) *the sequence of numbers* $\left(\sum_{i=1}^{\infty} |\xi_i^{(n)}|^p\right)$ *be bounded*

and

(15) $$\lim_{n\to\infty} \xi_i^{(n)} = \xi_i \text{ for every } i = 1, 2, \ldots$$

The proof follows from the remark on p. 79.

The space l^1. *For a sequence* (x_n) *where* $x_n = \left(\xi_i^{(n)}\right) \in l^1$ *to converge weakly to* $x = (\xi_i) \in l^1$, *it is necessary and sufficient that*

$$\lim_{n\to\infty} \|x_n - x\| = 0, \text{ i.e. } \lim_{n\to\infty} \sum_{i=1}^{\infty} |\xi_i^{(n)} - \xi_i| = 0.$$

Consequently:

In the space l^1, *weak convergence is equivalent to norm convergence.*

Proof. Suppose that (x_n) converges weakly to x. Putting $\eta_i^{(n)} = \xi_i^{(n)} - \xi_i$, we thus have that the sequence (y_n) where $y_n = \left(\eta_i^{(n)}\right)$ converges weakly to Θ as $n \to \infty$. Consequently, for every bounded sequence of numbers (c_i) we have

(16) $$\lim_{n\to\infty} \sum_{i=1}^{\infty} c_i \eta_i^{(n)} = 0.$$

Letting

$$c_i = \begin{cases} 1 & \text{for } j = i, \\ 0 & \text{for } j \neq i, \end{cases}$$

we thus have

(17) $$\lim_{n\to\infty} \eta_j^{(n)} = 0 \text{ for every } j = 1,2,\ldots$$

We need to show that

(18) $$\lim_{n\to\infty} \sum_{i=1}^{\infty} |\eta_i^{(n)}| = \lim_{n\to\infty} \|y_n\| = 0.$$

Suppose, on the contrary, that

(19) $$\overline{\lim_{n\to\infty}} \sum_{i=1}^{\infty} |\eta_i^{(n)}| > \varepsilon > 0.$$

By induction, let us define two increasing sequences of natural numbers (n_k) and (r_k) as follows:

1° n_1 is the least n such that $\sum_{i=1}^{\infty} |\eta_i^{(n)}| > \varepsilon$

2° r_1 is the least r such that $\sum_{i=1}^{r} |\eta_i^{(n_1)}| > \frac{\varepsilon}{2}$ and $\sum_{i=r+1}^{\infty} |\eta_i^{(n_1)}| < \frac{\varepsilon}{5}$,

3° n_k is the least natural number exceeding n_{k-1} and such that $\sum_{i=1}^{\infty} |\eta_i^{(n_k)}| > \varepsilon$ and $\sum_{i=1}^{r_{k-1}} |\eta_i^{(n_k)}| < \frac{\varepsilon}{5}$,

4° r_k is the least natural number exceeding r_{k-1} and such that $\sum_{i=r_{k-1}+1}^{r_k} |\eta_i^{(n_k)}| > \frac{\varepsilon}{2}$ and $\sum_{i=r_k+1}^{\infty} |\eta_i^{(n_k)}| < \frac{\varepsilon}{5}$.

The sequences (n_k) and (r_k) thus defined exist by virtue of (17) and (19).

Now let

(20) $$c_i = \begin{cases} \operatorname{sign} \eta_i^{(n_1)} & \text{for } 1 \leqq i \leqq r_1, \\ \operatorname{sign} \eta_i^{(n_{k+1})} & \text{for } r_k < i \leqq r_{k+1}. \end{cases}$$

We thus have $|c_i| = 1$ for every $i=1,2,\ldots$, whence by (16)

(21) $$\lim_{k\to\infty} \sum_{i=1}^{\infty} c_i \eta_i^{(n_k)} = 0.$$

But, by (20), we have

$$\left|\sum_{i=1}^{\infty} c_i \eta_i^{(n_k)}\right| \geqq \sum_{i=r_{k-1}+1}^{r_k} |\eta_i^{(n_k)}| - \sum_{i=1}^{r_{k-1}} |\eta_i^{(n_k)}| - \sum_{i=r_k+1}^{\infty} |\eta_i^{(n_k)}|,$$

whence, by 3° and 4°,

$$\left|\sum_{i=1}^{\infty} c_i \eta_i^{(n_k)}\right| \geqq \frac{\varepsilon}{2} - \frac{\varepsilon}{5} - \frac{\varepsilon}{5} = \frac{\varepsilon}{10}$$

for every $k=1,2,\ldots$, which contradicts (21). We therefore have (18), q.e.d.

§3. The relationship between weak and strong (norm) convergence in the spaces L^p and l^p for $p > 1$.

As far as the connection between weak and norm convergence in the spaces L^p and l^p, $p > 1$, is concerned, we have the following more general theorems:

If the sequence $(x_n(t))$, where $x_n(t) \in L^p$ and $p > 1$, converges weakly to $x(t) \in L^p$ and if, further,

$$\lim_{n\to\infty} \int_0^1 |x_n(t)|^p dt = \int_0^1 |x(t)|^p dt,$$

then the sequence $(x_n(t))$ converges to $x(t)$ in norm, i.e.

$$\lim_{n\to\infty} \int_0^1 |x_n(t) - x(t)|^p dt = 0.$$

We are going to prove the analogous theorem for the l^p spaces, $p > 1$, the case $p = 1$ having already been discussed in the preceding section §2.

If the sequence (x_n), *where* $x_n = \left(\xi_i^{(n)}\right) \in l^p$ *and* $p \geq 1$, *converges weakly to* $x = (\xi_i) \in l^p$ *and if*

$$\lim_{n\to\infty} \|x_n\| = \|x\|,$$

then

(22) $$\lim_{n\to\infty} \|x_n - x\| = 0.$$

Proof. We have by (15), p. 83,

(23) $$\lim_{n\to\infty} \xi_i^{(n)} = \xi_i$$

and

(24) $$\left(\sum_{i=1}^{\infty} |\xi_i^{(n)} - \xi_i|^p\right)^{\frac{1}{p}} \leq \left(\sum_{i=1}^{N-1} |\xi_i^{(n)} - \xi_i|^p\right)^{\frac{1}{p}} + \left(\sum_{i=N}^{\infty} |\xi_i^{(n)} - \xi_i|^p\right)^{\frac{1}{p}},$$

for any natural number N. Now,

$$\left(\sum_{i=N}^{\infty} |\xi_i^{(n)} - \xi_i|^p\right)^{\frac{1}{p}} \leq \left(\sum_{i=N}^{\infty} |\xi_i^{(n)}|^p\right)^{\frac{1}{p}} + \left(\sum_{i=N}^{\infty} |\xi_i|^p\right)^{\frac{1}{p}},$$

whence by hypothesis, together with (23) and (24)

$$\overline{\lim_{n\to\infty}} \sum_{i=1}^{\infty} |\xi_i^{(n)} - \xi_i|^p \leq \left[2\left(\sum_{i=N}^{\infty} |\xi_i|^p\right)^{\frac{1}{p}}\right]^p = 2^p \sum_{i=N}^{\infty} |\xi_i|^p.$$

Since $\lim_{N\to\infty} \sum_{i=N}^{\infty} |\xi_i|^p = 0$ and N is arbitrary, this implies (22), q.e.d.

§4. Weakly complete spaces.

If (x_n) is a sequence of elements of a Banach space E such that $\lim_{n\to\infty} f(x_n)$ exists for every bounded linear functional f on E, it is not necessarily the case that there exists an element $x_0 \in E$ to which the sequence (x_n) converges weakly, i.e. such that $\lim_{n\to\infty} f(x_n) = f(x_0)$ for every bounded linear functional $f \in E^*$.

Here is an example of this situation in the space C. Let $(x_n(t))$, $0 \leq t \leq 1$, be a norm-bounded sequence of functions which converges everywhere to a function $z(t)$ which is not continuous. The limit $\lim_{n\to\infty} \int_0^1 x_n(t)dg$ then exists for every function $g(t)$ of bounded variation (cf. Introduction §5, p. 4), but the sequence $(x_n(t))$ does not converge weakly to any continuous function.

Nevertheless, we have the following theorem:

In the spaces L^p *and* l^p *for* $p \geq 1$, *the existence of* $\lim_{n\to\infty} f(x_n)$, *for a sequence* (x_n), *for any bounded linear functional* f, *implies that the sequence* (x_n) *converges weakly to some element* x_0.

Proof for L^1. If $\lim_{n\to\infty} \int_0^1 x_n(t)\alpha(t)dt$, where $(x_n(t)) \subseteq L^1$, exists for every function $\alpha(t) \in M$, we must have

$$\lim_{p,q\to\infty} \int_0^1 [x_p(t) - x_q(t)]\alpha(t)dt = 0 \text{ for every } \alpha(t) \in M.$$

We shall show that there exists, for every $\varepsilon > 0$, an $\eta > 0$ and a natural number N such that

(25) $$\int_H |x_N(t) - x_n(t)|dt < \varepsilon,$$

for every $n \geqq N$ and every subset H of $[0,1]$ of measure $<\eta$.

Indeed, if this were not so, there would exist two strictly increasing sequences of natural numbers (p_k) and (n_k) and a sequence of subsets (H_k) whose measures tend to 0 such that $\int_{H_k} |x_{p_k}(t) - x_{n_k}(t)|dt \geqq \varepsilon$, whence

$$\lim_{k\to\infty} \int_0^1 [x_{p_k}(t) - x_{n_k}(t)]\alpha(t)dt = 0 \text{ for every } \alpha(t) \in M,$$

would contradict the theorem of Lebesgue (see Introduction, §6, p. 5).

This established, we therefore have in particular, if η is sufficiently small, $\int_H |x_n(t)|dt < \frac{1}{2}\varepsilon$ for every $n=1,2,\ldots,N$, whence by (25)

(26) $$\int_H |x_n(t)|dt < \frac{3}{2}\varepsilon \text{ for every } n=1,2,\ldots,$$

provided that the measure of H is $<\eta$.

Put

(27) $$\lim_{n\to\infty} \int_0^t x_n(u)du = \beta(t).$$

We are going to show that the function $\beta(t)$ is absolutely continuous.

In fact, for every $\varepsilon > 0$ there exists by (26) an $\eta > 0$ such that $\int_H |x_n(t)|dt < \varepsilon$ for $n=1,2,\ldots$, and for every set H of measure $<\eta$. In particular, if H consists of a finite number of non-overlapping intervals with end-points t_i and t_i', we therefore have

$$\lim_{n\to\infty} \int_H x_n(t)dt = \lim_{n\to\infty} \sum_i \int_{t_i}^{t_i'} x_n(t)dt = \sum_i [\beta(t_i') - \beta(t_i)],$$

whence $|\sum_i [\beta(t_i') - \beta(t_i)]| \leqq \varepsilon$, which gives the absolute continuity of the function $\beta(t)$.

This being so, we have only to put $\beta'(t) = x_0(t)$ to conclude from (27) together with the conditions for weak convergence established on p. 77, that the sequence $(x_n(t))$ converges weakly to $x_0(t)$.

Proof for L^p, $p > 1$. Suppose that $\lim_{n\to\infty} \int_0^1 x_n(t)y(t)dt$, where $x_n(t) \in L^p$ for $n=1,2,\ldots$, exists for every $y(t) \in L^{p/(p-1)}$. The f_n, given by $f_n(y) = \int_0^1 x_n(t)y(t)dt$ are clearly bounded linear functionals on $L^{p/(p-1)}$ and since, by hypothesis, $\lim_{n\to\infty} f_n(y)$ exists for every $y(t) \in L^{p/(p-1)}$, $\lim_{n\to\infty} f_n(y) = f(y)$ also defines a bounded linear functional f on $L^{p/(p-1)}$ by theorem 4 (Chapter I, §3, p. 15); this f is therefore (cf. Chapter IV, §4, p. 39) of the form $f(y) = \int_0^1 x_0(t)y(t)dt$, for $y \in L^{p/(p-1)}$, where $x_0 \in L^p$.

It follows from this that

$$\lim_{n\to\infty} \int_0^1 x_n(t)y(t)dt = \int_0^1 x_0(t)y(t)dt \text{ for every } y \in L^{p/(p-1)},$$

i.e. that (x_n) converges weakly to x_0, q.e.d.

The proof for l^1 is similar to that of the theorem proved in §2, pp. 83-84, and consists of showing that the sequence (x_n) converges in norm to an element x_0.

The proof for l^p where $p > 1$ is similar to that for L^p.

§5. A theorem on weak convergence.

We conclude this chapter with the following general theorem.

THEOREM 3. *Let U be a bounded linear operator from one Banach space E to another E_1. If a sequence (x_n) converges weakly to x in E, then the sequence $(U(x_n))$ converges weakly to $U(x_0)$ in E_1.*

Proof. Let Y be any bounded linear functional on E_1. Then $X = U^*(Y)$, given by $X(x) = Y[U(x)]$, is a bounded linear functional on E, as $|X(x)| = |Y[U(x)]| \leq \|Y\| . \|U(x)\| \leq \|Y\| . \|U\| . \|x\|$.
The weak convergence of (x_n) to x_0 thus implies that

$$\lim_{n\to\infty} Y[U(x_n)] = \lim_{n\to\infty} X(x_n) = X(x_0) = Y[U(x_0)],$$

i.e. that $(U(x_n))$ converges weakly to $U(x_0)$, q.e.d.

Remark. With the additional hypothesis that *the operator U is compact*, the weak convergence of (x_n) to x_0 implies that $(U(x_n))$ *converges to $U(x_0)$ in norm*, i.e. that

$$\lim_{n\to\infty} \|U(x_n) - U(x_0)\| = 0.$$

In fact, if this were not the case, there would exist an $\varepsilon > 0$ and a subsequence (x_{n_i}) such that

(28) $$\|U(x_{n_i}) - U(x_0)\| > \varepsilon \text{ for every } i=1,2,\ldots,$$

with the sequence $(U(x_{n_i}))$ converging in norm to an element $y' \in E_1$. Now as the weak convergence of (x_{n_i}) to x_0 implies, by the preceding theorem 3, that of $(U(x_{n_i}))$ to $U(x_0)$, we would have $y' = U(x_0)$, which is impossible by (28).

CHAPTER X

Linear functional equations

§1. Relations between bounded linear operators and their adjoints.

In this chapter we shall concern ourselves with equations of the form $y = U(x)$ where U is a bounded linear operator whose domain is a Banach space E and whose codomain is a subspace E_1 of another Banach space E'.

Bounded linear functionals on E, i.e. elements of the dual of E, will be denoted by X and those on E' by Y.

If the bounded linear operator U defines a bijective transformation from E to E_1, the inverse operator U^{-1} is clearly linear (although not necessarily continuous). It is easy to see that for the inverse operator to exist, it is necessary and sufficient that

$$U(x) = \Theta \text{ implies } x = \Theta.$$

If the inverse operator is continuous, there exists an $M > 0$ such that, if $y = U(x)$, $\|x\| \leq M.\|y\|$.

Conversely, if there exists a number $m > 0$ such that $m\|x\| \leq \|U(x)\|$, then U has a continuous inverse.

If the *inverse operator is continuous, then the codomain* E_1 *is closed.*

Indeed, putting $\lim_{n\to\infty} y_n = y$, where $y_n = U(x_n)$, we have

$$\lim_{p,q\to\infty} \|x_p - x_q\| \leq M. \lim_{p,q\to\infty} \|y_p - y_q\| = 0,$$

whence, putting $\lim_{n\to\infty} x_n = x$, we conclude that $U(x) = y$.

If the functional Y_0 is a transfinite limit of the sequence (Y_ξ) of type θ, then the conjugate functional $X_0 = U^*(Y_0)$ is a transfinite limit of the sequence $(X_\xi) = (U^*(Y_\xi))$ of type θ.

In fact, for every x we have $X_\xi(x) = Y_\xi[U(x)]$ where $1 \leq \xi < \theta$.

LEMMA. *If the adjoint operator* U^* *has a continuous inverse and* Γ_1 *denotes any regularly closed linear subspace of the dual of* E', *then the corresponding set* $\Gamma = U^*(\Gamma_1)$ *is also regularly closed.*

Proof. By hypothesis there exists a number $M > 0$ such that $\|U^*(Y)\| \geq M.\|Y\|$ for every Y. Consequently, if $X_\xi \in U^*(\Gamma_1)$ and $\|X_\xi\| \leq C$ for every $1 \leq \xi < \theta$, where $X_\xi = U^*(Y_\xi)$, we will also have $Y_\xi \in \Gamma_1$ and $\|Y_\xi\| \leq \frac{1}{M}C$ for every $1 \leq \xi < \theta$. Since, by hypothesis the set Γ_1 is regularly closed, there exists by lemma 3 (Chapter VIII, §3, p. 75) a transfinite limit $Y_0 \in \Gamma_1$ of the sequence (Y_ξ). The functional $X_0 = U^*(Y_0)$ clearly therefore belongs to $U^*(\Gamma_1)$ and is a transfinite limit of the sequence (X_ξ). The set $\Gamma = U^*(\Gamma_1)$ is thus transfinitely closed and therefore, by the same lemma, regularly closed, q.e.d.

THEOREM 1. *If the adjoint operator* U^* *has a continuous inverse, the equation* $y = U(x)$ *has a solution for every* y.

Proof. For any given non-zero $y_0 \in E'$, let Γ_1 denote the set of all bounded linear functionals Y such that $Y(y_0) = 0$ and let $\Gamma = U^*(\Gamma_1) = \{X\colon X = U^*(Y),\ Y \in \Gamma_1\}$.

The set Γ_1 is regularly closed; it follows from this by the preceding lemma that the set Γ is also regularly closed. Moreover, if Y_0 is a bounded linear functional such that $Y_0(y_0) = 1$, the functional $X_0 = U^*(Y_0)$ does not belong to Γ. Hence by theorem 1 (Chapter VIII, §3, p. 75) there exists an element $x_0 \in E$ such that

(1) $$X_0(x_0) = 1 \text{ and } X(x_0) = 0 \text{ for every } X \in \Gamma.$$

Putting

(2) $$y_1 = U(x_0),$$

we have $Y_0(y_1) = X_0(x_0)$ and $Y(y_1) = X(x_0)$, whence by (1)

(3) $$Y_0(y_1) = 1 \text{ and } Y(y_1) = 0 \text{ for every } Y \in \Gamma_1.$$

Now, for any bounded linear functional Y, the functional $\overline{Y} = Y - [Y(y_0)].Y_0$ clearly belongs to Γ_1, because $\overline{Y}(y_0) = Y(y_0) - [Y(y_0)].Y_0(y_0) = 0$. Consequently, we have, by (3), $\overline{Y}(y_1) = Y(y_1) - [Y(y_0)].Y_0(y_1) = Y(y_1) - Y(y_0) = 0$, so that $Y(y_1 - y_0) = 0$ for every Y. It follows that $y_1 - y_0 = 0$, and so, by (2), x_0 satisfies $y_0 = U(x_0)$, and is therefore the required solution for the arbitrarily chosen element y_0, q.e.d.

Conversely, we have

THEOREM 2. *If the equation $X = U^*(Y)$ admits a solution for every X, then*

1° *the operator U has a continuous inverse,*

2° *the codomain of U is the set of y which satisfy the condition*

(4) $$Y(y) = 0 \text{ if } U^*(Y) = 0.$$

Proof. 1°. If the operator U did not admit a continuous inverse, there would exist a sequence $(x_n) \subseteq E$ such that

(5) $$\lim_{n\to\infty} \|x_n\| = \infty$$

and $\lim\limits_{n\to\infty} \|y_n\| = 0$ where $y_n = U(x_n)$.

Now, as the equation $X = U^*(Y)$ has a solution for any X by hypothesis, we have $\lim\limits_{n\to\infty} X(x_n) = \lim\limits_{n\to\infty} Y(y_n) = 0$ for every bounded linear functional X defined in E, which, by theorem 6 (Chapter V, §1, p. 50), implies that the sequence of norms $(\|x_n\|)$ is bounded, contradicting (5).

2°. Suppose that for some element $y_0 \in E^*$,

(6) $$U^*(Y) = 0 \text{ implies } Y(y_0) = 0.$$

Since the codomain E_1 of the operator U is closed by 1° above, if y_0 did not belong to E_1, there would exist (cf. Chapter IV, §3, p. 35, lemma) a bounded functional Y_0 such that

(7) $$Y_0(y_0) = 1$$

and $Y_0(y) = 0$ for every $y \in E_1$. Putting $X_0 = U^*(Y_0)$, we would thus have $X_0(x) = Y_0(y) = 0$ where $y = U(x) \in E_1$, whence $U^*(Y_0) = 0$, which by (6) would imply that $Y_0(y_0) = 0$, contradicting (7). Consequently, $y_0 \in E_1$.

Conversely, if $U^*(Y) = X = 0$, we have for every $y \in E_1$ the equality $Y(y) = X(x) = 0$, q.e.d.

Replacing, in the preceding theorems 1 and 2, x, y, X, Y, U and U^* by Y, X, y, x, U^* and U respectively and using theorems about functionals instead of those involving elements in the arguments, we obtain the following theorems.

THEOREM 3. *If the operator U admits a continuous inverse, the equation $X = U^*(Y)$ has a solution for every bounded linear functional X defined on E.*

THEOREM 4. *If the equation $y = U(x)$ has a solution for every y, then*

1° *the operator U admits a continuous inverse,*

2° *its codomain is the set of X satisfying, for every $x \in E$, the condition:*

(8) $$X(x) = 0 \text{ if } U(x) = 0.$$

Theorems 1–4 lead easily to the following theorems.

THEOREM 5. *If the equation $y = U(x)$ admits exactly one solution for every y, then the equation $X = U^*(Y)$ also admits exactly one solution for every X and conversely.*

THEOREM 6. *If the operators U and U^* admit continuous inverses, then for every y and for every X there exist exactly one x and one Y such that $y = U(x)$ and $X = U^*(Y)$.*

THEOREM 7. *If the equations $y = U(x)$ and $X = U^*(Y)$ admit solutions for every y and for every X, then these solutions are unique.*

Furthermore, we shall prove the three theorems that now follow.

THEOREM 8. *If the codomain of a bounded linear operator U is closed, that of the adjoint operator U^* is the set of X which satisfy condition* (8): $X(x) = 0$ *if* $U(x) = 0$.

Proof. The derived set E_1' of the codomain $E_1 \subseteq E'$ of the operator U, being a closed linear subspace, is itself a Banach space.

Now if Z denotes an arbitrary bounded linear functional on E_1' and $U_1^*(Z)$ denotes the bounded linear functional X satisfying the equation

$$Z[U(x)] = X(x) \text{ for every } x \in E,$$

it is easily verified that the codomains of the operators U_1^* and U^* are the same. Indeed, for every bounded linear functional Y on E' and satisfying the condition

(9) $$Z(y) = Y(y) \text{ for every } y \in E_1',$$

we have $Z[U(x)] = Y[U(x)]$ for every $x \in E$, whence

(10) $$U_1^*(Z) = U^*(Y)$$

and, by definition of Z, there exists, by theorem 2 (Chapter IV, §2, p. 34), a bounded linear functional Y on E' satisfying condition (9) and therefore (10). Condition (8) follows from this by theorem 4, 2°, above, on replacing E' by E_1.

THEOREM 9. *If the codomain of the bounded linear operator U^* is closed, that of the operator U is the set of all y which satisfy condition* (4): $Y(y) = 0$ *if* $U^*(Y) = 0$.

Proof. The functionals Z and $U_1^*(Z)$ being defined as in the proof of the preceding theorem 8, observe that $U_1^*(Z) = \Theta$ implies $Z(y) = 0$ for every $y \in E_1'$; hence $Z = \Theta$.

Now as the sets of Z and of X are Banach spaces, the operator U_1^*, where $X = U_1^*(Z)$, admits a continuous inverse by theorem 5 (Chapter III, §3, p. 26). It follows from this by theorem 1, p. 89, that the equation $y = U(x)$ possesses a solution for every $y \in E_1'$. The codomain $E_1 = E_1'$ of the operator U is therefore closed.

As condition (4) is plainly satisfied, when $y \in E_1$, it only remains to establish the converse, i.e. to show that every $y_0 \in E'$ which satisfies (4) belongs to E_1.

In fact, as E_1 is a closed linear subspace, in the contrary case (cf. Chapter IV, §3, p. 35, lemma) there would exist a bounded linear functional Y_0 such that $Y_0(y_0) = 1$ and $Y_0(y) = 0$ for every $y \in E_1$. Therefore, putting $X_0 = U^*(Y_0)$, we would have $X_0(x) = Y_0(y) = 0$ for $x \in E$, whence $X_0 = 0$, and consequently $U^*(Y_0) = 0$, contradicting condition (4) which y_0 is assumed to satisfy.

THEOREM 10. *If the codomain E_1 of the bounded linear operator U is closed, there exists a number $m > 0$ such that for every $y \in E_1$ one can find a corresponding $x \in E$ satisfying the conditions*

$$y = U(x) \text{ and } \|x\| \leq m\|y\|.$$

Proof. In the course of the proof of theorem 3 (Chapter III, §3, p. 25) we established proposition (1) which, under the hypothesis of the theorem to be proved, yields the existence for every $\varepsilon > 0$ of an $\eta > 0$ such that, given an arbitrary y satisfying the inequality $\|y\| < \eta$, one can find a corresponding x satisfying the conditions $y = U(x)$ and $\|x\| < \varepsilon$.

We easily deduce from this the existence, for every y, of an x meeting the requirements of the theorem with $m = \frac{\varepsilon}{\eta}$.

§2. Riesz' theory of linear equations associated with compact linear operators.

We are going to concern ourselves here with equations of the form $y = x - U(x)$, where U is a compact linear operator from the space E into itself.

LEMMA. *If the linear operator U is compact, then the operator T given by $T(x) = x - U(x)$ transforms every bounded closed set $G \subseteq E$ into a closed set.*

Proof. Put

(11) $$x_n \in G \text{ for } n = 1,2,\ldots, \text{ and } \lim_{n\to\infty} T(x_n) = y_0,$$

so that the sequence $(U(x_n))$ thus forms a relatively compact set and there exists a subsequence $(U(x_{n_i}))$ which converges to some element $x_0 \in E$. As $x_{n_i} = U(x_{n_i}) + T(x_{n_i})$, we have, by (11), that $\lim_{n\to\infty} x_{n_i} = x_0 + y_0$, whence $T(y_0 + x_0) = y_0$.

THEOREM 11. *If U is a compact operator, the codomains of the operators T and T^*, given by*

$$T(x) = x - U(x) \text{ and } T^*(X) = X - U^*(X)$$

are closed.

Proof. With G denoting the set of solutions of the equation $T(x) = 0$, let $y_0 \neq \Theta$ be a point of accumulation of the codomain of T, so that there exists a sequence $(x_n) \subseteq E$ such that $y_0 = \lim_{n\to\infty} T(x_n)$.

If the sequence $(\|x_n\|)$ were bounded, the element y_0 would belong to the codomain by the lemma just proved.

Letting d_n denote the distance between x_n and the set G, there thus exists a $w_n \in G$ such that

(12) $$d_n \leq \|x_n - w_n\| \leq \left(1 + \frac{1}{n}\right)d_n.$$

We have

(13) $$\lim_{n\to\infty} T(x_n - w_n) = y_0.$$

If the sequence $(\|x_n - w_n\|)$ were bounded, the proof would be complete by the preceding lemma.

Suppose therefore that $\lim_{n\to\infty} \|x_n - w_n\| = \infty$, whence, putting $z_n = \frac{x_n - w_n}{\|x_n - w_n\|}$, we have, by (13), $\lim_{n\to\infty} T(z_n) = \Theta$ and $\|z_n\| = 1$. By the lemma we can thus extract, from the sequence (z_n), a subsequence (z_{n_i}) convergent to an element w_0 such that $T(w_0) = 0$, whence $w_0 \in G$. Putting $z_n - w_0 = \varepsilon_n$, we have

(14) $$\lim_{i\to\infty} \|\varepsilon_{n_i}\| = 0,$$

so that $z_n - w_0 = \frac{x_n - w_n}{\|x_n - w_n\|} - w_0 = \varepsilon_n$ and consequently $x_n - w_n - w_0 . \|x_n - w_n\| = \varepsilon_n \|x_n - w_n\|$, whence by (12)

(15) $$\|x_{n_i} - w_{n_i} - w_0\|x_{n_i} - w_{n_i}\|\| \leq \|\varepsilon_{n_i}\|\left(1 + \frac{1}{n_i}\right)d_{n_i}.$$

Now, by (14) and (15) there exists an n_i such that

$$\|x_{n_i} - w_{n_i} - w_0\|x_{n_i} - w_{n_i}\|\| \leq \frac{d_{n_i}}{2};$$

but this is impossible because $w_{n_i} + w_0\|x_{n_i} - w_{n_i}\| \in G$ and d_{n_i} is the distance between x_{n_i} and G.

Thus the codomain of T is closed. The argument for T^* is similar.

THEOREM 12. *If U is a compact linear operator, the equations*

$$x - U(x) = \Theta \text{ and } X - U^*(X) = \Theta$$

have at most a finite number of linearly independent solutions.

Proof. Suppose, on the contrary, that there exists an infinite sequence (x_n) of linearly independent elements of E satisfying the equations $x_n - U(x_n) = \Theta$ for $n=1,2,\ldots$ Let E_n be the set $\left\{\sum_{i=1}^{n} h_i x_i : h_i \text{ arbitrary real numbers}\right\}$. Clearly

(16) $$x \in E_n \text{ implies } x - U(x) = \Theta$$

and it is easy to see that for every $n=1,2,\ldots$, the set E_n is a closed linear subspace not containing x_{n+1} and thus a *proper* subset of E_{n+1}.

By the lemma of Chapter V, §3, p. 52, there therefore exists a sequence (y_n) such that

(17) $$y_n \in E_n,\ \|y_n\| = 1 \text{ and } \|y_n - x\| > \frac{1}{2} \text{ for every } x \in E_{n-1},$$

whence, by (16), $y_n - U(y_n) = \Theta$ and consequently $y_n = U(y_n)$. The sequence (y_n) is, therefore, a relatively compact set, which contradicts (17).

For the equation $X - U^*(X) = \Theta$ the reasoning is similar, applied to the dual space, of all bounded linear functionals on E, which is itself a Banach space.

THEOREM 13. *If, for a compact linear operator U, the equation $y = x - U(x)$, respectively $Y = X - U^*(X)$, has a solution for every y or Y respectively, then the equation $x - U(x) = \Theta$, respectively $X - U^*(X) = \Theta$, has exactly one solution, namely $x = \Theta$ or $X = \Theta$ respectively.*

Proof. Put

$$T^{(1)}(x) = x - U(x) = T(x) \text{ and } T^{(n)}(x) = T[T^{(n-1)}(x)].$$

Let E_n denote the set of all $x \in E$ satisfying the equation $T^{(n)}(x) = \Theta$ and suppose that there exists an $x_1 \neq \Theta$ such that $T(x_1) = \Theta$. Letting x_n denote the element satisfying the equation $x_{n-1} = T(x_n)$, we therefore have

$$T^{(n)}(x_{n+1}) = x_1 \neq \Theta \text{ and } T^{(n+1)}(x_{n+1}) = T(x_1) = \Theta,$$

whence

$$x_{n+1} \in E_{n+1} \setminus E_n.$$

The set E_n is plainly a closed linear subspace, and is a proper subset of E_{n+1}. Hence by the lemma on p. 92, there exists a sequence (y_n) satisfying condition (17).

Now, as $y_n \in E_n$, we have, by definition of T and of E_n, the equality $T(y_n) = y_n - U(y_n)$, whence

(18) $$U(y_p) - U(y_q) = y_p - [y_q + T(y_p) - T(y_q)] = y_p - x$$

and $p > q$ implies $T^{(p-1)}(x) = T^{(p-1)}(y_q) + T^{(p)}(y_p) - T^{(p)}(y_q) = \Theta$.

Consequently $x \in E_{p-1}$, whence, by (17), $\|y_p - x\| > \frac{1}{2}$, so that, by (18), $\|U(y_p) - U(y_q)\| > \frac{1}{2}$ for $p > q$, which is impossible, as the sequence $(U(y_n))$ has convergent subsequences. It must therefore be the case that $x_1 = \Theta$, q.e.d.

For the equation $X - U^*(X) = \Theta$, the proof is similar, again working with the dual space of E.

THEOREM 14. *If, for a compact linear operator U, the equation $x - U(x) = \Theta$, respectively $X - U^*(X) = \Theta$, has the unique solution $x = \Theta$ or $X = \Theta$ respectively, then the equation $y = x - U(x)$, respectively $Y = X - U^*(X)$, has a solution for every y or for every Y respectively.*

Proof. As the codomain of the operator $I - U$, where I is the *identity* operator on E, $I(x) = x$ for every $x \in E$, is closed by theorem 11, p. 92, the hypothesis implies, by theorem 3, p. 91, that the equation $Y = X - U^*(X)$ has a solution for every Y. Hence by the preceding theorem 13, the only solution of the equation $X - U^*(X) = \Theta$ is given by $X = \Theta$ and consequently, by theorem 5, p. 91, the equation $y = x - U(x)$ is soluble for every y.

The proof for $Y = X - U^*(X)$ is similar.

THEOREM 15. *If U is a compact linear operator, the equations*

$$x - U(x) = \Theta \text{ and } X - U^*(X) = \Theta$$

have the same number of linearly independent solutions.

Proof. As before, put

(19) $$T(x) = x - U(x) \text{ and } T^*(X) = X - U^*(X).$$

Let

(20) $$T(x_i) = \Theta \text{ for } i=1,2,\ldots,n \text{ and } T^*(X_i) = \Theta \text{ for } i=1,2,\ldots,\nu,$$

where the terms of the sequence (x_i) and equally those of the sequence (X_i) are assumed to be linearly independent and the numbers n and ν denote, respectively, the largest possible numbers of linearly independent solutions of the equations $T(x) = \Theta$ and $T^*(X) = \Theta$.

Denote by z_i, for $i=1,2,\ldots,\nu$, any element such that

(21) $$X_j(z_i) = \begin{cases} 1 \text{ for } i = j, \\ 0 \text{ for } i \neq j. \end{cases}$$

Such z_i exist, as the linear subspace of functionals of the form

$$\sum_{j=1}^{i-1} \alpha_j X_j + \sum_{j=i+1}^{\nu} \beta_j X_j$$

is weakly closed and does not contain X_i.

Similarly, let Z_i, for $i=1,2,\ldots,n$, denote a bounded linear functional such that

(22) $$Z_j(x_i) = \begin{cases} 1 \text{ for } i = j, \\ 0 \text{ for } i \neq j. \end{cases}$$

Such functionals Z_i exist, because x_i does not belong to the closed linear subspace of elements of the form

$$\sum_{j=1}^{i-1} \alpha_j x_j + \sum_{j=i+1}^{n} \beta_j x_j.$$

Having said this, *suppose, to begin with, that* $\nu > n$. Let

(23) $$R(x) = U(x) + \sum_{i=1}^{n} Z_i(x).z_i \text{ and } W(x) = x - R(x).$$

It is easy to see that the operator R thus defined is compact. We shall show that the equation $W(x) = \Theta$ has exactly one solution, namely $x = \Theta$.

In fact, suppose that $W(x_0) = \Theta$. We need to prove that $x_0 = \Theta$. Now, we have, by (19) and (23):

(24) $$W(x_0) = x_0 - R(x_0) = T(x_0) - \sum_{i=1}^{n} Z_i(x_0).z_i = 0$$

and by (20)

(25) $$X_i T(x) = \Theta \text{ for every } x \text{ and } i=1,2,\ldots,\nu;$$

we deduce from (21) and (24) that

(26) $$X_i W(x_0) = Z_i(x_0) = 0 \text{ for } i=1,2,\ldots,n,$$

whence $T(x_0) = \Theta$, which implies by (20) and the definition of n that $x_0 = \sum_{i=1}^{n} \alpha_i x_i$ where α_i are suitable real numbers. By (26) and (22) we therefore have $Z_i(x_0) = \alpha_i = 0$ for every $i=1,2,\ldots,n$, whence, finally, $x_0 = \Theta$.

This established, we conclude, by theorem 14, p. 94, that the equation $x - R(x) = T(x) - \sum_{i=1}^{n} Z_i(x).z_i - z_{n+1}$ has a solution. However we immediately see, by (21) and (25), that $X_{n+1}[x - R(x)] = 0$ and moreover, by (21), that $X_{n+1}(z_{n+1}) = 1$. The assumption that $\nu > n$ is thus untenable.

Now suppose that $\nu < n$. Let $R(x) = \sum_{i=1}^{\nu} Z_i(x).z_i$, whence $R^*(X) = \sum_{i=1}^{\nu} X(z_i).Z_i$. Proceeding as above, one would then show that the equation $T^*(x) - \sum_{i=1}^{\nu} X(z_i).Z_i = \Theta$ (the adjoint of the equation $T(x) - \sum_{i=1}^{\nu} Z_i(x).z_i = \Theta$) has exactly one solution, namely $X = \Theta$. The equation $T^*(X) - \sum_{i=1}^{\nu} X(z_i).Z_i = Z_{\nu+1}$ would therefore have a solution by theorem 14, p. 94, and this is, however, impossible, because we have $T^*(X)x_{\nu+1} = X[T(x_{\nu+1})] = 0$ for every X, whence, by (22), $Z_i(x_{\nu+1}) = 0$ for $i=1,2,\ldots,n$ and moreover $Z_{\nu+1}(x_{\nu+1}) = 1$. Thus the assumption that $\nu < n$ also leads to a contradiction.

§3. Regular values and proper values in linear equations.

Suppose now that U, still a bounded linear operator, maps E into itself.

If I is the identity operator on E, then $I - hU$ is a bounded linear operator for every real number h and its adjoint is $I - hU^*$ where U^* is the adjoint of U.

Having said this, we are going to study the equations

(27) $$x - hU(x) = y \text{ and } X - hU^*(X) = Y.$$

If, for a given h_0, the first or second, respectively, of the equations (27) admits exactly one solution for every y or for every Y, respectively, then h_0 is called a *regular value* of this equation; otherwise, h_0 is called a *proper value*. The set of all such proper values is called the *spectrum*.

If x or X, respectively, satisfies the first or second, respectively, of the equations

$$x + hU(x) = \Theta \text{ and } X + hU^*(X) = \Theta, \tag{28}$$

it is known as a *proper element* (*vector*) or *functional* respectively.

By theorem 5, p. 91, the two equations (27) have the same set of regular values, and therefore also of proper values.

Theorems 1-9, established on p. 90 - 91, are easily seen to apply to equations of the form (27). These theorems enable one to deduce the behaviour of one of the two equations from that of the other and conversely.

THEOREM 16. *The set of regular values is open.*

Proof. If h_0 is a regular value, there exists a number $m > 0$ satisfying the conditions

$$\|x - h_0U(x)\| \geqq m.\|x\| \text{ and } \|X - h_0U^*(X)\| \geqq m.\|X\|.$$

Consequently, for every ε we have:

$$\|x - (h_0 + \varepsilon)U(x)\| \geqq \|x - h_0U(x)\| - |\varepsilon|.\|U(x)\| \geqq \left(m - |\varepsilon|.\|U\|\right)\|x\|$$

and, similarly,

$$\|X - (h_0 + \varepsilon)U^*(X)\| \geqq \left(m - |\varepsilon|.\|U^*\|\right).\|X\|.$$

It follows that, for $|\varepsilon|$ sufficiently small, the operators

$$I - (h_0 + \varepsilon)U \text{ and } I - (h_0 + \varepsilon)U^*$$

have bounded inverses, from which it follows, by theorem 6, p. 91, that $h_0 + \varepsilon$ is also a regular value.

THEOREM 17. *If* $|h| < 1/\|U\|$, *then* h *is a regular value.*

Proof. If $|h| < 1/\|U\|$, the solutions can be written in the form

$$x = y + \sum_{n=1}^{\infty} h^n U^{(n)}(y) \text{ and } X = Y + \sum_{n=1}^{\infty} h^n U^{*(n)}(Y), \tag{29}$$

where

$$U^{(1)} = U \text{ and } U^{*(1)} = U^*,$$
$$U^{(n)} = U[U^{(n-1)}] \text{ and } U^{*(n)} = U^*[U^{*(n-1)}].$$

The series (29) are convergent, because we have

$$\sum_{n=1}^{\infty} \|h^n U^{(n)}(y)\| \leqq \sum_{n=1}^{\infty} \left[|h|.\|U\|\right]^n.\|y\|$$

and

$$\sum_{n=1}^{\infty} \|h^n U^{*(n)}(Y)\| \leqq \sum_{n=1}^{\infty} \left[|h|.\|U^*\|\right]^n.\|Y\|.$$

From (29) we obtain

$$U(x) = U(y) + \sum_{n=1}^{\infty} h^n U^{(n+1)}(y) = \frac{1}{h}.\sum_{n=1}^{\infty} h^n U^{(n)}(y) = \frac{1}{h}.(x - y),$$

whence $x - hU(x) = y$. Similarly, we have $X - hU^*(X) = Y$. The equations (27) thus admit solutions for every y and for every Y respectively. By theorem 7, p. 91, these solutions are therefore unique and consequently h is a regular value, q.e.d.

THEOREM 18. *If* $h \neq h'$ *and*

$$x - hU(x) = \Theta \quad \text{and} \quad X - h'U^*(X) = \Theta,$$

then $X(x) = 0$.

In other words: *any proper vector of the value h is orthogonal to every proper functional of a value h' not equal to h.*

Proof. We have $X(x) = hX[U(x)] = hU^*(X)x$ and as $U^*(X) = \frac{1}{h'}X$, it follows that $X(x) = \frac{h}{h'}X(x)$. If $h \neq h'$, we therefore have $X(x) = 0$.

§4. Theorems of Fredholm in the theory of compact operators.

If, under the hypotheses of the previous section, the operator is further supposed to be compact, one can state, for the equations (28), the following theorems which constitute a generalisation of Fredholm's theorems on integral equations.

THEOREM 19. *The equations* (28) *have the same finite number* $d(h)$ *of linearly independent solutions.*

This is just a restatement of theorem 15, p. 94.

THEOREM 20. *If* $d(h) = 0$, *h is a regular value.*

This is a consequence of theorems 14, p. 94, and 19, above.

THEOREM 21. *If* $d(h) > 0$ *and if*

$$\{x_i\}, \{X_i\} \text{ respectively, for } i=1,2,\ldots,d(h),$$

denote (linearly independent) solutions of the equations (28), *then the equations* (27) *admit solutions for every y such that* $X_i(y) = 0$ *and for every Y such that* $Y(x_i) = 0$ *respectively,* $(i=1,2,\ldots,d(h))$.

This is a consequence of theorems 8 and 9, respectively, p. 91, and 11, p. 92.

We now prove the

THEOREM 22. *If U is a compact linear operator, the proper values of the first equation* (27)

$$y = x - hU(x)$$

constitute an isolated (discrete) set.

Proof. Let (h_n) be an infinite sequence of proper values where $h_i \neq h_j$ for $i \neq j$. Put

(30) $$x_n = h_n U(x_n) \text{ and } x_n \neq \Theta.$$

We first show that the vectors x_n are linearly independent.

In fact, if $x_1, x_2, \ldots, x_{n-1}$ were linearly independent, but $x_n = \sum_{i=1}^{n-1} \alpha_i x_i$, we would have $x_n = h_n U(x_n) = \sum_{i=1}^{n-1} h_n \alpha_i U(x_i)$, whence $x_n = \sum_{i=1}^{n-1} h_n \frac{\alpha_i}{h_i} x_i$ and consequently $\sum_{i=1}^{n-1} \alpha_i \left(1 - \frac{h_n}{h_i}\right) x_i = 0$. Since, by hypothesis, $\frac{h_n}{h_i} \neq 1$ for $n > i$, it is plain that the vectors $x_1, x_2, \ldots, x_{n-1}$ could not be linearly independent.

This established, for each $n=1,2,\ldots$, let E_n be the linear subspace of elements y of the form $y = \sum_{i=1}^{n} \alpha_i x_i$; it is closed and forms a proper subset of E_{n+1}. For every $y \in E_n$, we have, by (30),

$$y - h_n U(y) = \sum_{i=1}^{n} \alpha_i x_i - \sum_{i=1}^{n} h_n \alpha_i \frac{x_i}{h_i} = \sum_{i=1}^{n-1} \alpha_i \left(1 - \frac{h_n}{h_i}\right) x_i,$$

whence $y - h_n U(y) \in E_{n-1}$. By the lemma on p. 92, there therefore exists a sequence of elements (y_n) satisfying the conditions (17), p. 93.

Now suppose that the sequence (h_n) were convergent. Since the operator U is compact, the sequence $(U(h_n y_n))$ would constitute a relatively compact set. Moreover, for $p > q$ we have

(31) $$\|U(h_p y_p) - U(h_q y_q)\| = \|y_p - [y_p - h_p U(y_p) + U(h_q y_q)]\|$$

and, by (17), $y_p \in E_p$, which implies, as we have seen, that $y_p - h_p U(y_p) \in E_{p-1}$; similarly $h_q U(y_q) \in E_q \subseteq E_{p-1}$, whence, by (17) and (31), $\|U(h_p y_p) - U(h_q y_q)\| > \frac{1}{2}$ for every $p > q$, from which it follows that the set $\{U(h_n y_n): n=1,2,\ldots\}$ would not be relatively compact.

This contradiction implies that no sequence (h_n) of distinct proper values can be convergent. Hence they form a discrete set.

§5. Fredholm integral equations.

We now discuss several applications of the theorems just proved.

In the L^p spaces, the equations of the form $x - hU(x) = y$ are the so-called Fredholm integral equations, which have the following general form

(32) $$x(s) - h\int_0^1 K(s,t)x(t)\,dt = y(s),$$

where the function $K(s,t)$ satisfies certain conditions.

The adjoint equation $X - hU^*(X) = Y$ takes the form

(33) $$X(t) - h\int_0^1 K(s,t)X(s)\,ds = Y(t).$$

It is easy to see how the preceding theorems may be interpreted in the context of these integral equations.

If $K(s,t)$ satisfies the appropriate conditions, the operator which maps $x(t)$ to the function $\int_0^1 K(s,t)x(t)\,dt$ is compact and so the theorems of §§2,3 and 4 of this chapter may be applied to the equations (32) and (33). In particular, theorems 19-21 then become the theorems of Fredholm mentioned, although, of course, they also hold outside the field of integral equations.

§6. Volterra integral equations.

Equations of the form

(34) $$x(s) - \int_0^s K(s,t)x(t)\,dt = y(s),$$

where $K(s,t)$ is a continuous function, are called *Volterra equations*.

The operator $\int_0^s K(s,t)x(t)\,dt$ is then compact, as an operator in the spaces C and L^p, $p > 1$.

We now show that the equation

(35) $$x(s) - \int_0^s K(s,t)x(t)\,dt = 0$$

admits the unique solution $x(s) = 0$.

Indeed, suppose that $x(s)$ satisfies this equation; clearly $x(s)$ is a continuous function. Put

$$m = \max_{0 \leqq s \leqq 1} |x(s)| \text{ and } M = \max_{\substack{0 \leqq s \leqq 1 \\ 0 \leqq t \leqq 1}} |K(s,t)|.$$

We therefore have by (35)

(36) $$|x(s)| \leqq M.\int_0^s |x(t)|\,dt,$$

whence $|x(s)| \leqq M.m.s$ for $0 \leqq s \leqq 1$, which, replacing $x(t)$ by $M.m.s$ in (36), yields the inequality $|x(s)| \leqq M^2.m.s^2/2$. Iterating this procedure, we therefore obtain $|x(s)| \leq \frac{(M.s)^n}{n!}.m$ for every $n=1,2,\ldots$, whence, clearly, $x(s) = 0$.

This established, let us return to equation (34). Since for $x,y \in C$, and similarly for $x,y \in L^p$, the operator $\int_0^s K(s,t)x(t)dt$ is compact, by theorem 14, p. 94, the equation (34) possesses exactly one solution $x \in C$ or $x \in L^p$ respectively for every $y \in C$ or $y \in L^p$ respectively

§7. Symmetric integral equations.

If U is a bounded linear operator from L^2 to itself, its adjoint U^* can also be regarded as such an operator.

This is because the dual space of L^2 can also be regarded as L^2 (cf. Chapter IV, §4, p. 39).

The operator U is called *symmetric*, when

$$\int_0^1 yU(x)dt = \int_0^1 xU(y)dt \text{ for } x,y \in L^2. \tag{37}$$

Since $\int_0^1 yU(x)dt = \int_0^1 xU^*(y)dt$, *every symmetric operator coincides with its own adjoint.*

When the function $K(s,t)$ is symmetric (i.e. $K(s,t) = K(t,s)$ for all s,t) and, further, the double integral

$$\int_0^1\int_0^1 K(s,t)x(t)y(s)dsdt$$

exists for all $x,y \in L^2$, the operators U and V given by

$$U(x) = \int_0^1 K(s,t)x(t)dt = y(s),$$

(38)

$$V(x) = x(s) - h\int_0^1 K(s,t)x(t)dt = y(s),$$

are bounded linear operators which are symmetric because they satisfy the condition (37).

Equations of the form (38) are known as *symmetric integral equations.*

THEOREM 23. *If U is a symmetric operator, the number h is a regular value of the operator $I - hU$ when this operator admits a continuous inverse or when the equation $x - hU(x) = y$ is soluble for each y.*

The proof follows from theorems 3 and 4, p. 91, due to the fact that in these circumstances, the relevant equation and its adjoint are one and the same.

CHAPTER XI

Isometry, equivalence, isomorphism

§1. Isometry.

Let E and E_1 be metric spaces (see Introduction, §7, p. 5) and let U be a bijective mapping from E onto E_1. This mapping is said to be *isometric* or is called an *isometry* if it does not change distances, i.e. when

$$d(x_1, x_2) = d(U(x_1), U(x_2))$$

for every pair x_1, x_2 of elements of E.

Since normed vector spaces are metric spaces (cf. Chapter IV, §1, p. 33), it makes sense to consider isometric transformations between them.

§2. The spaces L^2 and l^2.

THEOREM 1. *The spaces L^2 and l^2 are isometric.*

Proof. Let $(x_i(t))$, $0 \leq t \leq 1$, be any complete orthonormal sequence of functions in L^2. If $x \in L^2$, we know that

$$(1) \qquad \sum_{i=1}^{\infty}\left[\int_0^1 x_i(t)x(t)\,dt\right]^2 = \int_0^1 \big(x(t)\big)^2\,dt.$$

If $U(x)$ denotes the sequence $y = (\eta_i)$ where $\eta_i = \int_0^1 x_i(t)x(t)\,dt$, we have, by (1), that $y \in l^2$ and $\|U(x)\| = \|x\|$. As U is additive and does not alter the norms of elements, it is a bounded linear operator. Moreover, it follows from the theory of orthogonal series that, for each $y \in l^2$, there exists one and only one function $x(t) \in L^2$ such that $y = U(x)$.

The bounded linear operator U thus maps L^2 bijectively onto l^2 without changing norms, and so distances are also unchanged. Consequently the spaces L^2 and l^2 are isometric.

Remark. We shall later see that the spaces L^p and L^q are only isometric in the case $p = q = 2$. This is a consequence of the corollary (Chapter XII, §3, p. 119).

§3. Isometric transformations of normed vector spaces.

THEOREM 2. *Every isometry U of one normed vector space to another such that $U(\Theta) = \Theta$ is a bounded linear operator.*

Proof. First let E be an arbitrary metric space and x_1, x_2 any pair of points of E.

Let H_1 denote the set of points $x \in E$ such that

$$(2) \qquad d(x, x_1) = d(x, x_2) = \tfrac{1}{2}d(x_1, x_2)$$

and, for $n = 2, 3, \ldots$, let H_n denote the set of points $x \in H_{n-1}$ such that, for every $z \in H_{n-1}$,

$$(3) \qquad d(x, z) \leq \tfrac{1}{2}\delta(H_{n-1})$$

where $\delta(H_{n-1}) = \sup\{d(x,y): x,y \in H_{n-1}\}$, is the *diameter* of the set H_{n-1}.

For the sequence (H_n) so defined, we have

$$(4) \qquad \lim_{n\to\infty} \delta(H_n) = 0.$$

Indeed, if the sets H_n are non-empty, we have for every pair x',x'' of points of H_n, $x'' \in H_{n-1}$, since, by definition, $H_1 \supseteq H_2 \supseteq \ldots \supseteq H_n \supseteq \ldots$, so that, by (3) $d(x',x'') \leq \frac{1}{2}\delta(H_{n-1})$. Consequently $\delta(H_n) \leq \frac{1}{2}\delta(H_{n-1})$, whence $\delta(H_n) \leq \frac{1}{2^{n-1}}\delta(H_1)$, and moreover, we have by (2) for each pair x',x'' of points of H_1, the inequality $d(x',x'') \leq d(x',x_1) + d(x'',x_1) = d(x_1,x_2)$, so that $\delta(H_1) \leq d(x_1,x_2)$ and consequently $\delta(H_n) \leq \frac{1}{2^{n-1}}d(x_1,x_2)$, whence (4).

It follows from this that the intersection of the sets H_n, if non-empty, reduces to a single point. We shall call this point the *centre* of the pair x_1,x_2.

Having said this, let E be a normed vector space, so that

$$d(x',x'') = \|x' - x''\| \text{ for all } x',x'' \in E.$$

Put $\bar{x} = x_1 + x_2 - x$ for $x \in E$. We easily see by induction that

$$(5) \qquad x \in H_n \text{ implies } \bar{x} \in H_n \text{ for each } n=1,2,\ldots$$

Indeed, if $x \in H_1$ we have $\|\bar{x} - x_1\| = \|x - x_2\|$ and $\|\bar{x} - x_2\| = \|x - x_1\|$, so that $\|\bar{x} - x_1\| = \|\bar{x} - x_2\| = \frac{1}{2}\|x_1 - x_2\|$, whence by (2) $\bar{x} \in H_1$ and, assuming (5) holds for $n-1$, we have, consequently, for $x' \in H_{n-1}$, $x_1 + x_2 - x' \in H_{n-1}$. If $x \in H_n$, we therefore have, by (3), $\|\bar{x} - x'\| = \|(x_1 + x_2 - x') - x\| \leq \frac{1}{2}\delta(H_{n-1})$, whence $\bar{x} \in H_n$.

We are going to show that the point $\xi = \frac{1}{2}(x_1 + x_2)$ is the centre of the pair x_1,x_2. In fact, we have $\xi \in H_1$, because $\|x_1 - \xi\| = \|x_2 - \xi\| = \frac{1}{2}\|x_1 - x_2\|$. Suppose, therefore, that $\xi \in H_{n-1}$. For every $x \in H_{n-1}$, we have, by (5), $x_1 + x_2 - x = \bar{x} \in H_{n-1}$ and as $2\|\xi - x\| = \|x_1 + x_2 - 2x\| = \|\bar{x} - x\| \leq \delta(H_{n-1})$, we conclude that $\|\xi - x\| \leq \frac{1}{2}\delta(H_{n-1})$, whence $\xi \in H_n$. Since it belongs to H_n for each natural number n, the point ξ is therefore the centre of x_1,x_2.

This established, let E_1 be another normed vector space and let U be an isometry of E onto all of E_1 such that $U(\Theta) = \Theta$. Since the notion of centre is a metric space one, it is easily seen that the centre of any pair x_1,x_2 of points of E will be mapped to the centre of the pair $U(x_1),U(x_2)$ of E_1. We therefore have

$$U[\tfrac{1}{2}(x_1 + x_2)] = \tfrac{1}{2}[U(x_1) + U(x_2)] \text{ for } x_1,x_2 \in E,$$

whence, putting $x_1 = x$ and $x_2 = \Theta$, we obtain, because of the hypothesis $U(\Theta) = \Theta$:

$$U(\tfrac{1}{2}x) = \tfrac{1}{2}U(x) \text{ for every } x \in E.$$

It follows from this that, for any points x_1 and x_2 of E:

$$U(x_1 + x_2) = U[\tfrac{1}{2}(2x_1 + 2x_2)] = \tfrac{1}{2}U(2x_1) + \tfrac{1}{2}U(2x_2) = U(x_1) + U(x_2).$$

The operator U is thus additive, and, as it is continuous, in fact a bounded linear operator.

§4. Spaces of continuous real-valued functions.

For any compact metric space Q, (cf. Introduction, §7, p. 6), the set E of continuous real-valued functions $x(q)$ defined on Q may be regarded as a Banach space, when addition and scalar multiplication are defined in the usual (pointwise) way and the norm is taken to be the maximum of the absolute value of the function.

LEMMA. *Let $x(q) \in E$, $q \in Q$. For a given element $q_0 \in Q$, the inequality*

(6) $$|x(q_0)| > |x(q)| \text{ for every } q \neq q_0,$$

holds if and only if

(7) $$\lim_{h\to 0} \frac{\|x+hz\| - \|x\|}{h}$$

exists for every $z(q) \in E$.

Furthermore, if the function $x(q)$ satisfies the inequality (6), *we have*

$$\lim_{h\to 0} \frac{\|x+hz\| - \|x\|}{h} = z(q_0).\text{sign } x(q_0) \text{ for every } z(q) \in E.$$

Proof. The condition is necessary. In fact, we have $\|x\| = |x(q_0)|$ and as the continuous function $|x + hz|$ attains its maximum, we obtain

(8) $$|x(q_0) + hz(q_0)| - |x(q_0)| \leq \|x + hz\| - \|x\| = |x(q_h) + hz(q_h)| - |x(q_0)|,$$

where q_h is a point of Q which depends on h. Now, we deduce from (8) that $|x(q_0) + hz(q_0)| \leq |x(q_h) + hz(q_h)|$ and consequently $0 \leq |x(q_0)| - |x(q_h)| \leq |h|.|z(q_0)| + |h|.|z(q_h)| \leq 2|h|.\|z\|$, whence $\lim_{h\to\infty} |x(q_h)| = |x(q_0)|$. This implies, by the compactness of Q that

(9) $$\lim_{h\to 0} q_h = q_0.$$

This established, first consider the case where $x(q_0) > 0$. There then exists an $\varepsilon > 0$ such that, for $|h| < \varepsilon$, we have

$$|x(q_0) + hz(q_0)| - |x(q_0)| = x(q_0) + hz(q_0) - x(q_0) = hz(q_0)$$

and, by (9),

$$|x(q_h) + hz(q_h)| - |x(q_0)| = x(q_h) + hz(q_h) - x(q_0) \leq hz(q_h),$$

whence, by (8), $hz(q_0) \leq \|x + hz\| - \|x\| \leq hz(q_h)$ and consequently, again by (9) together with the continuity of $z(q)$,

$$\lim_{h\to 0} \frac{\|x+hz\| - \|x\|}{h} = z(q_0).$$

In the case where $x(q_0) < 0$, we would obtain, proceeding similarly,

$$\lim_{h\to 0} \frac{\|x+hz\| - \|x\|}{h} = -z(q_0).$$

We have thus proved the necessity of the condition (the existence of the limit (7)), and at the same time, the second part of the lemma.

To show that the condition is sufficient, suppose that the modulus of the function $x(q)$ attains its maximum at two distinct points q_0 and q_1 of Q, i.e. that

$$|x(q_0)| = |x(q_1)| \geq |x(q)| \text{ for every } q \in Q.$$

In the case where $x(q_0) > 0$, put $z(q) = d(q, q_1)$. We then have: $\|x + hz\| - \|x\| \geq x(q_0) + h.d(q_0, q_1) - x(q_0)$, whence

(10) $$\liminf_{h\to 0+} \frac{\|x+hz\| - \|x\|}{h} \geq d(q_0, q_1) > 0.$$

At the same time we have $\|x + hz\| - \|x\| \geq |x(q_1) + hd(q_1, q_1)| - |x(q_1)| = 0$, whence

(11) $$\limsup_{h\to 0-} \frac{\|x+hz\| - \|x\|}{h} \leq 0,$$

and it follows from the inequalities (10) and (11) that the limit (7) cannot exist.

In the case where $x(q_0) < 0$, the same conclusion is reached on putting $z(q) = -d(q,q_1)$, q.e.d.

Two sets are said to be *homeomorphic* when there exists a bijection from one to the other such that both it and its inverse are continuous, and such a bijection is called a *homeomorphism*.

THEOREM 3. *For two compact metric spaces Q and Q_1 to be homeomorphic, it is necessary and sufficient that the spaces E and E_1 of continuous real-valued functions on the two spaces be isometric.*

Proof. Necessity. It is easily verified that if f is a homeomorphism of Q onto Q_1, the transformation of E_1 to E under which, to each function $y \in E_1$ there corresponds the function $x \in E$ given by $x(q) = y[f(q)]$ is an isometry of E_1 onto the whole of E.

Sufficiency. Assuming that the spaces E and E_1 are isometric, let V be an isometry of E onto E_1, i.e. $\|V(x_1) - V(x_2)\| = \|x_1 - x_2\|$ for all $x_1, x_2 \in E$.

Putting $U(x) = V(x) - V(\Theta)$, it is easily seen that the operator U has the same properties as V and, further, that $U(\Theta) = \Theta$. By theorem 2, p.101, U is therefore a bounded linear operator.

Let q_0 be a given point of Q and let $x \in E$ be a function satisfying the inequality (6) of the lemma on p.103. As the operator U leaves norms unaltered, we have, for every number h, putting $U(z) = t$ for $z \in E$:

$$\frac{\|x+hz\| - \|x\|}{h} = \frac{\|y+ht\| - \|y\|}{h},$$

whence, by the preceding lemma,

$$(12) \qquad z(q_0).\text{sign } x(q_0) = \lim_{h\to 0} \frac{\|y+ht\| - \|y\|}{h}.$$

Now, as the operator U maps E onto the whole of E_1, the limit (12) exists for every $t \in E_1$. Consequently there exists, by the lemma, a $q_0' \in Q_1$ such that $|y(q_0')| > |y(q')|$ for every point $q' \neq q_0'$ of Q_1 and

$$\lim_{h\to 0} \frac{\|y+ht\| - \|y\|}{h} = t(q_0').\text{sign } y(q_0') \text{ for every } t \in E_1.$$

It follows from this by (12) that $z(q_0).\text{sign } x(q_0) = t(q_0').\text{sign} y(q_0')$ whence, putting $\varepsilon(q_0') = \text{sign } x(q_0).\text{sign } y(q_0')$, we obtain the following relation between $q_0 \in Q$ and $q_0' \in Q_1$:

$$(13) \qquad t(q_0') = z(q_0).\varepsilon(q_0') \text{ where } |\varepsilon(q_0')| = 1,$$

which holds whenever $z \in E$ and $t = U(z)$.

Consider the function

$$q_0' = f(q_0),$$

defined by this relation, from Q to Q_1.

Firstly, it is one-to-one. Indeed, if $q_1' = f(q_1) = q_2' = f(q_2)$, then by (13) $|z(q_1)| = |z(q_2)|$ for every function $z \in E$, which implies that $q_1 = q_2$, on taking z to be the particular function given by $z(q) = d(q,q_1)$.

Furthermore, f maps Q onto *all* of Q_1, since for any $\bar{q}' \in Q_1$ we have by (13), putting $t(q') = \dfrac{1}{1+d(q',\bar{q}')}$,

$$(14) \qquad |z(q_0)| = \frac{1}{1+d(q_0',\bar{q}_0')} \text{ for every } q_0 \in Q.$$

Now as $\|z\| = \|t\| = 1$, there exists a $q_0 \in Q$ such that $|z(q_0)| = 1$. For the point $q_0' = f(q_0)$ we therefore have, by (14),

$\frac{1}{1+d(q'_0,\bar{q}')} = 1$, whence $d(q'_0,\bar{q}') = 0$ and consequently $\bar{q}' = q'_0$.

Finally, the mapping f is *continuous*. In fact, let $q_0 = \lim_{n\to\infty} q_n$ and put $q'_n = f(q_n)$ for $n=1,2,\ldots$ By (13), we must have $\lim_{n\to\infty} |t(q'_n)| = |t(q'_0)|$ for every $t \in E_1$, and so, in particular, for $t(q') = d(q',q'_0)$, we have $\lim_{n\to\infty} d(q'_n,q'_0) = d(q'_0,q'_0) = 0$. Consequently $\lim_{n\to\infty} q'_n = q'_0$.

Since Q and Q_1 are compact, it now follows that they are homeomorphic.

Remark. This proof shows that if U is an isometry of E onto E_1 and $U(\Theta) = \Theta$, then there exists a homeomorphism f from Q onto Q_1 and a continuous real-valued function ε on E_1 such that

$$y(q') = x[f^{-1}(q')].\varepsilon(q')$$

where $y = U(x)$, $q' \in Q_1$ and $|\varepsilon(q')| = 1$ for all q'.

Applications. The above theorem 3 implies, in particular, that the space C of continuous real-valued functions $x(t)$, $0 \leq t \leq 1$, is not isometric with the space of continuous real-valued functions $x(u,v)$ of two variables u and v, defined on the unit square, $0 \leq u \leq 1$, $0 \leq v \leq 1$.

Nevertheless, the space L^p of pth-power summable functions on the interval $0 \leq t \leq 1$ *is* isometric to the space of pth-power summable functions defined on the unit square. In fact, there exists a bijection $t = \phi(u,v)$ which maps this square (less a set of measure zero) onto the interval $[0,1]$ (again excluding a set of measure zero) in a measure-preserving fashion, i.e. measurable sets are mapped to measurable sets of equal measure.

Thus, if to each function $x(t) \in L^p$, one makes correspond the function $y(u,v) = x[\phi(u,v)]$, it is easy to see that one obtains a bijection between these two function spaces under which distances are unchanged.

§5. Rotations.

A *rotation* of a Banach space E *about the point* $x_0 \in E$ is, by definition, any isometric bijection of E onto itself which maps the point x_0 to itself.

By theorem 2, p. 101, every rotation about Θ is a bounded linear operator.

We are going to study rotations in some particular Banach spaces.

The space C. In C the most general rotation about Θ is given by operators of the form

$$y(t) = \varepsilon.x[\alpha(t)],$$

where $x(t) \in C$, $\varepsilon = +1$ or -1, independently of x, and $\alpha(t)$ is any homeomorphism of the closed unit interval $[0,1]$ *onto itself.*

The proof follows from the remark above, using the fact that if $\varepsilon(t)$ is a continuous real-valued function on $[0,1]$ such that $|\varepsilon(t)| = 1$, then $\varepsilon(t)$ is constant.

The space c. We can regard this space as the space of continuous real-valued functions defined on a bounded closed set of real numbers with exactly one accumulation point. By the remark above, we can easily deduce from this the following theorem.

In c, the most general rotation about Θ is given by $y = U(x)$ where

$$x = (\xi_n) \in c,\ y = (\eta_n) \in c \text{ and } \eta_n = \varepsilon_n.\xi_{\phi(n)},$$

where (ε_n) is any convergent sequence such that $|\varepsilon_n| = 1$ for $n=1,2,\ldots$ and $\phi(n)$ is an arbitrary bijection of the natural numbers onto them-

selves, i.e. a permutation of the natural numbers.

The space L^2. Every rotation of L^2 about Θ is of the form

(15) $$y(t) = \sum_{n=1}^{\infty} \beta_n(t) \int_0^1 \alpha_n(t)x(t)\,dt,$$

where $x(t) \in L^2$ and $(\alpha_n(t))$, $(\beta_n(t))$ for $0 \leq t \leq 1$ are arbitrary complete orthonormal sequences of functions in L^2.

Proof. From (15), we have

$$\int_0^1 (y(t))^2\,dt = \sum_{n=1}^{\infty}\left[\int_0^1 \alpha_n(t)x(t)\,dt\right]^2 = \int_0^1 (x(t))^2\,dt,$$

whence $\|y\| = \|x\|$. Every transformation of the form (15) is thus a rotation about Θ.

Conversely, let U be a rotation about Θ in L^2 and let $(\alpha_n(t))$ be any complete orthonormal sequence in L^2. Putting $\beta_n(t) = U[\alpha_n(t)]$, for $n=1,2,\ldots$, we therefore have

$$x(t) = \sum_{n=1}^{\infty} \alpha_n(t) \int_0^1 \alpha_n(t)x(t)\,dt$$

and consequently $y(t) = U[x(t)]$ is of the form (15). Furthermore

(16) $$\int_0^1 (\beta_n(t))^2\,dt = \int_0^1 \left[U(\alpha_n(t))\right]^2 dt = \int_0^1 (\alpha_n(t))^2\,dt = 1$$

and as $\beta_i(t) + \beta_j(t) = U[\alpha_i(t) + \alpha_j(t)]$, we have for $i \neq j$

$$\int_0^1 [\beta_i(t) + \beta_j(t)]^2\,dt = \int_0^1 [\alpha_i(t) + \alpha_j(t)]^2\,dt = 2,$$

whence by (16)

(17) $$\int_0^1 \beta_i(t)\beta_j(t)\,dt = 0 \text{ for } i \neq j.$$

Consequently, if for some function $\beta(t) \in L^2$ we have $\int_0^1 \beta_n(t)\beta(t)\,dt = 0$, for any $n=1,2,\ldots$, we will have, by (15), $\int_0^1 y(t)\beta(t)\,dt = 0$ for every function $y(t) \in L^2$, from which it follows that $\beta(t) = 0$. It follows from this together with (16) and (17) that $(\beta_n(t))$ is a complete orthonormal sequence of functions in L^2.

The space l^2. A completely analogous theorem can be stated for l^2. This is a consequence of the isometry of the spaces L^2 and l^2 (cf. theorem 1, p. 101).

The spaces L^p and l^p where $1 \leq p \neq 2$. We have the following lemmas:

1. *Let U be a rotation of L^p about Θ, where $1 \leq p \neq 2$. If for a pair $x_1(t), x_2(t)$ of functions belonging to L^p we have*

(18) $$x_1(t) \cdot x_2(t) = 0 \text{ almost everywhere in } [0,1],$$

then for the pair $y_1(t), y_2(t)$, where $y_1 = U(x_1)$ and $y_2 = U(x_2)$, we also have

(19) $$y_1(t) \cdot y_2(t) = 0 \text{ almost everywhere in } [0,1].$$

Proof. For each pair of numbers α, β we have by the hypothesis (18), $\|\alpha x_1 + \beta x_2\|^p = |\alpha|^p \cdot \|x_1\|^p + |\beta|^p \cdot \|x_2\|^p$, whence, by definition of y_1 and y_2, it follows that $\|\alpha y_1 + \beta y_2\|^p = |\alpha|^p \cdot \|y_1\|^p + |\beta|^p \cdot \|y_2\|^p$ and consequently

(20) $$\int_0^1 |\alpha y_1(t) + \beta y_2(t)|^p\,dt = |\alpha|^p \int_0^1 |y_1(t)|^p\,dt + |\beta|^p \int_0^1 |y_2(t)|^p\,dt.$$

In the case $p = 1$, this yields on putting first $\alpha = \beta = 1$ and then $\alpha = -\beta = 1$, the relation

$$\int_0^1 |y_1(t) + y_2(t)|dt = \int_0^1 |y_1(t) - y_2(t)|dt = \int_0^1 [|y_1(t)| + |y_2(t)|]dt,$$

which is only possible when condition (19) is satisfied.

In the case $p > 2$, we obtain from (20), denoting by H the set of $t \in [0,1]$ for which $y_1(t).y_2(t) \neq 0$, the relation

(21) $$\int_H |\alpha y_1(t) + \beta y_2(t)|^p dt = |\alpha|^p \int_H |y_1(t)|^p dt + |\beta|^p \int_H |y_2(t)|^p dt,$$

which gives, putting $\phi(\alpha,t) = |\alpha y_1(t) + \beta y_2(t)|^p$, the equalities

(22) $$\frac{\partial\phi}{\partial\alpha} = p|\alpha y_1(t) + \beta y_2(t)|^{p-1}.\text{sign}[\alpha y_1(t) + \beta y_2(t)].y_1(t)$$

and

(23) $$\frac{\partial^2\phi}{\partial\alpha^2} = p(p-1)|\alpha y_1(t) + \beta y_2(t)|^{p-2}.[y_1(t)]^2.$$

Now, as $|\alpha y_1(t) + \beta y_2(t)|^{p-1} \in L^{p/(p-1)}$ and $y_1(t) \in L^p$, we can assert that the integral $\int_0^\infty \int_H \left|\frac{\partial\phi}{\partial\alpha}\right| d\alpha dt$ exists, whence by (22)

(24) $$\int_H \frac{\partial\phi}{\partial\alpha} = \frac{d}{d\alpha}\int_H \phi(\alpha,t)dt = p.\text{sign }\alpha.|\alpha|^{p-1}\int_H |y_1(t)|^p dt$$

and consequently $\int_H \left(\frac{\partial\phi}{\partial\alpha}\right)_{\alpha=0} dt = 0$; it immediately follows from this, since $\frac{\partial^2\phi}{\partial\alpha^2} \geqq 0$ by (23), that

$$\int_0^\infty \int_H \frac{\partial^2\phi}{\partial\alpha^2} d\alpha dt = \int_H \frac{\partial\phi}{\partial\alpha} dt,$$

whence by (24)

$$\int_H \frac{\partial^2\phi}{\partial\alpha^2} dt = p(p-1)|\alpha|^{p-2}\int_H |y_1(t)|^p dt$$

and consequently by (23)

(25) $$\int_H |\alpha y_1(t) + \beta y_2(t)|^{p-2}.(y_1(t))^2 dt = |\alpha|^{p-2}\int_H |y_1(t)|^p dt.$$

From (25), on putting $\alpha = 0$ and $\beta = 1$, we obtain the equality

(26) $$\int_H |y_2(t)|^{p-2}.|y_1(t)|^2 dt = 0,$$

which implies, by the definition of H, that the measure of H, $m(H) = 0$.

Finally, in the case where $1 < p < 2$, consider for $i = 1$ and 2 the functional Y_i where $Y_i(y) = \int_0^1 Y_i(t)y(t)dt$ for $y(t) \in L^p$ and $Y_i(t) = |y_i(t)|^{p-1}.\text{sign } y_i(t)$. The adjoint of U, U^* is a rotation of the space $L^{p/(p-1)}$ about Θ (for a proof of this see the proof of the subsequent theorem 11, p.113). Put $X_i = U^*(Y_i)$ and $X_i(x) = \int_0^1 X_i(t)x(t)dt$ where $x \in L^p$. We have $X_i(x_i) = Y_i(y_i) = \|Y_i\|.\|y_i\| = \|X_i\|.\|x_i\|$, whence by Riesz' inequality, $X_i(t) = 0$ for the same values of t for which $x_i(t) = 0$. Consequently $X_1(t).X_2(t) = 0$ and as $\frac{p}{p-1} > 2$, we conclude from the case previously considered that $Y_1(t).Y_2(t) = 0$, so that $y_1(t).y_2(t) = 0$. Condition (19) is thus proved.

2. *Let U be a rotation of l^p, where $1 \leq p \neq 2$, about Θ. If, for two sequences $x_1 = (\xi_n^{(1)})$ and $x_2 = (\xi_n^{(2)})$ belonging to l^p we have*

$$\xi_n^{(1)}.\xi_n^{(2)} = 0 \text{ for } n=1,2,\ldots,$$

then for the sequences $y_1 = U(x_1) = (\eta_n^{(1)})$ *and* $y_2 = U(x_2) = (\eta_n^{(2)})$ *we have, equally*

$$\eta_n^{(1)} \cdot \eta_n^{(2)} = 0 \text{ for } n=1,2,\ldots$$

The proof is similar to that of the preceding lemma for the L^p spaces with the appropriate obvious modifications.

The two lemmas yield, respectively, the following theorems on the general form of rotations.

I. *Let* U *be a rotation of the space* L^p, $1 \leqq p \neq 2$, *about* Θ. *Then there exist two functions* $\phi(t)$ *and* $\psi(t)$ *defined for* $0 \leqq t \leqq 1$ *and such that the following conditions are satisfied:*

(a) *the function* $\phi(t)$ *maps almost all of the closed interval* $[0,1]$ *bijectively onto (almost all of) itself in such a way that measurable sets are mapped to measurable sets and conversely,*

(b) *for almost every* $t \in [0,1]$, *we have*

$$\psi(t) = \left[\lim_{h\to 0+} \frac{m(\phi[t,t+h])}{h}\right]^{\frac{1}{p}}$$

where $\phi([t,t+h]) = \{\phi(s): t \leqq s \leqq t+h\}$ *is the image of the closed interval* $[t,t+h]$ *under the function* ϕ,

(c) *for every* $x \in L^p$

$$y(t) = x[\phi(t)] \cdot \psi(t)$$

where $y(t) = U[x(t)]$.

Conversely, if $\phi(t)$ *is a function satisfying condition* (a), *there exists a function* $\psi(t)$ *defined by* (b) *and the operator* U *defined by* (c) *is a rotation of* L^p *about* Θ.

II. *Let* U *be any rotation of the space* l^p, $1 \leqq p \neq 2$, *about* Θ. *Then there exists a function* $\phi(n)$ *and a sequence of numbers* (ε_n) *such that*

(a) ϕ *is a permutation of the natural numbers,*

(b) $|\varepsilon_n| = 1$ *for* $n=1,2,\ldots$,

(c) *for every pair of sequences* $x = (\xi_n) \in l^p$ *and* $y = (\eta_n) \in l^p$ *where* $y = U(x)$

$$\eta_n = \varepsilon_n \cdot \xi_{\phi(n)} \text{ for } n=1,2,\ldots$$

Conversely, for any ϕ *and* (ε_n) *satisfying the conditions* (a) *and* (b), *the operator* U *given by* $y = U(x)$ *as defined by condition* (c) *is a rotation.*

Proof. First let U be a rotation of l^p about Θ. Put

$$\xi_n^{(i)} = \begin{cases} 1 \text{ for } i = n, \\ 0 \text{ for } i \neq n, \end{cases} \tag{27}$$

and $x_i = (\xi_n^{(i)})$ for $i=1,2,\ldots$ We clearly have for each $x = (\xi_n) \in l^p$

$$x = \sum_{i=1}^{\infty} \xi_i x_i. \tag{28}$$

Putting $y_i = U(x_i) = (\eta_n^{(i)})$, we thus have, by (28), for $y = U(x) = (\eta_n)$, the equality $y = \sum_{i=1}^{\infty} \xi_i y_i$, whence

$$\eta_n = \sum_{i=1}^{\infty} \xi_i \eta_n^{(i)} \text{ for } n=1,2,\ldots \tag{29}$$

By (27) we have $\xi_n^{(i)} \cdot \xi_n^{(j)} = 0$ when $i \neq j$; we deduce from this by the second lemma (above) that

$$\eta_n^{(i)} \cdot \eta_n^{(j)} = 0 \text{ for } i \neq j \text{ and } n=1,2,\ldots \tag{30}$$

Since y can be any sequence belonging to l^p, by (29) and (30), for every natural number n, there exists just one natural number $\phi(n)$ such that $\eta_n^{\phi(n)} \neq 0$. It follows from this by (29) that we have

$$(31) \qquad \eta_n = \xi_{\phi(n)} \cdot \varepsilon_n \text{ for } \varepsilon_n = \eta_n^{\phi(n)} \text{ and } n=1,2,\ldots,$$

so that condition (c) is satisfied.

Furthermore, $n_1 \neq n_2$ implies $\phi(n_1) \neq \phi(n_2)$, because otherwise, by (31), we would have for each sequence $(\eta_n) \in l^p$ the equality $\varepsilon_{n_2}\eta_{n_1} - \varepsilon_{n_1}\eta_{n_2} = 0$ which is impossible; moreover, if there existed a natural number n_0 such that $\phi(n) \neq n_0$ for $n=1,2,\ldots$, we would have, by (31), for the sequence $x = (\xi_n)$ where

$$\xi_n = \begin{cases} 1 \text{ for } n = n_0, \\ 0 \text{ for } n \neq n_0, \end{cases}$$

the equality $\eta_n = 0$ for $n=1,2,\ldots$, which is also impossible. Condition (a) is thus proved as well.

Finally, by the definition of a rotation, we have $\|y\| = \|x\|$, which, by (31), yields

$$(32) \qquad \sum_{n=1}^{\infty} |\xi_{\phi(n)}|^p \cdot |\varepsilon_n|^p = \sum_{n=1}^{\infty} |\xi_n|^p \text{ for every } x = (\xi_n) \in l^p.$$

Consequently, if, given any natural number n_0, one chooses the sequence $x = (\xi_n)$ in such a way that

$$\xi_{\phi(n)} = \begin{cases} 1 \text{ for } n = n_0, \\ 0 \text{ for } n \neq n_0, \end{cases}$$

it follows from (32) that $|\varepsilon_{n_0}|^p = 1$, whence $|\varepsilon_{n_0}| = 1$, which proves condition (b).

The converse is obvious.

§6. Isomorphism and equivalence.

Two F-spaces E and E_1 are said to be *isomorphic* when there is a bijective bounded linear operator from E onto the whole of E_1.

Let U be such an operator; by theorem 5 (Chapter III, §3, p. 24) the inverse U^{-1} is also a bounded linear operator, from which it follows that U is a homeomorphism.

The spaces E and E_1 are said to be *equivalent* when there is a bijective bounded linear operator U from E onto E_1 such that $|U(x)| = |x|$ for every $x \in E$.

If two spaces are equivalent, then they are necessarily isomorphic, but, as we shall see, the converse is not true.

Consider two examples.

1° Let c_0 be the space of real sequences which converge to 0. We have the theorem:

The spaces c and c_0 are isomorphic.

In fact, if, for $x = (\xi_i) \in c$, we put

$$\eta_1 = \lim_{i\to\infty} \xi_i \text{ and } \eta_i = \xi_{i-1} - \eta_1 \text{ for } i > 1,$$

we clearly have $\lim_{i\to\infty} \eta_i = 0$, whence, putting $y = (\eta_i)$, we have $y \in c_0$ and it is easy to see that the operator $y = U(x)$ thus defined is additive and satisfies the condition $\|U(x)\| \leq 2\|x\|$; it is therefore a bounded linear operator.

Conversely, if $y = (\eta_i) \in c_0$, we need only put, with $x = (\xi_i)$,

$$\xi_i = \eta_{i+1} + \eta_1 \text{ where } i=1,2,\ldots,$$

to obtain $x \in c$, since $\lim_{i\to\infty} \xi_i = \eta_1$, and to see that $y = 0$ implies $x = 0$.

It follows that U is a bounded linear operator which maps c bijectively onto c_0.

2°. *The spaces of bounded linear functionals defined on*

$$L^p,\ l^p \text{ where } p > 1,\ L^1,\ l^1 \text{ and } c$$

are equivalent, respectively, to the spaces

$$L^q,\ l^q \text{ where } \frac{1}{p} + \frac{1}{q} = 1,\ M,\ m \text{ and } l^1.$$

This is nothing but a reformulation of the theorems on the general form of bounded linear functionals established in Chapter IV, §4 (see p. 36).

Theorem 2, p.101 , immediately implies the

THEOREM 4. *Two Banach spaces E and E_1 which are isometric are equivalent.*

§7. Products of Banach spaces.

Given two Banach spaces E and E_1, let $E \times E_1$ denote the space of all ordered pairs (x,y) where $x \in E$ and $y \in E_1$, with addition and scalar multiplication defined by putting

$$(x,y) + (x',y') = (x+x',y+y') \text{ and } h(x,y) = (hx,hy),$$

where, of course, $x,x' \in E$, $y,y' \in E_1$ and h is a number, and with the norm defined in such a way that the following condition is satisfied:

(33) $$\lim_{n\to\infty} x_n = x_0 \text{ and } \lim_{n\to\infty} y_n = y_0 \text{ if and only if } \lim_{n\to\infty} \|(x_n,y_n) - (x_0,y_0)\| = 0.$$

Thus defined, the space $E \times E_1$ is also a Banach space, which we will call the *product* of the spaces E and E_1.

It is easy to see that condition (33) will be satisfied, if, in particular, we take as norm of the pair $z = (x,y)$ one or other of the expressions

1) $\|z\| = [\|x\|^p + \|y\|^p]^{\frac{1}{p}}$ where $p \geqq 1$,

2) $\|z\| = \max [\|x\|,\|y\|]$,

and that these are not the only expressions that meet this condition.

Moreover, it is quite clear that whatever norms are chosen, provided they satisfy condition (33), isomorphic spaces will always be obtained.

To make clear which norm has been adopted, we shall denote the product of the spaces E and E_1 by $(E \times E_1)_{l^p}$, when endowed with the norm 1) and by $(E \times E_1)_\infty$, when endowed with the norm 2).

One similarly defines the product $E_1 \times E_2 \times \ldots \times E_n$ of a finite number of Banach spaces. It is plain that *the product of separable spaces is separable.*

The product $E \times E$ will be called the *square of E* and will be denoted by E^2.

THEOREM 5. *The spaces L^p, l^p, for $p \geqq 1$, and c are isomorphic with their respective squares.*

Proof. It is enough to associate with each function $x(t) \in L^p$ the pair of functions $(x_1(t),x_2(t))$ defined by the formulae

$$x_1(t) = x\left(\frac{t}{2}\right) \text{ and } x_2(t) = x\left(\frac{1}{2} + \frac{t}{2}\right) \text{ where } 0 \leqq t \leqq 1,$$

to set up a bijective bounded linear operator from L^p to $(L^p)^2$.

Similarly, it is enough to associate with each sequence $x=(\xi_n)\in l^p$ the pair of sequences $x_1=(\eta_n), x_2=(\zeta_n)$ defined by the formulae

$$\eta_n = \xi_{2n} \text{ and } \zeta_n = \xi_{2n-1} \text{ for } n=1,2,\ldots,$$

to see that l^p can be mapped bijectively onto $(l^p)^2$ by means of a bounded linear operator.

Finally, with each sequence $x=(\xi_n)\in c$ let us associate the pair $x_1=(\eta_n), x_2=(\zeta_n)$ defined by the formulae

$$\eta_n = \xi_{2n} - \xi_1 \text{ and } \zeta_n = \xi_{2n+1} - \lim_{n\to\infty}\xi_n + \xi_1 \text{ for } n=1,2,\ldots$$

We have

$$\xi_1 = \lim_{n\to\infty}\zeta_n,\ \xi_{2n} = \eta_n + \lim_{n\to\infty}\zeta_n \text{ and } \xi_{2n+1} = \zeta_n + \lim_{n\to\infty}\eta_n \text{ for } n=1,2,\ldots$$

and we see that we have a bounded linear bijection of c onto c^2.

THEOREM 6. *The space C is isomorphic with the product $C\times c$.*

Proof. Let E denote the subspace of C consisting of the functions $x(t)\in C$ which satisfy the condition

$$x\left(\frac{1}{n}\right) = 0 \text{ for } n=1,2,\ldots$$

For each function $x(t)\in C$, construct the function $\bar{x}(t)\in C$ such that $\bar{x}\left(\frac{1}{n}\right) = x\left(\frac{1}{n}\right)$ and which is linear in the intervals $\left[\frac{1}{n+1},\frac{1}{n}\right]$ for every natural number n.

With each $x(t)\in C$ we associate the pair (consisting of a function and a sequence of numbers)

$$\left(y(t),\left(x\left(\frac{1}{n}\right)\right)\right) \text{ where } y(t) = x(t) - \bar{x}(t).$$

We clearly have $y(t)\in E$ and $\left(x\left(\frac{1}{n}\right)\right)\in c$.

It is easy to see that this correspondence defines a bounded linear operator.

Equally, we see that for each pair $(y(t),(\xi_n))\in E\times c$ there exists a continuous function $x(t)$ such that $y(t) = x(t) - \bar{x}(t)$ and $\xi_n = x\left(\frac{1}{n}\right)$ for $n=1,2,\ldots$, from which it follows that the transformation under consideration is bijective between all of C and all of $E\times c$. These two spaces are therefore isomorphic.

Hence the spaces $C\times c$ and $E\times c\times c = E\times c^2$ are isomorphic. Now, as c^2 is isomorphic with c, by the preceding theorem 5, the space $C\times c$ is isomorphic with $E\times c$ and therefore with C, q.e.d.

THEOREM 7. *The space C is isomorphic with each of the spaces $C^{(p)}$ for $p=1,2,\ldots$*

Proof. With each function $x(t)\in C^{(p)}$ (cf. Introduction, §7, p. 7), associate the pair consisting of the function $y(t) = x^{(p)}(t)$ and the set of p numbers: $x(0), x'(0),\ldots,x^{p-1}(0)$. With R^p denoting p-dimensional space, $C^{(p)}$ is thus isomorphic with $C\times R_p$ and consequently, by the preceding theorem 6, with $C\times c\times R_p$.

Now, as $c\times R_p$ is isomorphic with c, the space $C^{(p)}$ is isomorphic with $C\times c$ and therefore, again by theorem 6, with the space C, q.e.d.

THEOREM 8. *The space C is isomorphic with the space C^2.*

Proof. With each pair $(x(t),y(t))$ of functions of C, associate the pair $(z(t),\xi)$ where $z(t)\in C$ is the function defined by the formulae

$$z(t) = \begin{cases} x(2t) & \text{for } 0 \leq t \leq \frac{1}{2}, \\ y(2t-1) - y(0) + x(1) & \text{for } \frac{1}{2} \leq t \leq 1, \end{cases}$$

and ξ is the number determined, for each $y(t) \in C$, by the equation $\xi = y(0)$.

The space C^2 is thus mapped to $C \times R$, where R is the real line. This transformation is a bounded linear operator and since, by definition, we have $x(t) = z\left(\frac{t}{2}\right)$ and $y(t) = z\left(\frac{1}{2} + \frac{t}{2}\right) - z\left(\frac{1}{2}\right) + \xi$, it is bijective. We have thus established the isomorphism of the spaces C^2 and $C \times R$ and since, by theorem 6 p. 111, C is isomorphic with $C \times c$, the space C^2 is isomorphic with $C \times c \times R$, and therefore, because $c \times R$ and c are isomorphic, with the space $C \times c$ and consequently (again by theorem 6) with the space C, q.e.d.

Remark. It is not known if the space C is isomorphic with the space of all continuous (real-valued) functions defined on the unit square.

§8. The space C as the universal space.

THEOREM 9. *Every separable Banach space E is equivalent to a closed linear subspace of the space C.*

Proof. Let Γ be the set of all bounded linear functionals on E of norm ≤ 1 and let (x_n) be a sequence in E, with $\|x_n\| \leq 1$ for $n=1,2,\ldots$, which is dense in the ball $\{x \in E : \|x\| \leq 1\}$.

Define a distance in Γ by putting, for each pair f_1, f_2 of functionals belonging to Γ

$$(34) \qquad d(f_1, f_2) = \sum_{n=1}^{\infty} \frac{1}{2^n} \cdot \frac{|f_1(x_n) - f_2(x_n)|}{1 + |f_1(x_n) - f_2(x_n)|}.$$

We shall show that, with this definition of distance, Γ is a compact metric space.

Consider a sequence $(f_i) \subseteq \Gamma$ such that $\lim_{p,q\to\infty} d(f_p, f_q) = 0$. By (34), the $\lim_{i\to\infty} f_i(x_n)$ then exists. As $\|f_i\| \leq 1$, it follows from theorem 3 (Chapter V, §1, p. 50) that the sequence $(f_i(x))$ is convergent for each $x \in E$; hence the sequence of functionals (f_i) is weakly convergent to a bounded linear functional f, say, and $\|f\| \leq 1$, whence $f \in \Gamma$. As $\lim_{i\to\infty} f_i(x_n) = f(x_n)$ for $n=1,2,\ldots$, we conclude from (34) that $\lim_{i\to\infty} d(f_i, f) = 0$. Thus Γ is complete.

Now, given any sequence $(f_i) \subseteq \Gamma$, we can, by a diagonal procedure, extract a subsequence (f_{i_k}) such that $\lim_{k\to\infty} f_{i_k}(x_n)$ exists for $n=1,2,\ldots$, whence, as above, we deduce the existence of a functional $f \in \Gamma$ such that $\lim_{k\to\infty} d(f_{i_k}, f) = 0$. Hence Γ is sequentially compact, and therefore a compact metric space.

Consequently there exists a continuous map of the (perfect, nowhere dense) Cantor set $P \subseteq [0,1]$ onto the space Γ. If $f_t \in \Gamma$ denotes the functional which is the image of the point $t \in P$ under this map, let $x \in E$ be an arbitrary element and define $y(t)$ as follows: for each $t \in P$ put

$$y(t) = f_t(x)$$

and for the points of the set $[0,1] \setminus P$, complete the definition of the function $y(t)$ in a linear manner, specifically, for $t \in [0,1] \setminus P$, putting

$$y(t) = \frac{y(t') - y(t'')}{t' - t''} \cdot (t - t'') + y(t''),$$

where t' and t'' denote the nearest points of P such that $t' < t < t''$.

Let us examine the properties of the function $y(t)$ thus defined.

If $\lim_{n\to\infty} t_n = t$ where $(t_n) \subseteq P$, the sequence (f_{t_n}) converges weakly to f_{t_0}, whence $\lim_{n\to\infty} f_{t_n}(x) = f_{t_0}(x)$, so that $\lim_{n\to\infty} y(t_n) = y(t_0)$. The function y is thus continuous in P. Since it is linear elsewhere, it is therefore continuous throughout $[0,1]$, hence $y(t) \in C$.

Moreover, by theorem 3 (Chapter IV, §2, p. 34) there exists a functional $f \in \Gamma$ such that $|f(x)| = \|x\|$. Let $t_0 \in [0,1]$ be the point such that $f = f_{t_0}$. We thus have $|y(t_0)| = |f_{t_0}(x)| = \|x\|$ and as

$$|y(t)| = |f_t(x)| \leq |f_t| \cdot \|x\| \leq \|x\| \text{ for every } t \in P,$$

we conclude from this, since the function $|y(t)|$ attains its maximum in the set P, that $\max_{0\leq t\leq 1} |y(t)| = \|x\|$.

We have thus associated with each element $x \in E$ an element $y = y(t) \in C$ and, putting $y = U(x)$, we see that we have defined an additive operator. As $\|y\| = \|U(x)\| = \|x\|$, it is actually a bounded linear operator and maps the space E isometrically onto a subspace E_1 of C. The spaces E and $E_1 \subseteq C$ are therefore equivalent, q.e.d.

THEOREM 10. *Every separable metric space E can be mapped isometrically onto a subset of C.*

Proof. By a remark due to M. Fréchet, every separable metric space E can be mapped isometrically onto a subset of m. Such a mapping may be obtained, as is easily verified, by associating with each $x \in E$ the sequence (ξ_n) defined by the formula

$$\xi_n = d(x,x_n) - d(x_0,x_n) \text{ for } n=1,2,\ldots,$$

where the sequence (x_n) forms a dense subset of E.

Consequently, it is enough to consider only the case where $E \subseteq m$. It is easily shown that the space consisting of all linear combinations of elements of E together with all limits of sequences thereof is a separable Banach space.

By the preceding theorem 9 there thus exists an isometric mapping of this space and *a fortiori* of its subset E, onto a subset of C, q.e.d.

Remark. By theorem 9 and 10 which have just been established, the space C can be regarded as the *universal* space for separable Banach (respectively metric) spaces. The study of separable Banach spaces thus reduces to that of closed linear subspaces of the space C.

§9. Dual spaces.

Given a Banach space E, the space E^* of all bounded linear functionals defined in E is clearly another Banach space. We shall call E^* the *dual* or *conjugate* space of E.

THEOREM 11. *If two Banach spaces E and E_1 are isomorphic or equivalent respectively, the spaces E^* and E_1^* are equally isomorphic or equivalent, respectively.*

Proof. In fact, if U is a linear homeomorphism of E onto E_1, it follows from theorem 5 (Chapter X, §1, p. 91) that the adjoint operator U^* is equally a linear homeomorphism of E_1^* onto E^*, so that these latter two spaces are isomorphic.

If, further, E and E_1 are equivalent, we have, whenever $X = U^*(Y)$:

$$\|X\| = \sup_{\|x\|\leq 1} |X(x)| = \sup_{\|x\|\leq 1} |Y(U(x))| = \sup_{\|y\|\leq 1} |Y(y)| = \|Y\|,$$

so that the spaces E^* and E_1^* are equivalent in this case, q.e.d.

Remark. Nevertheless, the equivalence of the spaces E^* and E_1^*

does not always imply that of the spaces E and E_1.

Consider, by way of example, the spaces $E = c$ and $E_1 = (c)^2_\infty$. The duals of these spaces are $E^* = l^1$ and $E_1^* = (l^1)^2_{l^1}$, which are easily shown to be equivalent.

However this is not true of the spaces E and E_1. We can regard E as the space of continuous real-valued functions defined on the set Q consisting of the numbers 0 and $\frac{1}{n}$, for $n=1,2,\ldots$, while the space E_1 can be regarded as the space of continuous real-valued functions defined on the set Q_1 consisting of the numbers $0, 1, \frac{1}{n}$ and $1+\frac{1}{n}$, for $n=1,2,\ldots$ Now as the sets Q and Q_1 in question are not homeomorphic, it follows from theorem 3, p. 104, that the spaces E and E_1 are not isometric, and therefore *a fortiori* not equivalent.

THEOREM 12. *If the dual space E^* is separable, so also is the space E.*

Proof. Let $\Gamma \subseteq E^*$ denote the set of all bounded linear functionals on E of norm 1, so that, by hypothesis, there exists a sequence $(X_n) \subseteq \Gamma$ which is dense in Γ.

Let (x_n) be a sequence of elements of E which satisfies the conditions

(35) $$\|x_n\| = 1 \text{ and } X_n(x_n) > \tfrac{1}{2} \text{ for } n=1,2,\ldots$$

If the space E is not separable, then the sequence (x_n) is not fundamental in E, and therefore, by theorem 7 (Chapter IV, §3, p.36), it is not total either. Consequently there exists a functional $X \in \Gamma$ such that

(36) $$\|X\| = 1 \text{ and } X(x_n) = 0 \text{ for } n=1,2,\ldots$$

Putting $Z_n = X_n - X$, we consequently have by (35) and (36) $Z_n(x_n) = X_n(x_n) - X(x_n) > \frac{1}{2}$, whence $\|Z_n\| > \frac{1}{2}$, so that $\|X_n - X\| > \frac{1}{2}$ for every natural number n, which is impossible, as the sequence (X_n) is assumed dense in Γ and X belongs to Γ.

THEOREM 13. *Let E be a separable Banach space such that every norm-bounded sequence (x_i) of elements of E contains a subsequence which is weakly convergent to an element of E. Then the space E is equivalent to the space E^{**} (the dual of E^*).*

Proof. Let G be the set of bounded linear functionals F defined on E^* which are of the form $F(X) = X(x_0)$, for every $X \in E^*$, for some $x_0 \in E$ independent of F. We thus have $|F(X)| \leq \|X\| . \|x_0\|$, whence $\|F\| \leq \|x_0\|$. By theorem 3 (Chapter IV, §2, p. 34) there exists, moreover, a functional $X_0 \in E^*$ such that $\|X_0\| = 1$ and $X_0(x_0) = \|x_0\|$, so that $F(X_0) = \|x_0\|$, whence $\|F\| \geq \|x_0\|$. These two inequalities give $\|F\| = \|x_0\|$.

G is a *total* subset of the space E^{**} of all bounded linear functionals on E^*.

In fact, if for some $X_0 \in E^*$ we have $F(X_0) = 0$ for any $F \in G$, we also have $X_0(x) = 0$, for any $x \in E$, and so $X_0 = 0$.

We now show that the set G is *transfinitely closed*.

To this end, let θ be any limit ordinal and $(F_\xi) \subseteq G$, for $1 \leq \xi < \theta$, a norm-bounded transfinite sequence of functionals. There therefore exists a number $M > 0$ such that $\|F_\xi\| < M$ for $1 \leq \xi < \theta$ and by definition of G every functional F_ξ is of the form $F_\xi(X) = X(x_\xi)$. Let (x_i) be a dense sequence in E which is separable by hypothesis.

For every natural number n let $x_\xi^{(n)}$ be any term of (x_i) which satisfies the inequality

(37) $$\|x_\xi^{(n)} - x_\xi\| < \frac{1}{n}$$

and put

$$F_\xi^{(n)}(X) = X(x_\xi^{(n)}) \text{ for } X \in E^*.$$

In the case where θ is cofinal with ω (therefore when there exists a sequence (ξ_i) of transfinite numbers such that $\lim_{i\to\infty} \xi_i = \theta$ and $\xi_i < \theta$ for $i=1,2,\ldots$), the sequence $\left(x_{\xi_i}^{(n)}\right)$ contains a subsequence which converges weakly to an element $x^{(n)} \in E$. Clearly we then have

$$\overline{\lim_{\xi\to\theta}}\, F_\xi^{(n)}(X) \geq \overline{\lim_{i\to\infty}}\, F_{\xi_i}^{(n)}(X) = \overline{\lim_{i\to\infty}}\, X\left(x_{\xi_i}^{(n)}\right) \geq X\left(x^{(n)}\right)$$

and consequently the functional $F^{(n)}(X) = X\left(x^{(n)}\right)$ is a transfinite limit of the sequence $\left(F_\xi^{(n)}\right)$.

In the case where the limit ordinal θ is not cofinal with ω, the transfinite sequence $\left(x_\xi^{(n)}\right)$, which, by definition, has at most countably many distinct terms, includes a term $x^{(n)}$ such that for every $\eta < \theta$ there exists an $\xi > \eta$ with $x_\xi^{(n)} = x^{(n)}$. We then have

$$\overline{\lim_{\xi\to\theta}}\, F_\xi^{(n)}(X) = \overline{\lim_{\xi\to\theta}}\, X\left(x_\xi^{(n)}\right) \geq X\left(x^{(n)}\right),$$

from which it follows that the functional $F^{(n)}(X) = X\left(x^{(n)}\right)$ is again a transfinite limit of the sequence $\left(F_\xi^{(n)}\right)$.

This established, consider the sequence $\left(x^{(n)}\right)$. This contains a subsequence which converges weakly to an $\bar{x} \in E$. Put $X(\bar{x}) = F_0(X)$. We thus have, on the one hand

$$(38) \qquad \overline{\lim_{n\to\infty}}\, F^{(n)}(X) \geq F_0(X) \text{ for every } X \in E^*$$

and, on the other hand, by definition of G, $F_0 \subset G$. Now, by (37) we have $X(x_\xi) \geq X\left(x_\xi^{(n)}\right) - \frac{1}{n}\|X\|$, whence, by definition of F_ξ and $F_\xi^{(n)}$,

$$\begin{aligned}\overline{\lim_{\xi\to\theta}}\, F_\xi(X) &= \overline{\lim_{\xi\to\theta}}\, X(x_\xi) \geq \overline{\lim_{\xi\to\theta}}\, X\left(x_\xi^{(n)}\right) - \frac{1}{n}\|X\| \\ &= \overline{\lim_{\xi\to\theta}}\, F_\xi^{(n)}(X) - \frac{1}{n}\|X\| \geq F^{(n)}(X) - \frac{1}{n}\|X\|\end{aligned}$$

and consequently, by (38), $\overline{\lim_{\xi\to\theta}}\, F_\xi(X) \geq \overline{\lim_{n\to\infty}}\, F^{(n)}(X) \geq F_0(X)$. The functional F_0 is therefore a transfinite limit of the sequence (F_ξ) and as $F_0 \in G$, the set G is indeed transfinitely closed.

Since it is both total and transfinitely closed, it follows from the remark (Chapter VIII, §2, p. 72) together with lemma 3 (Chapter VIII, §3, p. 75) that the set G is equal to the entire space E^{**}.

By definition of G, there therefore corresponds to every $F \in E^{**}$ an $x \in E$ such that, as was proved to begin with, $\|F\| = \|x\|$. The operator U defined by $U(x) = F$ is consequently a bounded linear bijection which maps E to E^{**} with no change in norm. The spaces E and E^{**} are thus equivalent, q.e.d.

Remark. Thus, for example, the spaces L^p and l^p, for $p > 1$, are equivalent to the dual spaces of the spaces of bounded linear functionals on them (cf. p. 110, 2°).

THEOREM 14. *The dual space of a product of Banach spaces is isomorphic to the product of their duals.*

Proof. If $E_1, E_2, \ldots, E_n$ are Banach spaces, we have to establish the isomorphism of the space E^* where $E = E_1 \times E_2 \times \ldots \times E_n$ with the space $E_1^* \times E_2^* \times \ldots \times E_n^*$. We need only consider the case where $n = 2$.

Let x_1, x_2 and z denote elements of E_1, E_2 and E respectively, and let X_1, X_2 and Z denote bounded linear functionals on these respective spaces.

Let H be the set of all pairs (x_1, Θ) where $x_1 \in E_1$. We can thus regard H as a subset of $E = E_1 \times E_2$ and consequently every bounded linear functional Z, restricted to the space H, determines a bounded linear functional X_1 on E_1. Put

$$Z(z) = X_1(x_1) \text{ for } z = (x_1, \Theta)$$

and, similarly,

$$Z(z) = X_2(x_2) \text{ for } z = (\Theta, x_2).$$

For $z = (x_1, x_2)$ we therefore have, as is easily verified,

(39) $$Z(z) = X_1(x_1) + X_2(x_2).$$

Conversely, given two bounded linear functionals $X_1 \in E_1^*$ and $X_2 \in E_2^*$, the formula (39) defines a functional $Z \in E^*$.

The correspondence is bijective and takes the form of a bounded linear operator from $E_1^* \times E_2^*$ onto the whole of E^*, so that these two spaces are isomorphic, q.e.d.

Remark. Putting $E = [E_1 \times E_2 \times \ldots \times E_n]_{l^p}$ or $E = [E_1 \times E_2 \times \ldots \times E_n]_\infty$ respectively, it is easily seen that the dual space E^* is *isometric*, for $p > 1$, with the space $[E_1^* \times E_2^* \times \ldots \times E_n^*]_{l^{p/(p-1)}}$ and, for $p = 1$, with the space $[E_1^* \times E_2^* \times \ldots \times E_n^*]_\infty$ or with the space $[E_1^* \times E_2^* \times \ldots \times E_n^*]_{l^1}$, respectively.

CHAPTER XII

Linear dimension

§1. Definitions.

Given two F-spaces E and E_1, we shall say that the *linear dimension of the space E does not exceed* that of the space E_1, or, symbolically:

(1) $$\dim_l E \leq \dim_l E_1,$$

if E is isomorphic with a closed linear subspace of E_1.

We will say that the spaces E and E_1 have the *same linear dimension*, symbolically:

$$\dim_l E = \dim_l E_1,$$

when both (1) and

(2) $$\dim_l E_1 \leq \dim_l E$$

hold simultaneously.

We will say that the linear dimension of E is *strictly less* than that of E_1, when (1) holds but (2) does not. Symbolically, we shall write:

$$\dim_l E < \dim_l E_1.$$

Finally, we shall say that the linear dimensions of the two spaces are *incomparable* when neither (1) nor (2) holds.

It follows that isomorphic spaces always have the same linear dimension. It is unknown whether or not the converse is true, but I think it very likely that there exist Banach spaces, even separable ones, which have equal linear dimensions without being isomorphic.

Every space which is isomorphic with n-dimensional Euclidean space will simply be called *n-dimensional*. A Banach space for which no such n exists will be said to be *infinite-dimensional*.

§2. Linear dimension of the spaces c and l^p, for $p \geq 1$.

THEOREM 1. *If, for a Banach space E, one has*

(3) $$\dim_l E < \dim_l c$$

or

(4) $$\dim_l E < \dim_l l^p \quad \textit{for some } p \geq 1,$$

then E is a finite-dimensional space.

Proof. As the space c is isomorphic with the space c_0 of sequences of numbers convergent to 0 (cf. Chapter XI, §6, p. 109, 1°), there exists, by (3), a closed linear subspace $G \subseteq c_0$ isomorphic with E. If E, and therefore G also, were infinite-dimensional, there would exist, for every natural number N, a sequence of $N+1$ elements $z_i \subset G$, $i=1,2,\ldots,N+1$ such that

$$\sum_{i=1}^{N+1} \alpha_i z_i = 0 \text{ implies } \alpha_1=\alpha_2=\ldots=\alpha_{N+1}=0.$$

Consequently, putting $z_i = (\beta_n^i)$, we would be able to find numbers α_i, $i=1,2,\ldots,N+1$, not all zero, satisfying the equations $\sum_{i=1}^{N+1}\alpha_i\beta_n^i = 0$ for $n=1,2,\ldots,N$. With (β_n) denoting the sequence $z = \sum_{i=1}^{N+1}\alpha_i z_i$, we would therefore obtain

(5) $$\|z\| > 0 \text{ and } \beta_n = 0 \text{ for } n=1,2,\ldots,N.$$

We have thus established the existence, for every natural number N, of an element $z = (\beta_n)$ of G satisfying (5).

Now define, by induction, a sequence (y_i) of elements of G, where $y_i = (\eta_n^i)$, by choosing for y_1 any element of G with $\|y_1\| = 1$ and for y_i, $i=1,2,\ldots$, an element of G such that

(6) $$\|y_i\| = 1 \text{ and } \eta_n^i = 0 \text{ for } n=1,2,\ldots,N_{i-1},$$

where N_{i-1} is the least natural number satisfying the inequality

(7) $$|\eta_n^{i-1}| < \frac{1}{3^{i-1}} \text{ for every } n \geqq N_{i-1}.$$

The existence of such a sequence (y_i) is an immediate consequence of the result just established above.

Let G_0 be the set consisting of all polynomials of the form $\sum_{i=1}^{r}\alpha_i y_i$ where $r=1,2,\ldots$ together with all limits of sequences thereof i.e. G_0 is the closure of the set of all such polynomials. G_0 is clearly a closed linear subspace of c_0.

This established, let $x = (\xi_i)$ be any bounded sequence and put

(8) $$\eta_n = \sum_{i=1}^{\infty}\xi_i\eta_n^i \text{ for } n=1,2,\ldots$$

We shall show that

(9) $$\frac{1}{6}\|x\| \leqq \sup_{n\geqq 1}|\eta_n| \leqq \frac{3}{2}\|x\|.$$

Indeed, given a suffix n, there exists, by (6), a natural number m_i such that

(10) $$|\eta_{m_i}^i| = 1 \text{ for } i=1,2,\ldots,$$

whence by definition of N_i

(11) $$N_{i-1} \leqq m_i < N_i$$

and consequently $\lim_{i\to\infty} N_i = \infty$; there therefore exists a natural number k such that, for the suffix n in question,

(12) $$N_{k-1} \leqq n < N_k,$$

where $N_0 = 1$.

For every $i > k$, we consequently have, by (11), $N_k \leqq N_{i-1}$, whence, by (12), $n < N_{i-1}$. We conclude, by virtue of (6), that $\eta_n^i = 0$ for every $i > k$, and therefore by (8) that

(13) $$\eta_n = \sum_{i=1}^{k}\xi_i\eta_n^i$$

For every $i < k$ we also have, by (11), $N_i \leqq N_{k-1}$, whence, by (12), $N_i \leqq n$, so that, by (7), $|\eta_n^i| < \frac{1}{3^i}$. Since $|\eta_n^k| \leqq 1$ and $|\xi_i| \leqq \|x\|$ for every i, it follows from this, by (13), that, on the one hand, we have

$$|\eta_n| \leqq \|x\|\sum_{i=1}^{k-1}\frac{1}{3^i} + \|x\| \leqq \frac{3}{2}\|x\|,$$

whence

$$\sup_{n\geq 1} |\eta_n| \leq \tfrac{3}{2}\|x\|, \tag{14}$$

and on the other hand, for every k satisfying (12),

$$|\eta_n| \geq |\xi_k|.|\eta_n^k| - \|x\| \sum_{i=1}^{k-1} \frac{1}{3^i} \geq |\xi_k|.|\eta_n^k| - \tfrac{1}{2}\|x\|. \tag{15}$$

Now there exists a k such that $|\xi_k| \geq \frac{2}{3}\|x\|$, so that, according to (10), we have $|\eta_{m_k}^k| = 1$. Consequently, as the relation (15) was proved for the arbitrarily chosen suffix n, we deduce from it, for $n = m_k$: $|\eta_n| \geq \frac{2}{3}\|x\| - \frac{1}{2}\|x\| = \frac{1}{6}\|x\|$, whence $\sup_{n\geq 1} |\eta_n| \geq \frac{1}{6}\|x\|$. Taking this inequality together with the inequality (14), we see that formula (9) has thus been established.

Now, with every $x = (\xi_i)$ let us associate the sequence $y = (\eta_n)$, defined by formula (8). By (9), the sequence y is bounded and we have, putting $y = U(x)$,

$$\tfrac{1}{6}\|x\| \leq \|U(x)\| \leq \tfrac{3}{2}\|x\|, \tag{16}$$

so that U is a bounded linear operator.

However, for $x_i = (\xi_n^i)$, where

$$\xi_n^i = \begin{cases} 1 & \text{for } i = n, \\ 0 & \text{for } i \neq n, \end{cases}$$

we have, by definition, $y_i = U(x_i)$ for $i=1,2,\ldots$ Consequently for $x = (\xi_i) \in c_0$ we have $x = \sum_{i=1}^{\infty} \xi_i x_i$ whence, by the continuity of the operator U, it follows that $y = U(x) = \sum_{i=1}^{\infty} \xi_i U(x_i) = \sum_{i=1}^{\infty} \xi_i y_i$, so that, this last series being convergent, we obtain $y \subset G_0$.

Conversely, let $y \in G_0$. By definition of G_0 we therefore have $y = \lim_{n\to\infty} s_n$ where $s_n = \sum_{i=1}^{r_n} \alpha_i^n y_i$; for $t_n = \sum_{i=1}^{r_n} \alpha_i^n x_i$ we consequently have $t_n \in c_0$ and $U(t_n) = s_n$. Now (16) yields $\frac{1}{6}\|t_p - t_q\| \leq \|U(t_p - t_q)\| = \|s_p - s_q\|$; the equality $\lim_{p,q\to\infty} \|s_p - s_q\| = 0$ thus implies $\lim_{p,q\to\infty} \|t_p - t_q\| = 0$. Hence the sequence (t_n) is convergent. Putting $x = \lim_{n\to\infty} t_n$, we therefore have $x \in c_0$ and $U(x) = y$, from which it follows that the operator U is *one-to-one* and maps c_0 onto *all of* G_0.

The spaces c_0 and G_0 are thus isomorphic and as $G_0 \subseteq G$, this implies that $\dim_l c_0 \leq \dim_l G$, from which it follows, due to the isomorphism of G with E and of c_0 with c, that $\dim_l c \leq \dim_l E$, which contradicts the hypothesis (3). E is therefore finite-dimensional, q.e.d.

The proof for l^p, $p \geq 1$, is similar.

§3. Linear dimension of the spaces L^p and l^p for $p > 1$.

THEOREM 2. *Every sequence of functions* $(x_i(t)) \subseteq L^p$ *which converges weakly to* 0 *contains a subsequence* $(x_{i_k}(t))$ *such that*

$$\Big\|\sum_{k=1}^{n} x_{i_k}\Big\| = \begin{cases} O(n^{\frac{1}{p}}) & \textit{for } 1 < p \leq 2, \\ O(n^{\frac{1}{2}}) & \textit{for } p \geq 2. \end{cases} \tag{17}$$

Proof. The proof will rest on the following inequality for $p > 1$:

$$|a+b|^p \leq |a|^p + p|a|^{p-1} b.\operatorname{sign} a + A|b|^p + B \sum_{j=2}^{E(p)} |a|^{p-j}|b|^j, \tag{18}$$

where a and b are arbitrary real numbers, A and B are constants which only depend on p and $E(p)$ denotes the integer part of p. The last term on the right hand side thus vanishes when $p \leqq 2$.

Define the sequence (x_{i_k}) by induction, putting $i_1 = 1$ and, for $n > 1$, letting i_n be any natural number such that the inequality

$$(19) \qquad p\left|\int_0^1 |s_{n-1}(t)|^{p-1}.\mathrm{sign}\ s_{n-1}(t).x_{i_n}(t)dt\right| \leqq 1,$$

where $s_{n-1}(t) = \sum_{k=1}^{n-1} x_{i_k}(t)$, is satisfied. Such an i_n exists since, by hypothesis, the sequence $(x_i(t))$ converges weakly to 0 and $|s_{n-1}(t)|^{p-1} \in L^q$ where $\frac{1}{p} + \frac{1}{q} = 1$.

Putting $a = s_{n-1}(t)$ and $b = x_{i_n}(t)$, the inequality (18) yields, on integration:

$$(20) \qquad \begin{aligned} \int_0^1 |s_n|^p dt \leqq \int_0^1 |s_{n-1}|^p dt + p\int_0^1 |s_{n-1}|^{p-1}.\mathrm{sign}\ s_{n-1}x_{i_n}dt \\ + A\int_0^1 |x_{i_n}|^p dt + B\sum_{j=2}^{E(p)}\int_0^1 |s_{n-1}|^{p-j}|x_{i_n}|^j dt. \end{aligned}$$

The weak convergence of the sequence $(x_n(t))$ implies, by virtue of theorem 1 (Chapter IX, §1, p. 89) that the sequence of numbers $(\|x_n\|)$ is bounded and, without loss of generality, we can assume that that

$$(21) \qquad \|x_n\| \leqq 1 \text{ for } n=1,2,...$$

Now, in the case $p > 2$, we have by (21), in view of Riesz' inequality (cf. Introduction, §2, p.1) for $2 \leqq j \leqq p$:

$$\int_0^1 |s_{n-1}|^{p-j}|x_{i_n}|^j dt \leqq \left[\int_0^1 |s_{n-1}|^p dt\right]^{(p-j)/p} \leqq 1 + \left[\int_0^1 |s_{n-1}|^p dt\right]^{(p-2)/p}$$

whence, by (19) and (20), $\|s_n\|^p \leqq \|s_{n-1}\|^p + 1 + A + Bp(1 + \|s_{n-1}\|^{p-2})$, which by iteration yields

$$(22) \qquad \|s_n\|^p \leqq C.n + D\sum_{k=1}^{n-1}\|s_k\|^{p-2}$$

where $C = 1 + A + Bp$ and $D = Bp$.

Let $M = C + D + 2$. We are going to show by induction that

$$(23) \qquad \|s_n\| \leqq M.n^{\frac{1}{2}} \text{ for } n=1,2,...$$

In fact, by definition of s_n and by (21) we have $\|s_1\| \leqq 1$ and, assuming that the inequality (23) holds for suffixes less than some given n, we have by (22) that $\|s_n\|^p \leqq D.M^{p-2}\sum_{k=1}^{n-1} k^{(p-2)/2} + C.n \leqq D.M^{p-2}.n^{p/2} + C.n \leqq M^p n^{p/2}(D.M^{-2} + n^{1-p/2}C.M^{-p})$, which implies (23) since, as is easily checked, the sum in brackets is <1 for $p > 2$.

By (23), the equality $\|s_n\| = O(n^{\frac{1}{2}})$ for $p > 2$ is thus established.

We now pass to the case where $1 < p \leqq 2$. By definition of s_n we deduce from (20) and (21) that $\int_0^1 |s_n|^p dt \leqq \int_0^1 |s_{n-1}|^p dt + 1 + A + B$, whence $\|s_n\|^p \leqq \|s_{n-1}\|^p + C$, where $C = 1 + A + B$, and consequently $\|s_n\|^p \leqq \|s_1\|^p + C(n-1) \leqq C.n$, so that, putting $M^p = C$ we obtain $\|s_n\| \leqq M.n^{1/p}$, from which the equality $\|s_n\| = O(n^{1/p})$ follows in this case also, q.e.d.

Remark. The above theorem is no longer true, for any $p > 1$, if the symbol O in the relations (17) is replaced by o.

Indeed for $p \geqq 2$ let $x_i(t) = \sin 2\pi it$. Since we have $\lim_{i\to\infty} \int_0^1 \alpha(t)\sin 2\pi it dt = 0$ for any integrable function $\alpha(t)$, the sequence $(x_i(t)) \subseteq L^p$ is weakly convergent. Putting $s_n(t) = \sum_{k=1}^{n} x_{i_k}(t)$ where $(x_{i_k}(t))$ denotes an arbitrary subsequence, we therefore have

$$\|s_n(t)\| = \left(\int_0^1 |s_n(t)|^p dt\right)^{\frac{1}{p}} \geq \left(\int_0^1 s_n^2(t)\,dt\right)^{\frac{1}{2}} = \frac{1}{\sqrt{2}}.n^{\frac{1}{2}},$$

which shows that O cannot be replaced by o.

For $1 < p \leq 2$, putting

$$x_i(t) = \begin{cases} 2^{i/p} & \text{for } \frac{1}{2^i} \leq t \leq \frac{1}{2^{i-1}} \\ 0 & \text{for } 0 \leq t < \frac{1}{2^i} \text{ and } \frac{1}{2^{i-1}} < t \leq 1, \end{cases}$$

we have, for any subsequence $(x_{i_k}(t))$, the equality

$$\|s_n\| = \left(\int_0^1 |s_n(t)|^p dt\right)^{\frac{1}{p}} = n^{\frac{1}{p}},$$

which demonstrates the impossibility of replacing O by o in this latter case also.

THEOREM 3. *Every sequence* (x_i) *of elements of* l^p, *where* $p > 1$, *which converges weakly to* 0, *contains a subsequence* (x_{i_k}) *such that*

$$(24) \qquad \left\| \sum_{k=1}^{n} x_{i_k} \right\| = O(n^{1/p}).$$

Proof. Let $x_i = (\xi_r^i)$. The weak convergence of (x_i) to 0 implies (cf. p. 83) that

$$(25) \qquad \lim_{i\to\infty} \xi_r^i = 0 \text{ for } r=1,2,\ldots$$

and that

$$(26) \qquad \|x_i\| \leq M \text{ for } i=1,2,\ldots$$

The sequence (x_{i_k}) is defined inductively in the following way: $x_{i_1} = x_1$ and x_{i_n}, for $n > 1$, is any term of the sequence (x_i) satisfying the inequality

$$(27) \qquad \sum_{j=1}^{N} |\xi_j + \xi_j^{i_n}|^p \leq \sum_{j=1}^{N} |\xi_j|^p + 1,$$

where $(\xi_j) = s_{n-1} = \sum_{k=1}^{n-1} x_{i_k}$ and N denotes a natural number such that

$$(28) \qquad \sum_{j=N}^{\infty} |\xi_j|^p \leq 1.$$

Such an x_{i_n} exists by virtue of (25). We have by definition

$$\|s_n\|^p = \|s_{n-1} + x_{i_n}\|^p \leq \sum_{j-1}^{N} |\xi_j - \xi_j^{i_n}|^p + \sum_{j=N}^{\infty} |\xi_j + \xi_j^{i_n}|^p,$$

whence by (27) and Hölder's inequality

$$\|s_n\|^p \leq \sum_{j=1}^{N} |\xi_j|^p + 1 + \left[\left(\sum_{j=N}^{\infty} |\xi_j|^p\right)^{\frac{1}{p}} + \left(\sum_{j=N}^{\infty} |\xi_j^{i_n}|^p\right)^{\frac{1}{p}}\right]^p$$

and consequently, by (26) and (28), $\|s_n\|^p \leq \|s_{n-1}\|^p + 1 + (1+M)^p = \|s_{n-1}\|^p + C$ where $C = 1 + (1+M)^p$. It follows from this that $\|s_n\|^p \leq C.n$, from which, by definition of s_n, the equality (24) follows, q.e.d.

Remark. The above theorem 3 no longer holds for every $p > 1$ if O is replaced by o in the formlula (24).

In fact, it is enough to put

$$\xi_r^i = \begin{cases} 1 \text{ for } i = r, \\ 0 \text{ for } i \neq r, \end{cases}$$

to have $\left\| \sum_{k=1}^{\infty} x_{i_k} \right\| = n^{1/p}$ for any subsequence (x_{i_k}).

We are going to deduce from theorems 2 and 3 just proved, several relationships, firstly between the linear dimensions of the spaces L^p and L^q, then between those of the spaces l^p and l^q and finally between the linear dimensions of the spaces L^p and those of the spaces l^q, with $p,q > 1$ throughout.

LEMMA. *If* $\dim_l L^p \leqq \dim_l L^q$, *where* $p,q > 1$, *then either* $q \leqq p \leqq 2$ *or* $2 \leqq p \leqq q$.

Proof. By hypothesis, there exists a bounded linear operator U which maps L^p injectively onto a closed subspace G of L^q. If the sequence $(x_n) \subseteq L^p$ is weakly convergent to Θ, the same is true of the sequence (y_n) where $y_n = U(x_n)$. By theorem 2, p. 119, there consequently exists a subsequence (y_{i_n}) such that

$$\left\| \sum_{k=1}^{\infty} y_{i_k} \right\| = O\left(n^{\phi(q)}\right) \text{ where } \phi(q) = \begin{cases} 1/q \text{ for } 1 < q \leqq 2, \\ 1/2 \text{ for } q \geqq 2. \end{cases} \tag{29}$$

As the inverse operator U^{-1} is continuous, there exists an $M > 0$ such that $\|x\| \leqq M\|y\|$, where $x = U^{-1}(y)$, for every $y \in G$, whence $\left\| \sum_{k=1}^{n} x_{i_k} \right\| \leqq M \left\| \sum_{k=1}^{n} y_{i_k} \right\|$ and consequently, by (29), $\left\| \sum_{k=1}^{n} x_{i_k} \right\| = O(n^{\phi(q)})$, so that, (x_i) being any sequence weakly convergent to Θ, we conclude from (29) that

$$\phi(p) \leqq \phi(q). \tag{30}$$

Now as the spaces of bounded linear functionals on L^p and L^q are (cf. Chapter XI, §6, p. 110, 2°) isometric with $L^{p/(p-1)}$ and $L^{q/(q-1)}$ respectively, we have that the adjoint operator U^* maps $L^{q/(q-1)}$ to $L^{p/(p-1)}$ and it follows from theorem 3 (Chapter X, §1, p. 91) that its codomain is all of the space $L^{p/(p-1)}$. By theorem 10 (Chapter X, §1, p.92), there therefore exists an $m > 0$ such that to each $X \in L^{p/(p-1)}$ there corresponds a $Y \in L^{q/(q-1)}$ in such a way that $X = U^*(Y)$ and $\|Y\| \leqq m\|X\|$.

Having said this, let (X_n) be any sequence of elements of $L^{p/(p-1)}$ which converges weakly to 0 and (Y_n) the sequence satisfying the conditions $X_n = U^*(Y_n)$ and $\|Y_n\| \leqq m\|X_n\|$ for every natural number n. Since the sequence of norms $(\|Y_n\|)$ is bounded, (Y_n) has a weakly convergent subsequence (Y_{n_i}), (see Chapter VIII, §7, p. 80). If Y_0 denotes the limit of this subsequence, we have $U^*(Y_0) = 0$, since the sequence (X_{n_i}) converges weakly to 0. We consequently have $X_{n_i} = U^*(Y_{n_i} - Y_0)$ and, further, the sequence $(Y_{n_i} - Y_0)$ converges weakly to 0. Putting $Y_i = Y_{n_i} - Y_0$ for $i=1,2,...$, we can therefore, by theorem 2, p. 119, extract a subsequence (Y_{i_k}) such that

$$\left\| \sum_{k=1}^{n} Y_{i_k} \right\| = O\left(n^{\phi\left(\frac{q}{q-1}\right)}\right) \tag{32}$$

whence, putting $X_{i_k} = U^*(Y_{i_k})$, we obtain $\|X_{i_k}\| \leqq \|U^*\| . \|Y_{i_k}\|$ and

$$\left\| \sum_{k=1}^{n} X_{i_k} \right\| = O\left(n^{\phi\left(\frac{q}{q-1}\right)}\right) \tag{33}$$

Since (X_{i_k}) is a subsequence of (X_n), we conclude from (32) and (33), in view of the remark on p. 121, that

$$(34) \qquad \phi\left(\frac{p}{p-1}\right) \le \phi\left(\frac{q}{q-1}\right)$$

whence, by (30) and the definition of the function ϕ, the desired inequalities follow without difficulty.

This lemma easily leads to the following theorems.

THEOREM 4. *If* $\dim_l L^p = \dim_l L^q$, *where* $p,q > 1$, *we have* $p = q$.

THEOREM 5. *If* $1 < p < 2 < q$, *the spaces* L^p *and* L^q *are of incomparable linear dimensions.*

THEOREM 6. *If* $1 < p \neq 2$, *we have* $\dim_l L^2 < \dim_l L^p$.

Proof. For $x(t) \in L^2$, let

$$y(t) = \frac{a_0}{2} + \sum_{i=1}^{\infty} (a_i \cos 2^i t + b_i \sin 2^i t)$$

where $a_i = \frac{1}{\pi}\int_0^{2\pi} x(t)\cos it\,dt$ and $b_i = \frac{1}{\pi}\int_0^{2\pi} x(t)\sin it\,dt$, for any $i = 0,1,2,\ldots$

As $\sum_{i=0}^{\infty}(a_i^2 + b_i^2) = \int_0^{2\pi} x^2(t)\,dt$, there exists a constant $M > 0$, depending only on p, such that

$$\left[\int_0^{2\pi} |y(t)|^p\right]^{\frac{1}{p}} \le M\left[\sum_{i=0}^{\infty}(a_i^2 + b_i^2)\right]^{\frac{1}{2}}.$$

Putting $y = U(x)$, we therefore have $y \in L^p$ and the above inequality can be written in the form

$$\|y\| \leqq M\|x\|,$$

from which it follows that U is a bounded linear operator.

Moreover there exists a constant K such that

$$\left[\sum_{i=0}^{\infty}(a_i^2 + b_i^2)\right]^{\frac{1}{2}} \le K\int_0^{2\pi} |y(t)|\,dt,$$

whence, by Riesz' inequality (see Introduction, §2, p. 1):

$$\left[\sum_{i=0}^{\infty}(a_i^2 + b_i^2)\right]^{\frac{1}{2}} \le K(2\pi)^{\frac{1}{p}}\left[\int_0^{2\pi} |y(t)|^p dt\right]^{\frac{1}{p}},$$

so that $\|x\| \leqq C\|y\|$ where $C = K(2\pi)^{1/p}$, from which it follows that U has a continuous inverse.

Consequently we have the relationship

$$\dim_l(L^2) \leqq \dim_l(L^p)$$

where the equality sign is excluded (since we would then have, by theorem 4 above, the equality $p = 2$, contrary to hypothesis), q.e.d.

It is worth noting that the following problem is still open: *is it true that for* $q < p < 2$, *just as for* $2 < p < q$, *we always have* $\dim_l L^p < \dim_l L^q$?

For the spaces l^p and l^q we have the

THEOREM 7. *The spaces* l^p *and* l^q *where* $1 < p \neq q > 1$ *are of incomparable linear dimensions.*

Proof. Putting $\dim_l l^p \leqq \dim_l l^q$ and proceeding as in the proof of the lemma, p. 122, one obtains the inequalities (which correspond to the formulae (30) and (34)):

$$\frac{1}{p} \leqq \frac{1}{q} \text{ and } \frac{p-1}{p} \leqq \frac{q-1}{q},$$

whence $p = q$, contrary to hypothesis.

We now pass to the relationships between the linear dimensions of L^p and l^q.

THEOREM 8. *If* $\dim_l L^p \leqq \dim_l l^q$ *where* $p,q > 1$, *we have* $p = q = 2$.

Proof. By the same procedure one obtains (in place of (30) and (34)):

$$\phi(p) \leqq \frac{1}{q} \text{ and } \phi\left(\frac{p}{p-1}\right) \leqq \frac{q-1}{q},$$

where

$$(35) \qquad \phi(n) = \begin{cases} \dfrac{1}{n} & \text{for } n \leqq 2, \\ \dfrac{1}{2} & \text{for } n \geqq 2. \end{cases}$$

It immediately follows from this that $p = q = 2$, q.e.d.

The above theorem 8 implies, by virtue of theorem 1 (Chapter XI, §2, p. 101) the

COROLLARY. *For* $\dim_l L^p = \dim_l l^q$, *it is necessary and sufficient that* $p = q = 2$.

THEOREM 9. *If* $1 < p \neq 2$, *we have* $\dim_l L^p > \dim_l l^p$.

Proof. Indeed, if, on the contrary, we had $\dim_l L^p \leqq \dim_l l^p$, we would have by theorem 8 above, putting $p = q$ there, the equality $p = 2$, contrary to hypothesis.

It therefore remains to show that the spaces in question are of comparable linear dimensions. To this end put

$$y_i(t) = \begin{cases} 2^{i/p} & \text{for } \dfrac{1}{2^i} \leqq t \leqq \dfrac{1}{2^{i-1}} \\ 0 & \text{for } 0 \leqq t < \dfrac{1}{2^i} \text{ and } \dfrac{1}{2^{i-1}} < t \leqq 1, \end{cases}$$

whence $\int_0^1 |y_i(t)|^p dt = 1$, so that $y_i(t) \in L^p$ for $i=1,2,\ldots$; for every $x = (\xi_i) \in l^p$, let

$$y(t) = \sum_{i=1}^{\infty} \xi_i y_i(t),$$

whence $\int_0^1 |y(t)|^p dt = \sum_{i=1}^{\infty} |\xi_i|^p$. Consequently, putting $y = U(x)$, we obtain $\|y\| = \|x\|$, which shows that U is a bounded linear operator that admits a continuous inverse. Moreover, it maps l^p isomorphically onto a subspace of L^p.

THEOREM 10. *For* $1 < q < p < 2$, *just as for* $2 < p < q$, *the spaces* L^p *and* l^q *are of incomparable linear dimensions.*

Proof. Assuming that $\dim_l L^p \geqq \dim_l l^q$, the argument used in the proof of the lemma, p. 122, leads to the inequalities (analogous to (30) and (34)):

$$\frac{1}{q} \leqq \phi(p) \text{ and } \frac{q-1}{q} \leqq \phi\left(\frac{p}{p-1}\right),$$

where the function ϕ is defined by the formula (35). We immediately deduce from this that either $p \leqq q \leqq 2$ or $2 \leqq q \leqq p$, contrary to hypothesis.

The following question nevertheless remains unsettled: *is it true that* $p < q < 2$, *or likewise* $2 < q < p$, *implies the inequality* $\dim_l L^p > \dim_l l^q$?

Appendix

Weak convergence in Banach spaces

We distinguish two notions of weak convergence in Banach spaces, namely: weak convergence of bounded linear functionals and that of elements (cf. Chapter VIII, §4 and Chapter IX, §1). The two notions are clearly different. We are here going to add several theorems connected with the study of these notions.

§1. The weak derived sets of sets of bounded linear functionals.

Given a *separable* Banach space, let Γ be an arbitrary set of bounded linear functionals defined on E.

Let us call a bounded linear functional X a *weak accumulation point* of the set Γ when there exists a sequence of bounded linear functionals (X_k) with $X_k \neq X$ and $X_k \in \Gamma$ for every $k=1,2,\ldots$ which converges weakly to the functional X.

The set of all weak accumulation points of the set Γ will be called the *weak derived set* (*of order* 1) *of* Γ, and the weak derived set of the weak derived set of order $n-1$ of Γ will be called the *weak derived set of order* n of Γ. The successive weak derived sets of Γ will be denoted by $\Gamma_{(1)}, \Gamma_{(2)}, \ldots, \Gamma_{(n)}, \ldots$

If Γ is a linear set, we evidently have

$$\Gamma \subseteq \Gamma_{(1)} \subseteq \Gamma_{(2)} \subseteq \ldots \subseteq \Gamma_{(n)} \subseteq \Gamma_{(n+1)} \subseteq \ldots$$

It is easy to give an example of a linear set Γ which is closed, without being weakly closed.

Indeed, take for Γ the set of bounded linear functionals X defined on the space c_0 of the form

$$(1) \qquad X(x) = \sum_{i=1}^{\infty} C_i \xi_i,$$

where $x = (\xi_i) \in c_0$ and $C_1 = \sum_{i=2}^{\infty} C_i$.

It is easy to see that the set Γ thus defined is linear, closed and that it does not contain the functional of the form (1) where $C_1 = 1$ and $C_i = 0$ for $i=2,3,\ldots$ Moreover, since this last functional (see the Remarks to Chapter VIII, §6, p. 148) is the weak limit of the sequence (X_k) of functionals of the form (1) where

$$C_i = \begin{cases} 1 \text{ for } i = 1 \text{ or } i = k, \\ 0 \text{ for } i \neq 1 \text{ and } i \neq k, \end{cases}$$

the set Γ is not weakly closed.

THEOREM 1. *For every natural number n there exists a linear set of bounded linear functionals defined on the space c_0 whose weak derived set of order n is not weakly closed.*

Proof. Every bounded linear functional X defined on c_0 being of the form (1) where $x = (\xi_i) \in c_0$ and $\sum_{i=1}^{\infty} |C_i| = \|X\|$, let Δ_1 be the set of those for which one has $C_{2i} = 0$ and Δ_2 the set of those where $C_{2i-1} = 0$ for $i=1,2,\ldots$

Set up a one-to-one correspondence between pairs r,s of natural numbers and even numbers $N(r,s)$ and denote by $Z_{r,s}$ the bounded linear functional on c_0 given by $Z_{r,s}(x) = \sum_{i=1}^{\infty} C_i\xi_i$ where $x = (\xi_i) \in c_0$ and

$$(2) \qquad C_i = \begin{cases} 1 \text{ for } i = N(r,s), \\ 0 \text{ for } i \neq N(r,s). \end{cases}$$

Let G be an arbitrary linear set of bounded linear functionals defined on c_0. Let H be the set of all functionals of the form (1) where $C_{2i} = 0$ for $i=1,2,\ldots$ and such that the functional $\sum_{i=1}^{\infty} C_{2i-1}\xi_i$ belongs to G. The set H thus defined is clearly linear and we have $H \subseteq \Delta_1$. Being a subspace of l^1, the set Δ_1 is separable. H therefore contains a sequence of functionals (Y_r) which is dense in the set of bounded linear functionals of norm $\leqq 1$ belonging to H and such that

$$(3) \qquad \|Y_r\| \leqq 1 \text{ for } r=1,2,\ldots$$

For r,s natural numbers, put:

$$(4) \qquad X_{r,s} = Y_r + rZ_{r,s}$$

and let Γ denote the linear set of functionals X of the form

$$(5) \qquad X = \sum_{r,s=1}^{\infty} a_{r,s}X_{r,s} = \sum_{r=1}^{\infty} Y_r \sum_{s=1}^{\infty} a_{r,s} + \sum_{r,s=1}^{\infty} ra_{r,s}Z_{r,s},$$

where at most finitely many of the $a_{r,s}$ are non-zero.

By virtue of (4) and (5) we therefore have, by definition of the sets Δ_1 and Δ_2

$$(6) \qquad \left\| \sum_{r,s=1}^{\infty} a_{r,s}X_{r,s} \right\| \geqq \left\| \sum_{r,s=1}^{\infty} ra_{r,s}Z_{r,s} \right\| = \sum_{r,s=1}^{\infty} |ra_{r,s}|.$$

Now let (X_k), where $X_k \in \Gamma$ for $k=1,2,\ldots$, be a sequence which is weakly convergent to X. By (5) we can put

$$(7) \qquad X_k = \sum_{r,s=1}^{\infty} a_{r,s}^{(k)} X_{r,s} = X_k' + X_k'',$$

where

$$(8) \qquad X_k' = \sum_{r=1}^{\infty} Y_r \sum_{s=1}^{\infty} a_{r,s}^{(k)} \text{ and } X_k'' = \sum_{r,s=1}^{\infty} ra_{r,s}^{(k)} Z_{r,s}.$$

Clearly $X_k' \in \Delta_1$ and $X_k'' \in \Delta_2$ for any k, from which it follows that the sequences (X_k') and (X_k'') converge weakly to some functionals $X' \in \Delta_1$ and $X'' \in \Delta_2$; consequently $X = X' + X''$.

With H' denoting, as usual, the derived set of H in the ordinary sense, we shall show moreover that

$$(9) \qquad X' \in H'.$$

In fact, due to the weak convergence of the sequence (X_k) to X, there exists a number $M > 0$ such that $\|X_k\| \leqq M$ for $k=1,2,\ldots$, whence, by (6)-(8), $\sum_{r,s=1}^{\infty} |ra_{r,s}^{(k)}| \leqq M$; therefore, putting $b_r^{(k)} = \sum_{s=1}^{\infty} a_{r,s}^{(k)}$, we can write

$$(10) \qquad \sum_{r=1}^{\infty} |rb_r^{(k)}| \leqq M \text{ for } k=1,2,\ldots$$

Hence there exists a subsequence (of indices) (k_j) such that the limit $b_r = \lim_{j\to\infty} b_r^{(k_j)}$ exists for every $r=1,2,\ldots$

We therefore have, by (10),

(11) $$\sum_{r=1}^{\infty} r|b_r| \leq M.$$

For each natural number m we consequently have

$$\sum_{r=1}^{\infty} |b_r^{(k_j)} - b_r| \leq \sum_{r=1}^{m-1} |b_r^{(k_j)} - b_r| + \sum_{r=m}^{\infty} |b_r^{(k_j)}| + \sum_{r=m}^{\infty} |b_r|,$$

which, by (11) and the definition of b_r, yields the inequality

$$\overline{\lim_{j\to\infty}} \sum_{r=1}^{\infty} |b_r^{(k_j)} - b_r| \leq 2M/m,$$

whence, as m is arbitrary,

$$\lim_{j\to\infty} \sum_{r=1}^{\infty} |b_r^{(k_j)} - b_r| = 0.$$

Observe that, by (3) and (11), the series $\sum_{r=1}^{\infty} b_r Y_r$ is convergent and the above equality implies, by (8), that X' is its sum. As $Y_r \in H$ for every natural number r and H is a linear set, we therefore have $X' \subset H'$.

It is thus proved that for $X = X' + X'' \in \Gamma_{(1)}$, where $X' \in \Delta_1$ and $X'' \in \Delta_2$, we have $X' \in H'$. Formula (9) is thereby established.

Moreover, it is easily shown that the sequence $(Z_{r,s})$ converges weakly to Θ as $s \to \omega$; hence, by (4), the sequence $(X_{r,s})$ converges weakly to Y_r as $s \to \infty$. We therefore have

(12) $$Y_r \subseteq \Gamma_{(1)} \quad \text{for } r=1,2,\ldots$$

Now let $(X_k) \subseteq \Gamma_{(1)}$ be a sequence which converges weakly to $X \in \Delta_1 \cap \Gamma_{(2)}$. We plainly have $X_k = X_k' + X_k''$, where $X_k' \in H'$ and $X_k'' \in \Delta_2$. It is easily seen that the sequence (X_k') converges weakly to X, whence $X \in H_{(1)}$. Conversely, for each $X \in H_{(1)}$, there exists a sequence $(X_k) \subseteq H$ which converges weakly to X. Without loss of generality we can assume that $\|X_k\| \leq 1$ for $k=1,2,\ldots$ By definition of the sequence (Y_r), there exists, for every k, an index r_k such that $\|X_k - Y_{r_k}\| \leq 1/k$ from which it follows that the sequence (Y_{r_k}) is also weakly convergent to X. It follows from this, by (12), that $X \in \Gamma_{(2)}$, whence $X \in \Delta_1 \cap \Gamma_{(2)}$, since $H_{(1)} \subseteq \Delta_1$ by definition of Δ_1.

Hence

(13) $$\Delta_1 \cap \Gamma_{(2)} = H_1$$

Continuing in this way, it is shown by induction that, in general, one has

(14) $$\Delta_1 \cap \Gamma_{(n+1)} = H_{(n)} \quad \text{for every } n=1,2,\ldots$$

This established, let us return to the given set G. If we assume that the derived set G' of G is not weakly closed, the same will clearly be true of the derived set H' of H, and, by (9) and (13), the same will hold for the weak derived set $\Gamma_{(1)}$ of Γ. Similarly, assuming that the weak derived set $G_{(n-1)}$ of G of order $n-1$ is not weakly closed, the same will evidently be true of the weak derived set $H_{(n)}$ of H, of order n, and so, by (14), also of the weak derived set $\Gamma_{(n+1)}$ of Γ of order $n+1$, q.e.d.

Remark. One can define the *weak derived sets* $\Gamma_{(\xi)}$ *of* Γ *of transfinite order* ξ for transfinite numbers ξ of the second class by putting $\Gamma_{(\xi)} = \bigcup_{\eta<\xi} \Gamma_{(\eta)}$ or $\Gamma_{(\xi)} = (\Gamma_{(\xi-1)})_{(1)}$, according as ξ is or is not a limit ordinal.

One can then establish, by induction, the following theorem, analogous to theorem 1:

For every transfinite number ξ *of the second class, there exists a* *linear set of bounded linear functionals on the space c_0 whose weak derived set of order ξ is not weakly closed.*

Nevertheless, one can show that, if E is a separable Banach space and Γ an arbitrary set of bounded linear functionals on E, there always exists a number ξ, finite or transfinite of the second class, such that the set $\Gamma_{(\xi)}$ is weakly closed. This is an easy consequence of theorem 4 (Chapter VIII, §5, p. 76).

THEOREM 2. *Let E be a separable Banach space and Γ a linear subspace of E^*, the dual space of E. A necessary and sufficient condition for $\Gamma_{(1)} = E^*$ is that there exist a number $M > 0$ such that, for each $x \in E$, Γ contains a functional X satisfying the conditions*

$$\|X\| \leq M \text{ and } |X(x)| = \|x\|. \tag{15}$$

Proof. Necessity. For each natural number n, let Δ_n be the set of bounded linear functionals X on E which are weak limits of sequences (X_k) contained in Γ satisfying the inequality $\|X_k\| \leq n$ for $k=1,2,\ldots$ We therefore have, by theorem 2 (Chapter VIII, §4, p. 75),

$\Gamma_{(1)} = \bigcup_{n=1}^{\infty} \Delta_n$, whence, by hypothesis

$$E^* = \bigcup_{n=1}^{\infty} \Delta_n. \tag{16}$$

Observe that Δ_n is a closed set. Indeed, let $(X_j) \subseteq \Delta_n$ be a sequence where $\lim_{j\to\infty}\|X_j - X\| = 0$. By definition of Δ_n, there therefore exists, for each j, a sequence (X_k^j) which converges weakly to X_j, where $X_k^j \in \Gamma$ and $\|X_k^j\| \leq n$ for $k=1,2,\ldots$ If (x_r) is a dense sequence in E, the equalities $\lim_{k\to\infty} X_k^j(x_r) = X_j(x_r)$ and $\lim_{j\to\infty} X_j(x_r) = X(x_r)$, which hold for any j and r, imply the existence of a sequence $(X_{k_j}^{j})$ such that $\lim_{j\to\infty} X_{k_j}^j(x_r) = X(x_r)$ for every $r=1,2,\ldots$ Since $\|X_{k_j}^j\| \leq n$, it follows, by theorem 2 (Chapter VIII, §4, p. 75) that the sequence $(X_{k_j}^j)$ converges weakly to X, whence $X \in \Delta_n$.

Thus, as every Δ_n is closed and E^* is itself a Banach space, the equality (6) implies the existence of an index n_0 such that Δ_{n_0} contains a ball $K \subseteq E^*$. Let X' denote the centre and ρ the radius of K.

Given an element $x \in E$, there exists, by theorem 3 (Chapter IV, §2, p. 34), a functional $X_0 \in E^*$ such that

$$X_0(x) = \|x\| \text{ and } \|X_0\| = 1. \tag{17}$$

Put

$$\lambda = \frac{\rho}{1+\|X'\|} \text{ and } X'' = \lambda X_0 + (1-\lambda)X'. \tag{18}$$

It easily follows that $\|X'' - X'\| \leq \rho$, whence $X'' \in K \subseteq \Delta_{n_0}$. There consequently exist two sequences (X_k') and (X_k'') of functionals belonging to Γ converging weakly to X' and X'' respectively; we therefore have both

$$\|X_k'\| \leq n_0 \text{ and } \|X_k''\| \leq n_0 \text{ for } k=1,2,\ldots \tag{19}$$

The sequence $\left\{\frac{1}{\lambda}X_k'' - \frac{(1-\lambda)}{\lambda}X_k'\right\}$ is contained in Γ and, by (18), converges weakly to X_0. By (17), there consequently exists an index k_0 such that

$$\frac{1}{\lambda}X_{k_0}''(x) - \frac{(1-\lambda)}{\lambda}X_{k_0}'(x) = \alpha\|x\| \text{ where } \tfrac{1}{2} < \alpha < 2. \tag{20}$$

Therefore, putting $X = \frac{1}{\alpha}\left(\frac{1}{\lambda}X''_{k_0} - \frac{(1-\lambda)}{\lambda}X'_{k_0}\right)$, we obtain $X \in \Gamma$, $X(x) = \|x\|$, and, by virtue of (18)-(20), $\|X\| \leqq M = \frac{2n_0}{\rho}(2 + 2\|X'\| + \rho)$, from which it follows that M is independent of x. Condition (15) is thus seen to be satisfied.

Sufficiency. Let Δ denote the set $\{X: X \in \Gamma \text{ and } \|X\| \leqq 1\}$. Then, by theorem 4 (Chapter VIII, §5, p. 76), replacing Γ by Δ and Δ by (X_r) therein, there exists a sequence $(X_r) \subseteq \Delta$ which is weakly dense in Δ.

Put, for each $x \in E$,

(21) $$y = (\eta_r) \text{ where } \eta_r = X_r(x) \text{ for } r=1,2,\ldots$$

We therefore have

(22) $$|\eta_r| \leqq \|X_r\|.\|x\| \leqq \|x\|,$$

whence $y \in m$, and

(23) $$\|y\| \leqq \|x\|,$$

where the norm of y is that of the space m.

Moreover, with $X \in \Gamma$ denoting a functional which, by hypothesis, satisfies the condition (15), put $X' = \frac{1}{M}X$. Then $\|X'\| \leqq 1$, so that $X' \in \Delta$.

There therefore exists a subsequence (X_{r_j}) which converges weakly to X', whence $\lim_{j\to\infty} |X_{r_j}(x)| = |X'(x)|$, which, by (15) and (21), yields $\overline{\lim_{r\to\infty}}\, |\eta_r| \geq |X'(x)| \geq \frac{1}{M}\|x\|$ and consequently

(24) $$\|y\| \geq \frac{1}{M}\|x\|.$$

Thus, putting $y = U(x)$, we see easily from (21) and (23) that U is a bounded linear operator; by (24), the same is true of the inverse operator U^{-1}. Since the space E is separable by hypothesis, the co-domain E_1 of U is also separable, as U is continuous.

Having said this, let X be any bounded linear functional on E and put

(25) $$Y(y) = X[U^{-1}(y)],$$

so that, the inverse U^{-1} being a bounded linear operator, Y is a bounded linear functional on E_1. By the theorem of S. Mazur (Chapter IV, §4, p. 44), replacing (ξ_j) by (η_r) therein, there therefore exists a double sequence of numbers (α_{nr}) such that

(26) $$Y(y) = \lim_{n\to\infty} \sum_{r=1}^{\infty} \alpha_{nr}\eta_r \text{ for } y \in E_1$$

and $\alpha_{nr} = 0$ for $r > k_n$, where (k_n) is a sequence of natural numbers. By (21), this leads to:

(27) $$\sum_{r=1}^{\infty} \alpha_{nr}\eta_r = \sum_{r=1}^{k_n} \alpha_{nr}\eta_r = \sum_{r=1}^{k_n} \alpha_{nr}X_r(x) = \overline{X}_n(x),$$

from which it follows that $\overline{X}_n \in \Gamma$ for $n=1,2,\ldots$, because Γ is a linear set and $X_r \in \Delta \subseteq \Gamma$.

Moreover, we have, by (26) and (27), $Y[U(x)] = \lim_{n\to\infty} \overline{X}_n(x)$, whence, by (25), $X(x) = \lim_{n\to\infty} \overline{X}_n(x)$ for every $x \in E$; the sequence $(\overline{X}_n)$ therefore converges weakly to X. Hence $X \in \Gamma_{(1)}$, which shows that the condition is indeed sufficient, q.e.d.

It is easy to see that *the set E of all bounded continuous real-valued functions $x(q)$, defined on any metric space Q, constitutes a Banach space*, when addition and scalar multiplication are defined in the usual (pointwise) way and the norm is given by

$$\|x\| = \sup_{q \in Q} |x(q)| \tag{28}$$

If, further, the space Q is *compact*, the space E in question is *separable*.

In these circumstances (with Q compact), we have the following

THEOREM 3. *Let (q_r) denote a sequence of points which is dense in Q. Then, for each bounded linear functional X defined on E, there exists an array of real numbers (α_{ir}) and a sequence of natural numbers (k_n) such that*

$$\lim_{i\to\infty} \sum_{r=1}^{k_i} \alpha_{ir} x(q_r) = X(x) \text{ for } x \in E.$$

The proof follows from the preceding theorem 2, due to the fact that, under these conditions, the set Γ of bounded linear functionals of the form $\sum_{i=1}^{m} a_i x(q_i)$, where the a_i are real numbers and m is an arbitrary natural number, satisfies the hypothesis of theorem 2.

Indeed, for each $x \in E$, there exists a $q_0 \in Q$ such that $x(q_0) \geq \frac{1}{2} \max_{q\in Q} |x(q)| = \frac{1}{2}\|x\|$ and as $X_0(x) = x(q_0)$ is a bounded linear functional of norm 1, one has only to put $M = 2$.

Theorem 3 can also be easily proved by direct application of the theorem of S. Mazur, p. 44.

§2. Weak convergence of elements.

Now let Q be a general abstract set, not necessarily a metric space, and E the Banach space of all *bounded* real-valued functions $x(q)$ defined on Q, with the norm (28).

A functional X defined on E will be called *non-negative* when, for any function $x \in E$, the condition $x(q) \geq 0$, for every $q \in Q$, implies that $X(x) \geq 0$.

THEOREM 4. *Every bounded linear functional X defined on E is the difference of two non-negative bounded linear functionals on E.*

Proof. For each subset S of Q, put

$$\mu(S) = \sup_{T \subseteq S} X(\phi_T) \tag{29}$$

where ϕ_T denotes the characteristic function of the set T. We thus have

$$0 \leq \mu(S) \leq \|X\| \tag{30}$$

and $\mu(S_1 \cup S_2) = \mu(S_1) + \mu(S_2)$ for disjoint sets S_1 and S_2.

By (29), we have, further

$$X(\phi_S) \leq \mu(S). \tag{31}$$

For every function $x \in E$ such that $\|x\| = 1$ let

$$x_n(q) = \frac{i}{n} \text{ for } \frac{i}{n} \leq x(q) < \frac{i+1}{n}, \text{ where } -n \leq i \leq n. \tag{32}$$

We clearly have $|x_n(q) - x(q)| \leq 1/n$ for every $q \in Q$, whence $\|x_n - x\| \leq 1/n$ and consequently

$$x = \lim_{n\to\infty} x_n. \tag{33}$$

With $S_{i,n}$ denoting the set $\{q \in Q\colon x_n(q) = i/n\}$, where $-n \leq i \leq n$, put

$$X'(x) = \lim_{n\to\infty} \sum_{i=-n}^{n} \frac{i}{n}\mu(S_{i,n}). \tag{34}$$

It is easily shown that, by (33), the limit (34) exists and that, by (30), $|X'(x)| \leq \|X\|$.

Now the functional X' is non-negative, because, supposing that

(35) $$x(q) \geqq 0 \text{ for every } q \in Q,$$

we obtain, from (30) and (34), the inequality

(36) $$X'(x) \geqq 0.$$

Observe, moreover, that (32) yields

$$x_n(q) = \sum_{i=0}^{n} \frac{i}{n} \phi s_{i,n}(q),$$

whence, by (31)

$$X(x_n) \leq \sum_{i=0}^{n} \frac{i}{n} \mu(S_{i,n})$$

and consequently by (33) and (34)

(37) $$X(x) \leq X'(x),$$

from which it follows that the functional

(38) $$X'' = X' - X$$

is also non-negative, because, by (37), we always have the inequality $X''(x) \geqq 0$ whenever condition (35) holds. Finally, $X = X' - X''$, by (38).

THEOREM 5. *For a norm-bounded sequence of functions $(x_n) \subseteq E$ to converge weakly to Θ, it is necessary and sufficient that*

(39) $$\lim_{n\to\infty} \overline{\lim_{i\to\infty}} |x_n(q_i)| = 0$$

for every sequence of points $(q_i) \subseteq Q$.

Proof. Necessity. Suppose, on the contrary, that for some sequence $(q_i) \subseteq Q$ we have $\overline{\lim_{n\to\infty}}\ \overline{\lim_{i\to\infty}} |x_n(q_i)| > \alpha > 0$. There therefore exists an increasing sequence (n_k) of natural numbers such that $\overline{\lim_{i\to\infty}} |x_{n_k}(q_i)| > \alpha > 0$ for every k and we can consequently extract, by a diagonal procedure, a subsequence (q_{i_j}) of (q_i) such that

(40) $$|\lim_{j\to\infty} x_{n_k}(q_{i_j})| > \alpha > 0 \text{ for } k=1,2,\ldots$$

Consider the bounded linear functional X defined by the formula

$$X(x) = \operatorname{Lim}_{j\to\infty} x(q_{i_j}) \text{ for every } x \in E,$$

where the symbol Lim has the meaning defined in Chapter II, §3, p.21. We then have, by (40), $|X(x_{n_k})| > \alpha$ for $k=1,2,\ldots$, whence

(41) $$\overline{\lim_{n\to\infty}} |X(x_n)| > \alpha > 0,$$

from which it follows that the sequence (x_n) does not tend weakly to Θ.

Sufficiency. In order to prove that a sequence of functions (x_n), where $\|x_n\| < M$ for $n=1,2,\ldots$, converges weakly to Θ, it is now enough to show, conversely, that there exists no non-negative bounded linear functional X which satisfies the inequality (41).

Suppose, on the contrary, that such a functional X exists; we can obviously assume that

(42) $$\|X\| = 1 \text{ and } \overline{\lim_{n\to\infty}} X(x_n) > \alpha > 0.$$

Put, for every $q \in Q$

$$s_n(q) = \begin{cases} x_n(q) & \text{if } x_n(q) \geqq 0, \\ 0 & \text{if } x_n(q) < 0, \end{cases}$$

and

$$t_n(q) = x_n(q) - s_n(q).$$

At least one of $\overline{\lim\limits_{n\to\infty}}\, X(s_n)$ and $\overline{\lim\limits_{n\to\infty}}\, X(t_n)$ must clearly exceed $\frac{1}{2}\alpha$. Assume, therefore, that

$$(43) \qquad \overline{\lim_{n\to\infty}}\, X(s_n) > \tfrac{1}{2}\alpha > 0.$$

Put, then, for each $q \in Q$

$$y_n(q) = \begin{cases} s_n(q) & \text{if } s_n(q) \geq \frac{1}{6}\alpha, \\ 0 & \text{if } s_n(q) < \frac{1}{6}\alpha. \end{cases}$$

Then $\|s_n - y_n\| \leq \frac{1}{6}\alpha$, whence, by (42) and (43)

$$(44) \qquad \overline{\lim_{n\to\infty}}\, X(y_n) > \frac{1}{3}\alpha > 0.$$

Let S_n denote the subset $\{q \in Q\colon |x_n(q)| \geq \frac{1}{6}\alpha\}$ of Q and let ϕ_n be its characteristic function. As $\|y_n\| \leq \|s_n\| \leq \|x_n\| < M$, we have $\phi_n(q) \geq \frac{1}{M} y_n(q)$ for every $q \in Q$ and $n=1,2,\ldots$, so that, as the functional X is non-negative, $X(M.\phi_n) \geq X(y_n)$, whence, by (44), putting $\beta = \alpha/3M$,

$$(45) \qquad \overline{\lim_{n\to\infty}}\, X(\phi_n) > \beta > 0.$$

Consider the set-function F defined for the subsets S of Q by

$$(46) \qquad F(S) = X(\phi_S)$$

where ϕ_S is the characteristic function of S. The inequality (45) can therefore be written in the form $\overline{\lim\limits_{n\to\infty}}\, F(S_n) > \beta > 0$. Let n_1 be the least natural number such that

$$(47) \qquad \overline{\lim_{n\to\infty}}\, F(S_{n_1} \cap S_n) > 0.$$

Such an n_1 exists.

Indeed, suppose, on the contrary, that $\lim\limits_{n\to\infty} F(S_k \cap S_n) = 0$ and consequently that

$$\lim_{n\to\infty} F\Big(\bigcup_{i=1}^{k} (S_i \cap S_n)\Big) = 0$$

for $k=1,2,\ldots$ There would thus exist two increasing sequences (k_j) and (n_j) such that for $j=1,2,\ldots$

$$k_j < n_j < k_{j+1}, \; F(S_{n_j}) > \beta \text{ and } F\Big(\bigcup_{i=1}^{k_j} (S_i \cap S_{n_j})\Big) < \frac{1}{2}\beta.$$

Putting $T_j = S_{n_j} \setminus \bigcup_{i=1}^{k_j} (S_i \cap S_{n_j})$, we would consequently have

$$(48) \qquad T_{j_1} \cap T_{j_2} = \emptyset, \text{ the empty set, for } j_1 \neq j_2$$

and

$$(49) \qquad F(T_j) > \frac{1}{2}\beta \text{ for } j=1,2,\ldots$$

Hence, with γ_j denoting the characteristic function of the set T_j, formulae (48) and (49) would yield

$$(50) \qquad X\Big(\sum_{j=1}^{n} \gamma_j\Big) > n.\frac{1}{2}\beta \text{ for } n=1,2,\ldots$$

However, by (48), we have $\Big\|\sum_{j=1}^{n} \gamma_j\Big\| \leq 1$, whence $X\Big(\sum_{j=1}^{n} \gamma_j\Big) \leq 1$, for

$n=1,2,\ldots$, contradicting (50).

Proceeding as for (47), one establishes the existence of an increasing sequence (n_j) satisfying the inequalities $\lim_{n\to\infty} F(S_{n_1} \cap S_{n_2} \cap \ldots \cap S_{n_j} \cap S_n) > 0$, from which it follows that none of the sets (S_{n_j}) is empty.

Now let q_i, for $i=1,2,\ldots$, be an arbitrary point of the set $S_{n_1} \cap S_{n_2} \cap \ldots \cap S_{n_i}$. We thus have $q_i \in S_{n_j}$ whenever $i \geq j$, whence, by definition of the set S_n, we have the inequality $|x_{n_j}(q_i)| \geq \frac{1}{6}\alpha$ for each $j=1,2,\ldots$ It follows from this that $\lim_{i\to\infty} |x_{n_j}(q_i)| \geq \frac{1}{6}\alpha$ and consequently that $\overline{\lim_{n\to\infty}} \lim_{i\to\infty} |x_n(q_i)| \geq \frac{1}{6}\alpha$, contrary to the hypothesis (39).

THEOREM 6. *For a norm-bounded sequence* (x_n) *in a Banach space* E *to converge weakly to* Θ, *it is necessary and sufficient that one has*

$$(51) \qquad \lim_{n\to\infty} \lim_{i\to\infty} |X_i(x_n)| = 0$$

for each sequence of functionals (X_i) *belonging to a set* Γ *of bounded linear functionals on* E *possessing the following properties:*

1° Γ *is a norm-bounded set of bounded linear functionals*

2° *there exists a number* $N > 0$ *such that, for each element* $x \in E$, *the set* Γ *contains a functional* X *satisfying the inequality*

$$(52) \qquad X(x) \geq N.\|x\|.$$

Proof. To show that the condition is sufficient, consider the space E_1 of all bounded real-valued functions defined on Γ. With each element $x \in E$ associate the function $f \in E_1$ given by the relation

$$(53) \qquad f(X) = X(x) \text{ for } X \in \Gamma.$$

Let $M = \sup_{X\in\Gamma} \|X\|$ and put $f = U(x)$. By (53) and (52), $N.\|x\| \leq \|f\| \leq M.\|x\|$; consequently, as U is additive, both U and its inverse are in fact bounded linear operators.

This established, if the sequence (x_n) satisfies the condition (51), it follows, by (53), putting $f_n(X) = X(x_n)$, that $\lim_{n\to\infty} \underline{\lim}_{i\to\infty} |f_n(X_i)| = 0$. It follows from this by theorem 5, p. 133, that the sequence (f_n) converges weakly to Θ. As U^{-1} is a bounded linear operator and $x_n = U^{-1}(f_n)$, it results, by theorem 3 (Chapter IX, §5, p. 87) that the sequence (x_n) converges weakly to Θ.

A similar argument shows that the condition is necessary.

THEOREM 7. *For a norm-bounded sequence* (x_n) *in a Banach space* E *to converge weakly to* Θ, *it is necessary and sufficient that one has*

$$(54) \qquad \lim_{n\to\infty} X(x_n) = 0 \text{ for every } X \in \Gamma,$$

where Γ *is a set of functionals possessing properties* 1° *and* 2° (*of theorem* 6) *and, further, is weakly compact.*

Proof. The condition is necessary by the definition of weak convergence of elements. To prove that it is sufficient, it is enough, by virtue of theorem 6, to show that (54) implies (51).

Suppose on the contrary, that there exists a subsequence (x_{n_k}) and a sequence $(X_i) \subseteq \Gamma$ such that

$$(55) \qquad \underline{\lim}_{i\to\infty} |X_i(x_{n_k})| > \alpha > 0 \text{ for } k=1,2,\ldots$$

Now, as the set Γ is, by hypothesis, weakly compact, there would exist a subsequence (X_{i_j}) weakly convergent to a functional $X_0 \in \Gamma$, whence, by (55), $|X_0(x_{n_k})| \geq \alpha > 0$ for $k=1,2,\ldots$, contradicting (54).

The following theorems are easily deduced from the theorems that have just been established.

THEOREM 8. *For a norm-bounded sequence (x_n) of continuous real-valued functions on a compact metric space Q to converge weakly to Θ, it is necessary and sufficient that one has*

$$\lim_{n\to\infty} x_n(q) = 0 \text{ for every } q \in Q.$$

The proof comes from theorem 7, with E denoting the space of continuous real-valued functions on Q and Γ the set of all bounded linear functionals X on E of the form $X(x) = x(q)$, for $x \in E$, for some $q \in Q$. We clearly then have $\|X\| = 1$ for every $X \in \Gamma$ and it is easy to see that Γ also satisfies the other hypotheses of theorem 7.

Remark. In particular, theorem 8 immediately yields conditions for the (weak) convergence of sequences of continuous functions on the closed unit interval and square, respectively.

THEOREM 9. *For a sequence of functions $(x_n) \subseteq M$, the space of (essentially) bounded real-valued functions on $[0,1]$, to converge weakly to Θ, it is necessary and sufficient that for every sequence of functions $(\alpha_i(t))$ such that*

$$\int_0^1 |\alpha_i(t)| = 1 \text{ for } i=1,2,\ldots,$$

we have

$$\lim_{n\to\infty} \overline{\lim_{i\to\infty}} \left|\int_0^1 \alpha_i(t)x_n(t)\,dt\right| = 0.$$

The proof follows from theorem 6, p. 135, with Γ denoting the set of all bounded linear functionals X on M of the form

$$X(x) = \int_0^1 x(t)\alpha(t)\,dt \text{ where } \int_0^1 |\alpha(t)|\,dt = 1.$$

We then have $\|X\| = 1$ for every $X \in \Gamma$ and, for each $x \in M$, there exists a function $\alpha(t)$ satisfying the conditions

$$\int_0^1 |\alpha(t)|\,dt = 1 \text{ and } \int_0^1 \alpha(t)x(t)\,dt \geqq \tfrac{1}{2}\|x\|.$$

It is thus enough to put $N = \frac{1}{2}$ in the theorem mentioned.

THEOREM 10. *For a sequence (x_n), where $x_n = (\xi_k^n)$, of elements of m, the space of bounded real sequences, to converge weakly to 0, it is necessary and sufficient to have, for every sequence of indices (k_i)*

$$\lim_{n\to\infty} \overline{\lim_{i\to\infty}} |\xi_{k_i}^n| = 0.$$

The proof follows from theorem 6, p. 135, with Γ denoting the sequence (X_j) of all functionals of the form

$$X_j(x) = \xi_j \text{ for } x = (\xi_j) \in m \text{ and } j=1,2,\ldots$$

We then have $\|X_j\| = 1$ for $j=1,2,\ldots$ and there further exists, for each $x \in m$, a j such that $|X_j(x)| \geqq \frac{1}{2}\|x\|$. We therefore put $N = \frac{1}{2}$.

Remarks

INTRODUCTION

§3. We write $\lim_{n\to\infty}$ as $x_n(t) = x(t)$ when the sequence of functions $(x_n(t))$ converges asymptotically to the function $x(t)$.

§5. The last theorem implies that if $(x_n(t))$ is a uniformly bounded sequence of functions and $(x_n(t))$ is everywhere convergent, then $\lim_{n\to\infty} \int_a^b x_n(t)\,d\alpha(t)$ exists for every function $\alpha(t)$ of bounded variation (cf. F. Riesz, *Sur le théorème de M. Egoroff et sur les opérations fonctionelles linéaires*, Acta Szeged 1 (1922), p. 18-26).

§6. The proof of Lebesque's theorem, from H. Lebesgue, Annales de Toulouse 1909, is also to be found in H. Hahn, *Über Folgen linearer Operationem*, Monatshefte für Math. u. Phys. 32 (1922), p. 1-88.

§7. The three conditions 1)-3) can be replaced by the following two: 1*) $d(x,y) = 0$ if and only if $x = y$, 2*) $d(x,z) \leq d(x,y) + d(y,z)$, cf. A. Lindenbaum, *Sur les espaces métriques*, Fundamenta Mathematicae 8 (1926) [p. 209-222] p. 211.

The distance between the elements x and y in S can also be defined by the formula $d(x,y) = \inf_{0\leq\omega<\infty} [\omega + m(\{t: |x(t) - y(t)| > \omega\})]$. The metric thus obtained is equivalent to that given in the text.

Similarly, in s the metric $d(x,y) = \inf_{1\leq n} \left[\frac{1}{n} + \max_{1\leq k\leq n} |\xi_k - \eta_k|\right]$ is equivalent to that given in the text (cf. M. Fréchet, *Les espaces abstraits*, Paris 1928, p. 82 and 92).

In the examples 1,3,5,7,8 and 10 one could take the functions to be defined on a more general set. Thus, for example, in 5, p. 6, the functions can be supposed to be defined on an arbitrary compact metric space, or even just a complete metric space, provided, in this last case, one only considers bounded continuous functions.

Many examples of metric spaces, interesting from the point of view of the theory of operators, can be found in the cited works of H. Hahn and M. Fréchet; with regard to its applications, the spaces considered in the works of J. Schauder, *Zur Theorie stetiger Abbildungen in Funktionalräumen* and *Bemerkungen zu meiner Arbeit...* Math. Zeitschr. 26 (1927), p. 47-65 and 417-431, deserve special attention. [Cf. also J. P. Schauder, *Oeuvres*, PWN. Editions Scientifiques de Pologne, Warsaw 1978, p. 63-82 and 83-98].

Among other examples, we note the following.

11. *The space Q of all almost periodic functions* with the metric $d(x,y) = \max_{-\infty<t<+\infty} |x(t) - y(t)|$.

12. *The space* H^p, *for* $p \geq 1$, *of all functions defined in the unit circle* $s^2 + t^2 \leq 1$ *and equivalent* (i.e. equal almost everywhere) *to harmonic functions.* The appropriate metric here is given by the formula

$$d(x,y) = \left(\iint_{s^2+t^2 \leq 1} |x(s,t) - y(s,t)|^p ds dt \right)^{\frac{1}{p}}$$

13. *The space* R *of functions defined in* $[0,1]$ *and equivalent to Riemann integrable functions* with the metric $d(x,y) = \operatorname{ess\,sup}_{0 \leq t \leq 1} |x(t) - y(t)|$. For a function $z(t)$, $0 \leq t \leq 1$, measurable and bounded above almost everywhere, $\operatorname{ess\,sup}_{0 \leq t \leq 1} z(t)$ denotes the infimum of the numbers ω such that $z(t) \leq \omega$ almost everywhere.

Examples 11 and 12 are to be found in the work of G. Ascoli, *Sugli spazi lineari metrici e le loro varieta lineari*, Annali di Mathematica X (1932), p. 33-81, and example 13 in that of W. Orlicz, *Beiträge zur Theorie der Orthogonalentwicklungen*, Studia Mathematica 1 (1929), p. 1-39 and 241-255.

Orlicz has further studied a class of spaces, which includes the L^p spaces, for $p > 1$, with which the other spaces in the class share many properties.

Specifically, let $M(u)$ be a convex function defined for all real values u and such that 1° $M(-u) = M(u)$, 2° $\lim_{u\to\infty} \frac{1}{u} M(u) = 0$, 3° $\lim_{u\to+\infty} \frac{1}{u} M(u) = +\infty$ and 4° $\overline{\lim}_{u\to+\infty} \frac{1}{M(u)} M(2u) < +\infty$.

Let $N(u)$ be the function defined for all real values of v by the relations: $N(v) = \max_{0<u<\infty} [uv - M(u)]$, when $v \geq 0$ and $N(v) = N(-v)$ for $v < 0$.

This done, *the set* O *of all functions* $x(t)$ *defined on* $[0,1]$ *for which the integral* $\int_0^1 M[x(t)]dt$ *exists, metrised as follows*

$$d(x,y) = \sup \int_0^1 [x(t) - y(t)]\omega(t)dt \text{ where } \int_0^1 N[\omega(t)]dt \leq 1,$$

constitutes a complete metric space.

In particular, for

$$M(u) = \frac{1}{p-1}\left(1 - \frac{1}{p}\right)^p \cdot |u|^p$$

where $p > 1$, one has $N(v) = |v|^{p/(p-1)}$ and

$$d(x,y) = \left(\int_0^1 |x(t) - y(t)|^p\right)^{\frac{1}{p}},$$

from which it follows that the space O in this case coincides with L^p.

Replacing, in the definition of $M(u)$, the condition 4° by $\overline{\lim}_{u\to\infty} \frac{1}{M(u)} M(2u) < +\infty$ without altering the definition of $N(v)$, *the space* o *of real sequences* (ξ_n) *such that the series* $\sum_{n=1}^{\infty} M(\xi_n)$ *is convergent, metrised by*

$$d(x,y) = \sup \sum_{n=1}^{\infty} (\xi_n - \eta_n)\omega_n \text{ where } x = (\xi_n),\ y = (\eta_n) \text{ and } \sum_{n=1}^{\infty} N(\omega_n) \leq 1,$$

also constitutes a complete metric space, of which the l^p spaces, $p > 1$, are particular cases (cf. W. Orlicz, *Über eine gewisse Klasse von Räumen vom Typus (B)*, Bull. de l'Acad. Polonaise des Sci. et des Lettres, February 1932).

We finally observe that none of the spaces 1-13, O and o is compact; further: in each of them, the compact sets are nowhere dense.

§8. The spaces 1,2,5-10 and 12, as well as the spaces O and o of W. Orlicz are separable. On the other hand the spaces 3,4,11 and 13 are not separable, while being of the power of the continuum, like the previous ones. In each of these latter spaces the sets of power less than that of the continuum are nowhere dense.

Theorem 6: see F. Hausdorff, *Mengenlehre*, Berlin and Leipzig 1927, p. 195, II.

§9. As K. Kuratowski has observed, if a B-measurable operator maps a separable metric space E bijectively to a metric space E_1, the inverse operator satisfies the Baire condition. The proof rests on theorem 7, p. 10, and on the following theorem: for an operator U from a metric space E to another metric space E_1 to satisfy the Baire condition, it is necessary and sufficient that, for each closed set $G_1 \subseteq E_1$, the set $G = U^{-1}(G_1)$ of elements $x \subset E$ such that $U(x) \in G_1$ satisfy the Baire condition (see K. Kuratowski, *La propriété de Baire dans les espaces métriques*, Fundamenta Mathematicae 16 (1930), p. 390-394.

Every analytic set satisfies the Baire condition: cf. O. Nikodym, *Sur une propriété de l'opération A*, Fundamenta Mathematicae 7 (1925), p. 149-154; the proof for Euclidean spaces, which is to be found there, can be applied to the general case without difficulty, bearing in mind the aforementioned theorem on sets of Category I.

CHAPTER I

§1. In view of the fact that F-spaces, studied in subsequent chapters, are particular examples of G-spaces, when one regards them as groups with respect to the *addition* operation defined on them, we have chosen to settle at the outset on the name *addition* for the fundamental group operation and to make the statements and notation comply with this.

All the metric spaces 1-13, O and o equally constitute G-spaces, as one sees immediately, the fundamental group operation being taken as the usual addition of functions or sequences, respectively. All these spaces are *abelian*, i.e. their addition is commutative, symbolically $x + y = y + x$.

Among other examples of G-spaces, one can mention the following:

14. The space of homeomorphisms of a compact metric space Q into itself, when the distance between two homeomorphisms x and y is defined by the formula $d(x,y) = \sup_{q\in Q} d\big(x(q),y(q)\big) + \sup_{q\in Q} d\big(x^{-1}(q),y^{-1}(q)\big)$, and 'addition' is taken to mean the usual composition of functions.

15. The space of isometric transformations of a ball (lying in a metric space) into itself, when distance and addition are defined as in the preceding example.

16. The space of all functions defined on a metric space Q and taking as values complex numbers of modulus 1 (one can further take them to be continuous or even uniformly continuous), when the distance between two functions x and y is defined by the formula $d(x,y) = \sup_{q\in Q} |x(q) - y(q)|$ and 'addition' is the usual multiplication of functions.

17. The space of bijective transformations of the set of natural numbers to itself, or *permutations* of the natural numbers, with the metric

$$d(x,y) = \sum_{n=1}^{\infty} \frac{1}{2^n} \cdot \frac{|x(n) - y(n)| + |x^{-1}(n) - y^{-1}(n)|}{(1 + |x(n) - y(n)| + |x^{-1}(n) - y^{-1}(n)|)},$$

(where $x(n)$, etc., denotes the image of n under the transformation x, etc.) and with 'addition' being composition of transformations.

Given an arbitrary G-space E, if a sequence (x_n) of elements of E is convergent, we plainly have

$$\text{(I)} \qquad \lim_{p,q\to\infty} d(x_p - x_q, \Theta) = 0,$$

but, in general, we do not know, conversely, if the condition I always implies the convergence of this sequence.

If, in a metric space E, addition of elements is defined in such a way that E, with this addition, becomes a group, and even the axioms II_1 and II_2 are satisfied, it is not sufficient that the condition I always implies the convergence of the sequence (x_n) to an element of E, for the space E to be complete. Nevertheless, it is not known whether there then exists in E another metric, equivalent to the given metric, which would make E a G-space. D. van Dantzig has shown that this is the case under the additional hypothesis that E is an abelian space; in this case, one can even find an equivalent translation-invariant metric, i.e. such that one has $d(x,y) = d(x+z,y+z)$ for every $z \in E$ (cf. D. van Dantzig, *Einige Sätze über topologische Gruppen*, Jahresber. d. Deutsch. Math. Ver. 41, 1932).

The definition of G-spaces, along with all the theorems of the text, is to be found in the note: S. Banach, *Über metrische Gruppen*, Studia Mathematica 3 (1931), p. 101-113 [Oeuvres II, p. 402-411]; cf. also F. Leja, *Sur la notion de groupe abstrait topologique*, Fundamenta Mathematicae 9 (1927), p. 37-44.

§2. The spaces 1-10 (Introduction, §7, p. 6), as well as the spaces 11-13, O and o defined here (see p. 138) are connected.

§3. As well as theorem 5, p. 15, we have the theorem: *the space E being connected, if* (U_n) *is a sequence of bounded linear functionals, the set of points where this sequence is bounded is either of category* I *or is the whole of E.*

§4. It follows from the previous remark that, *the space E being connected, if* $(U_{p,q})$ *is a double sequence of bounded linear functionals such that, for a sequence* $(x_p) \subseteq E$ *one has* $\overline{\lim_{q\to\infty}} |U_{p,q}(x_p)| = +\infty$ *for any p, the set of all* $x \in E$ *such that* $\overline{\lim_{q\to\infty}} |U_{p,q}(x)| = +\infty$ *for* $p=1,2,\ldots$ *is of category* II *and its complement is of category* I.

One can show that theorems 3-7 of Chapter III (p. 24-26) hold even for G-spaces E and E_1, *when the space E is assumed separable* (cf. S. Banach, loc. cit., Studia Math. 3, p. 101-113 [Oeuvres II, p. 402-411]). Theorem 5, p. 15 is also an immediate consequence of theorem 4, p. 15, and the remark, p. 139. The hypothesis that E be separable is essential; it would be interesting to know if the theorems 3-7 in question also hold for non-separable, but connected, G-spaces.

It is to be noted that the following two properties are equivalent for every G-space E:

(α) if U is a bounded linear operator which maps E bijectively to a G-space E_1, the inverse U^{-1} is also a bounded linear operator;

(β) given another metric $d^*(x,y)$ on E with respect to which E is equally a G-space, if $\lim_{n\to\infty} x_n = x_0$ always implies $\lim_{n\to\infty} d^*(x_n, x_0) = 0$, then one also has the reverse implication.

Moreover, it is unknown whether or not these properties hold, for example, for the function space E of example 16, p. 139, when Q there denotes the set of complex numbers of modulus 1.

CHAPTER II

§1. One can equally consider vector spaces with multiplication of elements, not only by real numbers, but also by complex numbers without modifying axioms 1)-7). These spaces form the point of departure of the theory of complex linear operators and of a class, much larger still, of analytic operators, which furnish a generalisation of ordinary analytic functions (cf. for example, L. Fantappié, *I funzionali analitici*, Città di Castello 1930). We intend to expound this theory in another volume.

A subset H of a vector space E is called a *Hamel base in* E when every element $x \in E$ is a (finite) linear combination of elements of H and no element of H is a linear combination of *other* elements of H, i.e. H is a linearly independent set. Every vector space admits Hamel bases and any two such are always of the same cardinality.

§2. The preceding remark implies, for every vector space E, the existence of non-zero additive, homogeneous functionals on E.

§3. The last theorem (see p. 21, 4) immediately implies that to each subset S of the set of natural numbers N one can assign a measure $m(S)$ in such a way that 1) $m(S) \geq 0$, 2) $m(S_1 \cup S_2) = m(S_1) + m(S_2)$ *for disjoint sets* S_1 *and* S_2, 3) $m(S_1) = m(S_2)$, *when* $S_1 \cong S_2$ and 4) $m(N) = 1$.

For any measure satisfying the conditions 1) to 4), the set of all numbers of the form $an + b$ for $n=1,2,\ldots$, with a and b fixed, has measure $1/a$; the set of all prime numbers has measure 0. A measure satisfying conditions 1)-4) does not always coincide with the density (when this is defined), but one can always arrange matters in such a way that this additional condition is also satisfied.

Regarding this theorem, cf. S. Mazur, *O metodach sumowalności*, Księga Pamiatkowa I Polskiego Zjazdu Matematycznego (in Polish), supplement to the Annales de la Société Polonaise de Math. (1929), p. 102-107, see p. 103.

CHAPTER III

§1. Regarding the definition of F-spaces, see M. Fréchet, *Les espaces abstraits topologiquement affines*, Acta. Math. 47 (1926), p. 25-52.

The spaces 11-13, O and o, defined on p. 138, are clearly also F-spaces. S. Mazur has observed that every F-space satisfies the condition

(1) *if* $\lim_{n\to\infty} x_n = x$, $\lim_{n\to\infty} y_n = y$, $\lim_{n\to\infty} h_n = h$ *and* $\lim_{n\to\infty} k_n = k$, *then* $\lim_{n\to\infty} (h_n x_n + k_n y_n) = hx + ky$.

It is unknown whether or not in every vector space, which is complete and satisfies condition (1), the metric can be replaced by an equivalent metric which makes the space an F-space.

§3. Theorems 3-9, p. 24-27, remain valid for every metric vector space E satisfying condition (1) and the following condition:

(2) *if* $\lim_{p,q\to\infty}(x_p - x_q) = 0$, *there exists an element* $x \in E$ *such that* $\lim_{n\to\infty} x_n = x$.

Mazur suggests that this last condition can be replaced by the hypothesis that the space E is complete. A simple proof of theorems 3-5 for the case of Banach spaces is to be found in the note of J. Schauder, *Über die Umkehrung linearer stetiger, Funktionaloperationen*, Studia Math. 2 (1930), p. 1-6 [*Oeuvres*, p. 162-167].

Now consider, in an F-space E, an arbitrary closed linear subspace G. It is clear that a partition of E into disjoint subsets is obtained if we agree that two elements x and y of E shall be in the same subset if and only if $x - y \in G$. The following theorem holds: the set E' of subsets of E thus obtained constitutes an F-space, when the distance and the basic operations are defined by the conditions, where X,Y and Z denote elements of E':

1° $d(X,Y) = \inf\{d(x,y): x \in X \text{ and } y \in Y\}$,

2° $X + Y = Z = \{x+y: x \in X \text{ and } y \in Y\}$,

3° $tX = Y = \{tx: x \in X\}$.

The proof of this theorem may be found in my book *Teorja operacyj*, Tom. I, Warsaw 1931, p. 47-49 (in Polish); cf. also F. Hausdorff, *Zur Theorie der linearen metrischen Räume*, Journ. f. reine u. angew. Math. 167 (1932), p. 294-311.

Using this theorem, one can show that, if U is a continuous linear operator from an F-space E into another F-space E_1, then *if* E *is separable, the codomain of* U *is B-measurable*. However, it is not known if the hypothesis that E be separable is essential.

§4. The method applied here has been further developed by S. Saks and H. Steinhaus who have used it to treat various problems in the theory of functions (cf. S. Saks, [*Sur les fonctionelles de M. Banach et leur application aux développements des fonctions*], Fundamenta Mathematicae 10 (1927), p. 186-196, and H. Steinhaus [*Anwendungen der Funktionenanalysis auf einige Fragen der reellen Funktionentheorie*], Studia Mathematica 1 (1929), p. 51-81).

The following works include applications of another method from the theory of operators to this and related problems:

S. Mazurkiewicz, *Sur les fonctions non dérivables*, Studia Math. 3 (1931), p. 92-94.

S. Mazurkiewicz, *Sur l'intégrale* $\int_0^1 \frac{f(x+t) + f(x-t) - 2f(x)}{t}dt$, ibid., p. 114-118.

S. Banach, *Über die Baire'sche Kategorie gewisser Funktionenmengen*, ibid., p. 173-179, [Oeuvres I, p. 218-222].

H. Auerbach and S. Banach, *Über die Holdersche Bedingung*, ibid., p. 180-184, [Oeuvres I, p. 223-227].

S. Kaczmarz, *Integrale vom Dinischen Typus*, ibid., p. 189-199.

§5. Other applications of the theory of operators to problems in differential equations are given in the following notes:

S. Banach, *Sur certains ensembles de fonctions conduisant aux équations partielles du second ordre*, Math. Zeitschr. 27 (1927), p. 68-75 [Oeuvres I, p. 169-177].

W. Orlicz, *Zur Theorie der Differentialgleichung* $\frac{dy}{dx} = f(x,y)$, Bull. Acad. Polon. Sci. et des Lett., February 1932.

§6. Case 4° of the final theorem of this section is already known (cf. F. Riesz, *Les systèmes d'équations linéaires à une infinité d'inconnues*, Paris 1913).

§7. Given a closed linear subspace $G \subseteq s$, for each element $x_0 \in s \setminus G$ there exists a bounded linear functional f on s such that $f(x) = 0$ for every $x \in G$ and $f(x_0) = 1$.

Theorem 12 implies that if the codomain of a bounded linear operator on s also lies in s, then it is closed. The reference for this theorem is O. Toeplitz, *Über die Auflösung undendlich vieler linearer Gleichungen mit unendlich vielen Unbekannten*, Rendiconti del Circ. Mat. di Palermo XXVIII (1909), p. 88-96.

CHAPTER IV

§1. Normed vector spaces have been discussed independently of me and almost simultaneously by N. Wiener, in his work, *Limit in terms of continuous transformations*, Bull. de la Soc. Math. de France 150 (1922), p. 124-134.

The general class of Banach spaces was studied for the first time in my work, *Sur les opérations dans les ensembles abstraits et leur application aux équations intégrales*, Ph. D. thesis, University of Leopol, June 1920; published in Fund. Math. 3 (1922), p. 133-181.

The spaces 11-13, O and o defined on p. 138, are Banach spaces. On the other hand, the space s of example 2, p. 6 (see also p. 31-32) is not a Banach space, nor, as S. Mazur has shown, is it even homeomorphic to any Banach space.

§2 and 3. Theorems 2-6 may be found in the note of H. Hahn, *Über linearer Gleichungen in linearen Räumen*, Journal für die reine und angewandte Mathematik 157 (1927) p. 214-229; cf. also S. Banach, *Sur les fonctionelles linéares*, Studia Math. 1 (1929) p. 211-216 [Oeuvres II, p. 375-380], in particular theorem 2 and the remark.

Theorem 4 was proved for certain special spaces by F. Riesz *Untersuchungen über Systeme integrierbarer Funktionen*, Mathematische Annalen 69 (1910), p. 449-497) and, in a more general form, by E. Helly (*Über lineare Funktionaloperationen*, Berichte der Wiener Akademie der Wissenschaften, IIa, 121 (1912), p. 265-297).

For F-spaces E, one can establish the equivalence of the following two properties:

(α) Given a continuous linear functional f defined on a linear subspace $G \subseteq E$, there exists a continuous linear functional F on the whole of E such that $F(x) = f(x)$ for every $x \in G$.

(β) Under the same conditions, if, further, G is closed, there exists, for every $x_0 \in E \setminus G$, a continuous linear functional F on E such that $F(x_0) \neq 0$ but $F(x) = 0$ for any $x \in G$.

However, these properties do not necessarily hold in all F-spaces. Thus, for example, any continuous linear functional on the space S must vanish identically.

Given two Banach spaces E and E_1 and a bounded linear operator U, defined on the linear subspace $G \subseteq E$ and whose codomain lies in E_1, we do not know if it may be *extended* from G to the whole of E, i.e. if there exists a bounded linear operator V defined on all of E, with codomain in E_1, such that $V(x) = U(x)$ for every $x \in G$.

This extension of U is always possible when E_1 is finite-dimensional, but even then the condition $\|V\|_E = \|U\|_G$ is not always realisable.

§4. The general form of bounded linear functionals on the space C was first established by F. Riesz (*Sur les opérations fonctionelles linéaires*, C. R. Acad. Sc. Paris 149 (1909), p. 947-977).

The general form of bounded linear functionals on the space L^r, $r > 1$, was proved, for $r = 2$, by M. Fréchet (*Sur les ensembles de fonctions et les opérations linéaires*, C. R. Acad. Sc. Paris 144 (1907), p. 1414-1416), and in the general case by F. Riesz, loc. cit., Math. Ann. 69 (1910), p. 449-497 (see p. 475).

The general form of bounded linear functionals on the space L^1 was first shown by H. Steinhaus (*Additive und stetige Funktionaloperationen*, Math. Zeitschr. 5 (1918), p. 186-221).

The conditions 1°-3°, p. 45, can be replaced by the following two:

1) $\alpha_{nj} = 0$ for every $j > n$ where $n=1,2,\ldots$

2) $\sum_{j=1}^{n} |\alpha_{nj}| = \|f\|$ for $n=1,2,\ldots$

This theorem is due to S. Mazur.

The general form of bounded linear functionals in the Orlicz spaces O and o (cf. p. 138) is established in his paper mentioned there. Thus, for example every bounded linear functional f on the space O is of the form $f(x) = \int_0^1 x(t)\alpha(t)\,dt$ where $\alpha(t)$ is a function such that $\int_0^1 N\big(k\alpha(t)\big)\,dt$ exists for some k lying between 0 and 1.

According to F. Riesz the norm of the bounded linear functional $f(x) = \int_0^1 x(t)\,dg(t)$ on C, where $g(t)$ is a function of bounded variation, is equal to the variation of the function $\bar{g}(t)$ defined as follows:

$$\bar{g}(0) = g(0),\ \bar{g}(1) = g(1) \text{ and } \bar{g}(t) = \lim_{h\to 0+} g(t+h) \text{ for } 0 < t < 1.$$

§6. See F. Riesz, *Sur l'approximation des fonctions continues et des fonctions sommables*, Bull. Calcutta Math. Soc. 20 (1928/9), p. 55-58.

§7. These two theorems are both due to F. Riesz (cf. the papers of Riesz and Helly already mentioned in the Remarks to §2 of this Chapter.

§8. See F. Riesz' book, *Les systèmes d'équations linéaires à une infinité d'inconnues*, Paris 1913.

CHAPTER V

§1. Theorem 3, p. 49, (cf. S. Banach and H. Steinhaus, *Sur le principe de la condensation de singularités*, Fund. Math. 9, (1927) p. 50-61) [Oeuvres II, p. 368], implies that the set Q of points of convergence of a norm-bounded sequence of bounded linear operators (U_n) is always closed. In the general case Q is an $F_{\sigma\delta}$.

In this connection, it should be noted that, as has been shown by S. Mazur and L. Sternbach, if (U_n) is a sequence of bounded linear *functionals* and the set Q of its points of convergence is not closed, there exists in Q a sequence of points (x_i) and a point $x_0 \in E \setminus Q$ such that $\lim_{i\to\infty} x_i = x_0$ and the double sequence $\big(U_n(x_i)\big)$ is bounded.

We deduce from this as a corollary that under these conditions Q is not an F_σ. Moreover, these statements may be extended to the case where (x_i) is, more generally, a sequence of bounded linear *operators*, provided that their codomains lie in a space E_1, also a Banach space and posessing the property:

(γ) for every sequence (y_n), where $y_n \in E_1$ and $\|y_n\| = 1$ for $n=1,2,\ldots$, there exists a sequence of numbers (t_n) such that the series $\sum_{n=1}^{\infty} t_n y_n$ is divergent while the sequence of norms of its partial sums is bounded.

Property (γ) is possessed, for example, by the space c as well as all finite-dimensional Banach spaces.

The above-mentioned corollary may be made more precise, due to a remark of S. Mazur and myself, in the sense that, under the stated conditions, the set Q is not a $G_{\delta\sigma}$. As an application, it yields the theorem that every infinite-dimensional Banach space E contains a linear subspace that is an $F_{\sigma\delta}$ without being a $G_{\delta\sigma}$. S. Mazur and L. Sternbach have further shown that every space of this kind contains a linear subspace which, without being an F_σ, is the intersection of an F_σ and a G_δ; nevertheless every G_δ linear subspace is in fact closed.

In certain Banach spaces one can establish the existence of linear subspaces which are $F_{\sigma\delta\sigma}$ sets without being $F_{\sigma\delta}$'s. Whether such subspaces always exist in infinite-dimensional Banach spaces remains an open problem. Further, it is not known if there exist F-spaces containing linear subspaces of higher Borel class or linear subspaces which are analytic but not B-measurable or, again, linear subspaces satisfying the Baire condition without being analytic. Every infinite-dimensional F-space contains linear subspaces failing to satisfy the Baire condition.

These problems are connected with certain questions concerning additive operators. If E and E_1 are F-spaces, every additive B-measurable operator U defined in a closed linear subspace $G \subseteq E$ and whose codomain lies in E_1 is, by virtue of theorem 4, p. 15, continuous. Moreover, if the set G is not closed, the operator U may not be continuous: we know of examples where, G being B-measurable, the operator U is discontinuous, of the first Baire class; however, we know of no example where it is of a higher Baire class. Similarly, it is not known if the operator U can satisfy the Baire condition without at the same time being B-measurable.

Inverting bounded linear operators leads to linear operators which are discontinuous. If E and E_1 are F-spaces and the bounded linear operator U maps E bijectively to a closed set $G_1 \subseteq E_1$, the inverse operator U^{-1} is continuous by theorem 5, p. 26. Moreover, if G_1 is not closed, the operator U^{-1} is not necessarily continuous, but if the space E is separable, it is always B-measurable. Thus, for example, in the case where $E = E_1 = L^2$, this operator is of Baire class I.

§3. The lemma and theorem 8 are to be found in the note of F. Riesz, loc. cit., p. 151. It is easily seen that the converse of theorem 8 is also true. Furthermore, theorem 8 may be generalised as follows: *every F-space containing a ball which is compact is finite-dimensional*; it is easy to see that the converse is also true.

§4. The theorem on L^r, proved on p. 52, is due, for $r > 1$, to F. Riesz. For $r = 1$, it is due to H. Lebesgue (see Annales de Toulouse, 1909). The theorem for l^r was discovered by E. Landau (*Über einen Konvergenzsatz*, Göttinger Nachrichten 1907, p. 25-27).

§6. All these examples are well known.

§7. The method A, which corresponds to the array A, is called *normal*, when $a_{ik} = 0$ for $i < k$ and $a_{ik} \neq 0$ for $i = k$. For $k > 0$, the C_k Cesàro methods and similarly the E_k Euler methods are examples of such methods. These last are perfect methods, according to S. Mazur (loc. cit. Studia Math. 2, p. 40-50).

Theorem 10 is due to O. Toeplitz (*Über allgemeine lineare Mittelbildungen*, Prace Mat.-Fiz. XXII, Varsovie (1911) p. 113-119).

We do not know if theorem 11, p. 58, holds when the method A is not reversible. For a special class of reversible methods, namely normal methods (see above), this theorem has been proved by S. Mazur (loc. cit. Math. Zeitschr. 28 (1928), p. 599-611, Satz VII).

Theorem 12, p. 97, can be completed as follows: if the method A is permanent, reversible and such that each sequence which is summable by A to a number is also summable to the same number by every permanent method weaker than A, then A is a perfect method. Also, regarding theorem 12, for normal methods, cf. S. Mazur, *Über eine Anwendung der Theorie der Operationen bei der Untersuchung der Toeplitzschen Limitierungsverfahren*, Studia Math. 2 (1930) p. 40-50.

The theorem on the general form of bounded linear functionals defined on a separable linear subspace E of m (see p. 44) shows that every bounded linear functional f agrees on E with a *generalised limit* obtained by a certain method A, i.e. there exists an array A such that every sequence $x \in E$ is summable to $f(x)$ by the method corresponding to this array. Moreover, if E is not separable, this theorem can fail to hold; furthermore, there can then exist, as S. Mazur has observed, a sequence (f_n) of bounded linear functionals on E, weakly convergent to a bounded linear functional f and such that f_n coincides, for every $n=1,2,\ldots$, with a generalised limit obtained by a suitable method, while f lacks this property.

CHAPTER VI

§1. The notion of a compact operator is due to D. Hilbert and F. Riesz who were also the first to show its utility.

According to a remark of S. Mazur, we have the following theorem: if (U_n) is a sequence of compact linear operators, defined in a Banach space E and such that $\lim_{n\to\infty} U_n(x) = x$ for every $x \in E$, a necessary and sufficient condition for a set $G \subseteq E$ to be compact is that the convergence of $\big(U_n(x)\big)$ on G be uniform. A space E for which such a sequence of operators exists is separable by theorem 1, p. 59. The question of if, conversely, every separable Banach space E admits such a sequence of operators, remains open.

On the subject of criteria for the compactness of a set $G \subseteq E$, cf. also A. Kolmogoroff, *Über Kompakheit der Funktionenmengen bei der Konvergenz im Mittel*, Göttinger Nachrichten 1931, pp. 60-63.

§2. All these examples are known.

§3. The notion of adjoint operator was first introduced in its full generality in my note *Sur les fonctionelles linéaires* II, Studia Math. 1 (1929), p. 223-239 [Oeuvres II, p. 381-395], which also includes theorem 3, p. 61. The proof of theorem 4, p. 62, is also to be found in the note of J. Schauder, *Über lineare vollstetige Funktionaloperationen*, Studia Math. 2 (1930), p. 185-196.

CHAPTER VII

§1. W. Orlicz has observed that for weakly complete spaces E theorem 2, p. 65, can be sharpened, namely, the series (2) is then convergent for *every* $x \in E$.

A biorthogonal system $(x_i),(f_i)$ is called *complete* if the sequences (x_i) and (f_i) are total sequences (see the definitions p. 27 and 36). One can show that there exist complete biorthogonal systems in every separable Banach space.

A biorthogonal system $(x_i),(f_i)$ is said to be *normalised* when we have $\|x_i\| = \|f_i\| = 1$ for $i=1,2,\ldots$ According to a remark of H. Auerbach, there exist complete normalised biorthogonal systems in every finite-dimensional Banach space. Nevertheless, we do not know if this is so in every separable Banach space or even if there always exists a complete biorthogonal system such that $\|x_i\| = 1$ for $i=1,2,\ldots$ and $\overline{\lim_{i\to\infty}} \|f_i\| < \infty$.

§2. The previous remark implies that, in theorem 5, p. 66, we can suppress the hypothesis that the sequences $(x_i(t))$ and $(y_i(t))$ are complete.

§3. The notion of base was first introduced in a general setting by J. Schauder in the paper *Zur Theorie stetiger Abbildungen in Funktionalräumen*, Math. Zeitschr. 26 (1927), p. 47-65, which also shows how a base may be constructed in the space C.

The theorem which states that the Haar system constitutes a base in L^p where $p \geq 1$ is to be found in the note of J. Schauder, *Eine Eigenschaft des Haarschen Orthogonalsystems*, Math. Zeitschr. 28 (1928), p. 317-320.

It can be shown that if a given sequence (x_n) in a Banach space E is such that for every $x \in E$ there exists exactly one sequence of numbers (t_n) with the property that the sequence

$$\left(\sum_{n=1}^{k} t_n x_n\right)$$

converges weakly to x, then the sequence (x_n) constitutes a base in E.

The space C^p (see example 7, p. 7) has a base for $p=1,2,\ldots$; however, we do not know if there is one in example 10, p. 7. Furthermore, we do not know if there is a base in, for example, the space of all real-valued functions $x(s,t)$ defined in the square $0 \leqq s \leqq 1, 0 \leqq t \leqq 1$ which admit continuous partial derivatives of the first order, where the algebraic operations are defined in the usual (pointwise) way and the norm is given by

$$\|x\| = \max_{0\leqq s,t\leqq 1} |x(s,t)| + \max_{0\leqq s,t\leqq 1} |x_s'(s,t)| + \max_{0\leqq s,t\leqq 1} |x_t'(s,t)|.$$

The existence of a base in every separable Banach space E is equivalent by theorem 9, established in Chapter XI, §8, p. 112, to the existence of a base in every closed linear subspace E_1 of C. Now we know of no example of a separable infinite-dimensional Banach space, not isomorphic with L^2 and such that each of its closed linear subspaces contains a base. At the same time, note that every infinite-dimensional Banach space contains an infinite-dimensional closed linear subspace which does have a base.

The notion of base can clearly be introduced, more generally, for F-spaces. In the space s a base is given, for example, by the sequence of elements

$$(x_i) \text{ where } x_i = (\xi_n^i) \text{ and } \xi_n^i = \begin{cases} 1 \text{ for } i = n, \\ 0 \text{ for } i \neq n. \end{cases}$$

The space S has no base; this is a consequence of the fact that there exists no non-zero continuous linear functional on S.

§4. Theorem 8, cf. S Banach and H. Steinhaus, *Sur quelques applications du calcul fonctionnel à la théorie des séries orthogonales*, Studia Math. 1 (1929), p. 191-200.

Theorem 10, cf. W. Orlicz, *Beiträge zur Theorie der Orthogonalentwicklungen*, Studia Math. 1 (1929), p. 1-39 and 241-255.

CHAPTER VIII

§4 and 5. According to a remark of S. Mazur, theorems 2-4, p. 75-76, also hold in F-spaces E, on replacing condition (20) in theorems 2 and 3 by the condition that the sequence of functionals (f_n) be bounded over a ball.

§6. The conditions for the weak convergence of functionals were given for the space c by H. Hahn and for the spaces l^p, $p \geq 1$, by F. Riesz.

Conditions (35) and (36) were discovered by H. Lebesgue.

Conditions (45) and (46), for the weak convergence of bounded linear functionals on the space c (see p. 79), for the case of the space c_0 take the form: 1° *the sequence* $\left(\sum_{i=1}^{\infty} |\alpha_{in}|\right)$ *is bounded* and 2° $\lim_{n\to\infty} \alpha_{in} = \alpha_i$ *for* $i=1,2,\ldots$

§7. The theorem on weak compactness in L^p, $p>1$, is due to F. Riesz, loc. cit., Math. Annalen 69 (1910), p. 466-467.

CHAPTER IX

§1. The notion of weak convergence (of elements) was first studied by D. Hilbert, in the space L^2, and by F. Riesz in the spaces L^p, for $p > 1$.

A subset G of a Banach space E is said to be (relatively) *weakly (sequentially* [trans.]) *compact*, when every sequence of elements of G has a weakly convergent subsequence. In the spaces L^p and l^p, for $p > 1$, every bounded set is relatively weakly sequentially compact (cf. p. 79-80). The same is true in c and c_0 while the spaces C, L^1, l^1 and m do not enjoy this property.

§2. The theorem on weak convergence of sequences in L^p, $p > 1$, was proved by F. Riesz, loc. cit., Math. Annalen 69 (1910), p. 465-466.

The conditions for the weak convergence of sequences in the space c were given by H. Hahn and in the spaces C and l^p, $p \geq 1$, by F. Riesz. The theorem on p. 83 on the equivalence of norm and weak convergence in the space l^1 is to be found in the note of J. Schur, *Über lineare Transformationen in der Theorie der unendlichen Reihen*, Journ. f. reine u. angew. Math. 151 (1921), p. 79-111.

It should be noted that the weak convergence of a sequence of bounded linear functionals on a Banach space E is not a sufficient condition for the weak convergence of the same sequence when regarded as a sequence of *elements* of the space E^*, (in modern terminology, weak* convergence of such a sequence does not imply its weak convergence [trans.]). Thus, for example, in l^1 the notion of weak convergence depends on whether l^1 is regarded as a space of bounded linear functionals (on e.g. c_0: this would give us a type of weak* convergence [trans.]), or not.

§3. The theorem stated here for L^p, $p > 1$, was first proved by M. Radon (Sitzungsberichte der Akad. für Wissensch. in Wien, 122 (1913), Abt. IIa, p. 1295-1438). Cf. also F. Riesz, Acta Litt. Ac. Scient. Szeged, 4 (1929), pp. 58-64 and 182-185.

§4. A Banach space E is said to be *weakly (sequentially) complete* when every weak Cauchy sequence $(x_n) \subseteq E$ (which implies that $\lim_{n\to\infty} f(x_n)$ exists for every bounded linear functional f on E) is weakly convergent to some element of E. The space c_0, and therefore also the spaces c and m, are not weakly sequentially complete. The properties of weak sequential completeness was established for the space L^1 by H. Steinhaus (see *Additive und stetige Funktionaloperationen*, Math. Zeitschr. 5 (1918), p. 186-221) and for the spaces L^p and l^p, for $p > 1$, by F. Riesz (see *Untersuchungen über Systeme integrierbarer Funktionen*, Math. Ann. 69 (1910), p. 449-497). According to a remark of W. Orlicz (loc. cit. Bull. de l'Acad. Polon. des Sci. et des Let., February 1932), the space O is weakly complete, if $\lim_{u\to\infty} \frac{1}{N(u)} N(2u) < +\infty$; the same is true of the space O.

A series of elements of a Banach space is called *unconditionally convergent* when it is always convergent no matter how the terms are ordered. Property 7° (Chapter III, §3, p. 24) established for F-spaces immediately implies that the absolute convergence of a series always implies its *unconditional convergence*, but it is not known if the converse is true in other than finite dimensional spaces. W. Orlicz has proved the following theorems:

(1) *The sum of an unconditionally convergent series is independent of the order of its terms,*

(2) *For a series to be unconditionally convergent, it is necessary and sufficient that every subseries of it be convergent,*

(3) *For the same conclusion, it is necessary and sufficient that every subseries be weakly convergent* (*to some element*).

It follows from this, under the hypothesis that the space E is weakly sequentially complete, that a necessary and sufficient condition for the unconditional convergence of a series $\sum_{n=1}^{\infty} x_n$ of elements of E is that the series $\sum_{n=1}^{\infty} |f(x_n)|$ be convergent for every bounded linear functional f on E. This last result enables one to establish, for weakly sequentially complete spaces, several important properties of unconditionally convergent series, entirely analogous to those of unconditionally convergent series of numbers. Thus, for example, a series $\sum_{n=1}^{\infty} x_n$ is unconditionally convergent when there exists a number $M > 0$ such that $\|x_{n_1} + x_{n_2} + \ldots + x_{n_k}\| < M$ for any set of distinct suffixes $n_1, n_2, \ldots, n_k$, or, again, when the series $\sum_{n=1}^{\infty} t_n x_n$ is convergent for any sequence of numbers (t_n) converging to 0. These theorems play a part in the theory of orthogonal series (cf. W. Orlicz, loc. cit., Studia Math. 1 (1929), p. 241-255).

CHAPTER X

§1. On the subject of the theory of linear equations, developed in this chapter, see F. Hausdorff, *Zur Theorie der linearen Räume*, Journ. f. reine u. angew. Math. 167 (1932), p. 294-311.

The theorems of this section were proved in the case $E = E' = L^2$ by E. Hellinger and O. Toeplitz (see *Integralgleichungen und Gleichungen mit unendlichvielen Unbekannten*, Encyklopedie der Math. Wiss., Leipzig 1923-1927). In the more general case where $E = E' = L^p$, with $p > 1$, theorems 1 and 3 were proved by F. Riesz, loc. cit., Math. Ann. 69 (1910), p. 449-497 and for $E = E' = l^p$ where $p > 1$ by the same author in his book *Les systèmes d'équations linéaires à une infinité d'inconnues*, Paris 1913. Theorems 2 and 4 for $E = E' = L^p$ and l^p respectively, where $p \geqq 1$, were proved by S. Saks, *Remarques sur les fonctionelles linéaires dans les champs* L^p, Studia Mathematica 1 (1929), p. 217-222.

§2. The theorems of this section, except those involving the notion of the adjoint operator, were first proved by F Riesz (*Über lineare Funktionalgleichungen*, Acta Mathematica 41 (1918), p. 71-98).

For certain special cases, theorem 15 was established by F. Riesz, loc. cit., Acta Mathematica 41 (1918), p. 96-98. In its full generality, but formulated differently, this theorem was proved by T. H. Hildebrandt (*Über vollstetige lineare Transformationen*, Acta Mathematica 51 (1928). p. 311-318) and in the version given here by J. Schauder, *Über lineare, vollstetige Funktionaloperationen*, Studia Math. 2 (1930), p. 183-196 [*Oeuvres*, p. 177-189]. If, in this same theorem, one suppresses the hypothesis that the operator U is compact, the equations in question cannot have the same number of linearly independent solutions. Nevertheless, one can show that for $U = I$ one has the inequality $n \leqq \nu$ and that it again becomes an equality when, further, the space E is weakly sequentially complete and such that its bounded subsets are relatively weakly compact (see S. Mazur, *Über die Nullstellen linearer Operationen*, Studia Math. 2 (1930), p. 11-20).

§4. For these theorems cf. J. Schauder, loc. cit., Studia Math. 2 (1930), p. 183-196 [*Oeuvres*, p. 177-189].

Theorem 22: cf. F. Riesz, loc. cit., Acta Mathematica 41 (1918), p. 90, Satz 12.

CHAPTER XI

§2. The oldest known example of an isometry between two Banach spaces is that of the isometry of L^2 to l^2 which is given by the Riesz-Fischer and Parseval-Fatou theorems.

§3. Theorem 2 was proved by S. Mazur and S. Ulam (see C. R. Acad. Sci. 194, Paris 1932, p. 946-948). It is not known if this theorem holds for F-spaces; according to a remark of Mazur and Ulam, it fails, however, for G-spaces. The same authors have, furthermore, pointed out the following corollary of this theorem 2: it is not possible to define, in a metric space E, operations (of addition and scalar multiplication) in two different ways, in such a way that in both cases E becomes a normed vector space with the same zero element Θ.

§4. We know of no example of a pair of separable infinite-dimensional Banach spaces which are not homeomorphic; however, we do not know how to prove that, for example, C is homeomorphic to c. Equally, we have not been able to establish the homeomorphism of C and l^1. The spaces L^p and l^q are, however, homeomorphic for any $p,q \geq 1$ (see S. Mazur, *Une remarque sur l'homéomorphie des champs fonctionelles*, Studia Math. 1 (1929). p. 83-85).

Of particular interest seems to be the question of whether C is homeomorphic with the space of continuous functions on the square. We know of no example of two compact metric spaces of finite, but different, dimensions (in the sense of Menger-Urysohn) such that the spaces of continuous functions defined on them are homeomorphic.

§5. The notion of rotation can be interpreted generally in G-spaces. It can happen that the only possible rotation about Θ is that given by the identity operator, i.e. $U(x) = x$; in F-spaces the transformation $V(x) = -x$ is also a rotation about Θ. There exist infinite-dimensional Banach spaces where these are the only two rotations about Θ. The general form (15) of rotations in L^2, established on p. 106, (actually known for a long time), shows that, for every pair of elements x and y of norm 1, there exists a rotation about Θ which maps x to y. S. Mazur has asked the question if every separable infinite-dimensional Banach space possessing this property is isometric with L^2.

§6. The notion of isomorphism also makes sense in G-spaces. Two G-spaces are equivalent when there exists an *additive* isometric transformation from one to the other.

For two isomorphic Banach spaces E and E_1, put

$$d(E,E_1) = \inf [\log(\|U\|.\|U^{-1}\|)]$$

where the inf is taken over all isomorphisms U from E to E_1. If $d(E,E_1) = 0$, the spaces E and E_1 will be called *almost isometric*. Isometric spaces are at the same time almost isometric. The converse is always true for finite-dimensional spaces, but we do not know how to refute the conjecture that, for example, the spaces c and c_0, which are not isometric, are almost isometric.

Consider the set J_E of all spaces which can be obtained from a given Banach space E by replacing the norm with any equivalent norm. It is plain that every space belonging to J_E is isomorphic with E and that every space isomorphic with E is *isometric* with a space belonging to the set J_E. Let us partition J_E into subsets, putting two spaces in the same subset ι when they are almost isometric. For two subsets ι_1 and ι_2 of J_E put $d(\iota_1,\iota_2) = d(E_1,E_2)$ where E_1 and E_2 are any two spaces belonging to ι_1 and ι_2 respectively. One can show that this distance is well-defined and that the set I_E of all the ι, thus metrised, constitutes a complete metric space. I have introduced these notions in collaboration with S. Mazur.

§7. Theorems 6,7, and 8 are due to K. Borsuk.

One can also study infinite products. Let us denote by $(E_1 \times E_2 \times \ldots)_{c_0}$, where $E_1, E_2, \ldots$ are Banach spaces, the Banach space E defined as follows: the elements of E are all sequences (x_n) where $x_n \in E_n$ for $n=1,2,\ldots$ such that $\lim_{n\to\infty}\|x_n\| = 0$; addition and scalar multiplication is defined termwise, and the norm in E is given by $\|(x_n)\| = \max_{n\geq 1} \|x_n\|$. In an analogous way, one can define, for example, the spaces $(E_1 \times E_2 \times \ldots)_c$, $(E_1 \times E_2 \times \ldots)_m$ and $(E_1 \times E_2 \times \ldots)_{l^p}$ where $p \geq 1$.

§8. The theorems of this section are due jointly to S. Mazur and myself.

P. Urysohn was the first to show the existence of a separable metric space containing subspaces isometric to every given separable metric space (see P. Urysohn, *Sur un espace métrique universel*, Bull. Sci. Math. 151 (1927), p. 1-38).

Theorem 10, proof: cf. M. Fréchet, *Les dimensions d'un ensemble abstrait*, Math. Annalen 68 (1910), p. 161.

§9. It is not known if the equivalence of the spaces E_1^* and E_2^* always implies the isomorphism of the spaces E_1 and E_2 (cf. theorem 11, p. 171). The converse of theorem 12, p. 114, is plainly false, but we do not know if the same is true of the converse of theorem 13, p. 114, in other words, if the equivalence of the separable Banach space E and the space E^{**} does or does not imply the existence, in every bounded sequence of elements of E, of a subsequence weakly convergent to an element of E. The following question also remains open: given a Banach space E whose dual space E^* is not separable, does there exist in E a bounded sequence with no weakly convergent subsequence?

CHAPTER XII

§3. The theorems of this section were discovered in collaboration with S. Mazur. For the proof of the inequality (18) see S. Banach and S. Saks, *Sur la convergence forte dans le champ* L^p, Studia Math. 2, p. 51-57 [Oeuvres II, p. 397].

Theorem 6, proof: the existence of the constant M follows from a theorem of A. Zygmund (see *Sur les séries trigonométriques lacunaires*, Proc. London Math. Soc. 5 (1930), p. 138-145).

We here list a series of, respectively, *isometric*, *isomorphic* and *dimensional invariants*, i.e. properties which, if they are possessed by some Banach space E, are also possessed by any space which is isometric with, isomorphic with or has the same linear dimension as E respectively.

Isometric invariants:

(1) The weak convergence of a sequence (x_n) to x_0, together with the condition $\lim_{n\to\infty} \|x_n\| = \|x_0\|$, implies that $\lim_{n\to\infty} \|x_n - x_0\| = 0$.

(2) $\|x_0\| = 1$ implies the existence of exactly one bounded linear functional f such that $f(x_0) = 1$ and $\|f\| = 1$.

(3) Isometry of the space with its dual space.

(4) Isometry of the space with every infinite-dimensional closed linear subspace.

(5) Isometry of every pair of linear subspaces of a given (finite) dimension $n \geqq 2$.

Isomorphic invariants:

(6) Existence of a base.

(7) Existence for every closed linear subspace S of a closed linear subspace T such that every element x can be written in exactly one way in the form $x = s + t$ where $s \in S$ and $t \in T$.

(8) Existence for every closed linear subspace S of a bounded linear transformation from the whole space onto all of S.

(9) Existence for every separable space E of a bounded linear transformation of the given space onto the whole of E.

(10) Isomorphism of the space with its dual space.

(11) Isomorphism of the space with its square.

Dimensional properties:

(12) The property of being weakly (sequentially) complete.

(13) Weak compactness of bounded subsets.

(14) Existence of a base in every closed linear subspace.

(15) Isomorphism of all infinite-dimensional closed linear subspaces.

(16) Equality of the linear dimension of all infinite-dimensional closed linear subspaces.

(17) Equivalence of weak and norm convergence of sequences of elements.

(18) Equality of the linear dimension of the space with that of its square.

In the table that follows, the presence and absence, where known, of these properties in various spaces is indicated by + and - respectively; the blank squares correspond to open (and difficult) problems while the symbols ⊕ and ⊖ correspond to results obtained since the book was first published: cf. the survey article of Cz. Bessaga and A. Pelczyński which follows.

As S. Mazur has observed, there exist separable infinite-dimensional spaces which, without being isomorphic with L^2, posess property (3), and therefore also property (10), whilst there exists no such space, at least among known spaces, which possesses property (4),(5) or (14). Moreover, Mazur has shown that every infinite-dimensional separable space which possesses property (5) for $n = 2$ is, conversely, isometric with L^2. Property (6) fails in all non-separable spaces, but we do not know if all separable spaces possess it. It is still not known if there exists a separable infinite-dimensional space which, without being isomorphic to L^2, L^1 or l^1 possesses property (8). Properties (11) and (18) hold for all known infinite-dimensional spaces; nevertheless, we do not know how to prove or disprove that every separable space possesses property (14). Finally, we know of no example of an infinite-dimensional space which, without being isomorphic with L^2, possesses property (15).

It should be noted that none of the isometric invariants considered here is an isomorphic invariant. Moreover, we do not know if all the isomorphic properties listed are not at the same time dimensional invariants. Among other open problems, we point out the following:

1° Let X_1 and X_2 be any two non-zero bounded linear functionals on an infinite-dimensional Banach space E. With G_1 and G_2 denoting the kernels (null-spaces) of these respective functionals, one can show that $\dim_l G_1 = \dim_l G_2$; is it true that $\dim_l G_1 = \dim_l E$?

2° If the linear dimensions of two Banach spaces are incomparable, is the same true of those of their squares?

SPACE			M	m	C	$C^{(p)}$ $p \geqq 1$	c	c_0	L^1	L^2	L^p $1<p\neq 2$	l^1	l^p $1<p\neq 2$
INVARIANTS	isometric	(1)	–	–	–	–	–	–	–	+	+	+	+
		(2)	–	–	–	–	–	–	–	+	+	–	+
		(3)	–	–	–	–	–	–	–	+	–	–	–
		(4)	–	–	–	–	–	–	–	+	–	–	–
		(5)	–	–	–	–	–	–	–	+	–	–	–
	isomorphic	(6)	–	–	+	+	+	+	+	+	+	+	+
		(7)	Θ	Θ	Θ	Θ	Θ	Θ	–	+	Θ	–	Θ
		(8)	Θ	Θ	Θ	Θ	Θ	Θ	+	+		+	
		(9)	Θ	Θ	Θ	Θ	–	–	+	–	–	+	–
		(10)	–	–	–	–	–	–	–	+	–	–	–
		(11)	+	+	+	+	+	+	+	+	+	+	+
	dimensional	(12)	–	–	–	–	–	–	+	+	+	+	+
		(13)	–	–	–	–	+	+	–	+	+	–	+
		(14)	–	–	Θ	Θ	Θ	Θ	Θ	+	Θ		Θ
		(15)	–	–	–	–	Θ	Θ	–	+	–		Θ
		(16)	–	–	–	–	+	+	–	+	–	+	+
		(17)	–	–	–	–	–	–	–	–	–	+	–
		(18)	+	+	+	+	+	+	+	+	+	+	+

We conclude by noting several results of S. Mazur concerning the geometry of normed vector spaces.

With E denoting such a space, a *translation* of E is any isometry of E to itself of the form $U(x) = x + x_0$, where $x_0 \in E$; the sets obtained by translation of linear subspaces will be called *linear varieties*. A linear variety $H \neq E$ will be known as a *hyperplane*, when there exists no closed linear variety G such that $H \subseteq G \subseteq E$ and $H \neq G \neq E$. We will say that a set *A lies on one side* of the hyperplane H, when every line segment joining two points of $A \setminus H$ is disjoint from H. A set C will be called a *convex body* when it is closed, convex and has interior points. A hyperplane H will be called a *supporting plane* of the convex body C when C lies on one side of H and is at distance 0 from H; in particular, H can therefore pass through frontier points of C.

With this terminology, we have the theorem: through each frontier point x_0 of a convex body C there passes a supporting plane H of C (cf. G. Ascoli, *Sugli spazi lineari* ..., Annali di Mathematica 10 (1932), p. 33-81). It follows from this that every closed convex set is weakly closed. In other words: given a sequence of points (x_n) of

E which is weakly convergent to $x_0 \in E$, there exist non-negative numbers $c_i^{(n)}$, both indices being natural numbers, such that, for every n, $c_i^{(n)} = 0$ for sufficiently large i and the sequence of points (y_n), where $y_n = \sum_{i=1}^{\infty} c_i^{(n)} x_i$, converges to x_0. This last convergence result was obtained by S. Mazur and myself, albeit by another method.

In particular, for the space C, it has also been established by D. C. Gillespie and W. A. Hurwitz (see *On sequences of continuous functions having continuous limits*, Trans. Amer. Math. Soc. 32 (1930), p. 527-543) and, independently, by Z. Zalcwasser (see *Sur une propriété du champ des fonctions continues*, Studia Math. 2 (1930), p. 63-67).

One can further show that a necessary and sufficient condition for the weak convergence of a (bounded) sequence (x_n) to a point x_0 is that every (bounded) convex body containing infinitely many of the points x_n contains x_0.

Index

SOME ASPECTS OF THE PRESENT THEORY OF BANACH SPACES

A. Pelczyński and Cz. Bessaga

Introduction

The purpose of this survey is to present some results in the fields of the theory of Banach spaces which were initiated in the monograph *Theory of linear operators*. The reader interested in the theory of functional analysis and the development of its particular chapters is referred to the Notes and Remarks in the monograph by Dunford and Schwartz [1], and to the Historical Remarks of Bourbaki [2](*).

The extensive bibliography at the end of this survey concerns only the fields which are discussed here, but even in this respect it is not complete. Large bibliographies of various branches of functional analysis can be found in the following monographs: Dunford and Schwartz [1], Köthe [1], Lacey [1], Lindenstrauss and Tzafriri [1], Semadeni [1], Singer [1].

Banach's monograph *Theory of linear operators* is quoted in this survey as [B]. When writing, for instance, [B], Rem. V, §2, we refer to "Remarks" to Chapter V, §2 of the monograph.

Some recent information is contained in the section "Added in proof".

Notation and terminology. We attempt to adjust our notation to that which is now commonly used (e.g. in Dunford and Schwartz [1]) and which differs to some extent from the notation of Banach.

We write L^∞ and l^∞ instead of M and m and we shall often deal with the following natural generalizations of L^p.

1. Let $1 \leq p \leq \infty$. Let μ be a non-trivial measure defined on a sigma-field Σ of subsets of a set S. For any μ-measurable scalar-valued function f defined on S, we let

$$\|f\|_p = \left(\int_S |f(s)|^p \mu(ds)\right)^{1/p} \text{ for } 1 \leq p < \infty;$$

$$\|f\|_\infty = \operatorname*{ess\,sup}_{s\in S} |f(s)|.$$

$L^p(\mu)$ is the Banach space (under the norm $\|\cdot\|_p$) of all classes of almost everywhere equal functions f defined on S such that $\|f\|_p < \infty$.

If S is an arbitrary non-empty set and μ is the measure defined for all subsets A of S by letting $\mu(A) = \infty$ if A is infinite and $\mu(A) =$ the cardinality of A otherwise, then the resulting space $L^p(\mu)$ will be denoted by $l^p(S)$.

In the case where S is finite and has n elements, the space $l^p(S)$ will be denoted by l^p_n.

2. By $c_0(S)$ we denote the closed linear subspace of $l^\infty(S)$ consisting of points $f \in l^\infty(S)$ such that, for every $\delta > 0$, the set $\{s\in S: |f(s)| > \delta\}$ is finite.

(*) Numbers in brackets refer to the "Bibliography" as well as to the "Additional bibliography".

3. By $C(K)$ we denote the Banach space of all continuous scalar-valued functions defined on a compact Hausdorff space K, with the norm $\|f\| = \sup_{k \in K} |f(k)|$.

We shall be concerned with the Banach spaces over the fields of both real and complex scalars.

By a *subspace* of a Banach space X we shall always mean a closed linear subspace of X.

For any Banach space X we denote by X^* and X^{**} the dual (conjugate) and the second dual (second conjugate) of X. If $T\colon X \to Y$ is a continuous linear operator, then T^* and T^{**} denote the conjugate (adjoint) and the second conjugate operator of T.

In the sequel we shall use the phrases "linear operator", "continuous linear operator" and "bounded linear operator" as synonyms; the same concerns "linear functionals", etc.

By a *projection* on a Banach space X we shall mean a bounded linear projection, i.e. a bounded linear operator $P\colon X \to X$ which is idempotent. A subspace of X which is a range of the projection is said to be *complemented in* X.

CHAPTER I

§1. Reflexive and weakly compactly generated Banach spaces. Related counter examples.

Theorem 13 in [B], Chap. XI, was a starting point for many investigations. In order to state the results let us recall several, already standard, definitions.

The *weak topology* of a Banach space X is the weakest topology in which all bounded linear functionals on X are continuous. A subset $W \subset X$ is said to be *weakly compact* if it is compact in the weak topology of X; W is said to be *sequentially weakly compact* if, for every sequence of elements of W, there is a subsequence which is weakly convergent to an element of W. The map $\varkappa: X \to X^{**}$ defined by $(\varkappa x)(x^*) = x^*(x)$ for $x \in X$, $x^* \in X^*$ is called the *canonical embedding* of X into X^{**}. A Banach space X is said to be *reflexive* if $\varkappa(X) = X^{**}$. Banach's Theorem 13, which we mentioned at the beginning, characterizes reflexive spaces in the class of separable Banach spaces. The assumption of separability turns out to be superfluous. This is a consequence of the following fundamental fact, discovered by Eberlein [1] and Šmulian [1].

1.1. *A subset W of a Banach space X is weakly compact if and only if it is sequentially weakly compact.*

A simple proof of 1.1 was given by Whitley [1]. For other proofs and generalizations see Bourbaki [1], Köthe [1], Grothendieck [1], Pták [1], Pełczyński [1].

From 1.1 we obtain the classical characterization of reflexivity generalizing Theorem 13 in [B], Chap. XI.

1.2. *For every Banach space X the following statements are equivalent:*

(i) *X is reflexive.*

(ii) *The unit ball of X is weakly compact.*

(iii) *The unit ball of X is sequentially weak compact.*

(iv) *Every separable subspace of X is reflexive.*

(v) *Every descending sequence of bounded nonempty convex closed sets has a nonempty intersection.*

(vi) *X^* is reflexive.*

Many interesting characterizations of reflexivity have been given by James [4],[5]. One of them, James [3], is theorem 1.3 below (see James [6] for a simple proof). For simplicity, we shall state this theorem only for real spaces.

1.3. *A real Banach space X is reflexive if and only if every bounded linear functional on X attains its maximum on the unit ball of X.*

It is interesting to compare 1.3 with the following theorem of Bishop and Phelps [1] (see also Bishop and Phelps [2]).

1.4. *For every real Banach space X, the set of bounded linear functionals which attain their least upper bounds on the unit ball is norm-dense in X*.*

The reader interested in other characterizations of reflexivity is referred to Day [1], to the survey by Milman [1], to Köthe [1] and the references therein.

James supplied counter-examples showing that the assumptions of Theorem 13 in [B], Rem. XI, in general cannot be weakened and answering questions stated in [B], Rem. XI, §9.

EXAMPLE 1 (James [2]). Let J be the space of real or complex sequences $x = x(j)_{1 \leq j < \infty}$ such that $\lim_j x(j) = 0$ and

$$\|x\| = \sup\left(|x(p_1)-x(p_2)|^2 + \ldots + |x(p_{n-1})-x(p_n)|^2 + |x(p_n)-x(p_1)|^2\right)^{\frac{1}{2}} < \infty,$$

where the supremum is extended over all finite increasing sequences of indices $p_1 < p_2 < \ldots < p_n$ $(n=1,2,\ldots)$.

It is easily seen that J under the norm $\|\cdot\|$ is a separable Banach space.

1.5. *The space J has the following properties:*

(a) *J is isometrically isomorphic to J^{**}.*

(b) *$x(J)$ has codimension* 1 *in J^{**}, i.e.* $\dim J^{**}/x(J) = 1$.

(c) *There is no Banach space X over the field of complex numbers which regarded as a real space, is isomorphic to the space J of real sequences* (Dieudonné [1]).

(d) *The space $J \times J$ is not isomorphic to any subspace of J* (Bessaga and Pelczyński [1]).

(e) *J is not weakly complete but has no subspace isomorphic to c_0.*

Statement (d) answers a question in [B], Rem. p. 153. Other examples of Banach spaces non-isomorphic to their Cartesian squares have been constructed by Semadeni [1] (cf. 11.20 in this article) and by Figiel [1]. Figiel's space is reflexive, while the dual of Semadeni's space is isomorphic to its Cartesian square.

In connection with question 1° in [B], Rem XII, p. 153, we shall mention that all subspaces of codimension one (i.e. kernels of continuous linear functionals) of a given Banach space are isomorphic to each other but it is not known whether there exists an infinite-dimensional Banach space which is not isomorphic to its subspace of codimension one. However, there exist infinite-dimensional normed linear spaces (Rolewicz [1] and Dubinsky [1]) and infinite-dimensional locally convex complete linear metric spaces (Bessaga, Pelczyński and Rolewicz [1]) with this property.

Now we shall discuss another example of James [8].

EXAMPLE 2. Let $I = \{(n,i) : n=0,1,2,\ldots;\ 0 \leq i < 2^n\}$. Call a *segment* any subset of I of the form $(n,i_1),(n+1,i_2),\ldots,(n+m,i_m)$ such that $0 \leq i_{k+1} - 2i_k \leq 1$ for $k=1,2,\ldots,m-1$ $(m,n = 0,1,\ldots)$. Let F denote the space of scalar-valued functions on I with finite supports. The norm on F is defined by the formula

$$\|x\| = \sup\left(\sum_{q=1}^{p}\left|\sum_{(n,i)\in S_q} x(n,i)\right|^2\right)^{\frac{1}{2}},$$

with the supremum taken over all finite systems of pair-wise disjoint segments $S_1, S_2, \ldots, S_p$. The completion of F in the norm $\|\cdot\|$ will be denoted by DJ.

1.6. *The Banach space DJ has the following properties* (James [8]):

(a) *DJ is separable and has a non-separable dual.*

(b) *The unit ball of DJ is conditionally weakly compact, i.e. every bounded sequence* (x_n) *of elements of DJ contains a subsequence* (x_{k_n}) *such that* $\lim_n x^*(x_{k_n})$ *exists for every* $x^* \in (DJ)^*$.

(c) *Every separable infinite-dimensional subspace E of the space* $(DJ)^*$ *contains a subspace isomorphic to the Hilbert space* l^2.

(d) *No subspace of DJ is isomorphic to* l^1.

(e) *If B is the closed linear subspace of* $(DJ)^*$ *spanned by the functionals* f_{ni} *for* $0 \le i < 2^n$; $n = 0, 1, \ldots,$ *where* $f_{ni}(x) = x(n,i)$ *for* $x \in DJ$, *then* $B^* = DJ$ *and the quotient space* $(DJ)^*/B$ *is isomorphic to a non-separable Hilbert space* (Lindenstrauss and Stegall [1]).

Property (b) of the space J and property (e) of DJ suggest the following problem. Given a Banach space X, does there exist a Banach space Y such that the quotient space $Y^{**}/x(Y)$ is isomorphic to X? This problem is examined in the papers by James [7], Lindenstrauss [5], Davis, Figiel, Johnson and Pełczyński [1]. The results already obtained in this respect concern an important class of WCG Banach spaces.

A Banach space X is said to be WCG (an abbreviation for *weakly compactly generated*) if there exists a continuous linear operator from a reflexive Banach space to X whose range is dense in X (cf. Amir and Lindenstrauss [1], Davis, Figiel, Johnson and Pełczyński [1]). Obviously every reflexive Banach space is WCG. We know that (Davis, Figiel, Johnson and Pełczyński [1]).

1.7. *For every* WCG *Banach space X there exists a Banach space Y such that the quotient space* $Y^{**}/x(Y)$ *is isomorphic to X.*

Setting $Z = Y^*$, we obtain

1.8. *If X is a* WCG *Banach space, then there exists a bounded linear operator* $T\colon Z^* \underset{\text{onto}}{\to} X$ *such that* Z^{**} *is a direct sum of* $x(Z)$ *and the subspace* $T^*(X^*)$ *which is isometrically isomorphic to* X^*.

Moreover, if X is separable, then the space Z above can be so constructed that Z^* *is separable and has a Schauder basis* (Lindenstrauss [5]).

The WCG spaces have been introduced by Amir and Lindenstrauss [1]. They share many properties of finite-dimensional Banach spaces. Amir and Lindenstrauss [1] proved the following:

1.9. *If X is a* WCG *Banach space, then for every separable subspace E of X there exists a projection* $P\colon X \to X$ *of norm* 1 *whose range* $P(X)$ *contains E and is separable.*

The last result is a starting point for several theorems on renorming WCG spaces. Recall that, if E is a normed linear space with the original norm $\|\cdot\|$, then a norm $p\colon E \to R$ is equivalent to $\|\cdot\|$ if there is a constant $a > 0$ such that $a^{-1}p(x) \le \|x\| \le ap(x)$ for $x \in X$. Troyansky [1] has proved the following:

1.10. *For every* WCG *Banach space X there exists an equivalent norm p which is locally uniformly convex, i.e. for every* $x \in X$ *with* $p(x) = 1$ *and for every sequence* (x_n) *in X, the condition* $\lim_n p(x_n) = 2^{-1} \lim_n p(x + x_n) = 1$ *implies* $\lim_n p(x - x_n) = 0$.

In particular, the norm p is strictly convex, i.e. $p(x) + p(y) = p(x + y)$ implies the linear dependence of x and y.

Assertion 1.10 for separable Banach spaces is due to Kadec [1],[2]. The existence of an equivalent strictly convex norm for WCG spaces has been established by Amir and Lindenstrauss [1].

In connection with 1.10 let us mention the following result of Day [2]:

1.11. *The space $l^\infty(S)$ with uncountable S admits no equivalent strictly convex norm.*

More information on renorming theorems can be found in Day [1] and papers by Asplund [1],[2], Lindenstrauss [6], Troyansky [1], Davis and Johnson [1], Klee [1], Kadec [2], Kadec and Pelczyński [2], Whitfield [1], Restrepo [1].

In contrast to the case of separable and reflexive Banach spaces we have (Rosenthal [1])

1.12. *There exists a Banach space X which is not* WCG *but is isomorphic to a subspace of a* WCG *space.*

Concluding this section, we shall discuss one more example.

EXAMPLE 3 (Johnson and Lindenstrauss [1]). Let S be an infinite family of subsets of the set of positive integers which have finite pair-wise intersections (cf. Sierpiński [1]). Let E_0 be the smallest linear variety in l^∞ containing all characteristic functions χ_A for $A \in S$ and all sequences tending to zero. It is easily seen that the formula

$$|||y||| = \Big\| x + \sum_{j=1}^{n} c_{A_j}\chi_{A_j} \Big\|_X + \Big(\sum_{j=1}^{n} |c_{A_j}|^2 \Big)^{\frac{1}{2}} \text{ for } y = \sum_{j=1}^{n} c_{A_j}\chi_{A_j},$$

where $x \in c_0$ and $A_1,\dots,A_n \in S$ $(n=1,2,\dots)$, defines a norm on E_0. The coefficient functionals $g_k(y) = y(k)$ for $y \in E_0$ are continuous in this norm. Let E be the Banach space which is the completion of E_0 in the norm $|||\cdot|||$ and let f_k be the continuous linear functional on E which extends g_k $(k=1,2,\dots)$. Then

1.13. *The space E has the following properties:*

(a) *The linear functionals $f_1, f_2, \dots$ separate points of E.*

(b) *E is not isomorphic to a subspace of any* WCG *space, in particular E is not isomorphically embeddable into l^∞.*

(c) *E^* is isomorphic to the product $l^1 \times l^2(S)$, hence it is* WCG.

CHAPTER II

Local properties of Banach spaces

§2. The Banach-Mazur distance and projection constants.

The distance between isomorphic Banach spaces introduced in [B], Rem. XI, §6, p. 151, plays an important role in the recent investigations of isomorphic properties of Banach spaces, and in particular in the study of the properties of finite-dimensional subspaces of a given Banach space X, which are customarily referred to as the "local properties" of the space X.

Let $a \geq 1$. Banach spaces X and Y are said to be *a-isomorphic* if there exists an isomorphism T of X onto Y such that $\|T\| \cdot \|T^{-1}\| \leq a$. The infimum of the numbers a for which X and Y are a-isomorphic is called the *Banach-Mazur distance* between X and Y and is denoted by $d(X,Y)$. Obviously 1-isomorphisms are the same as isometrical isomorphisms.

2.1. *There exist Banach spaces X_0, X_1 with $d(X_0,X_1) = 1$ which are not isometrically isomorphic.*

Proof. Consider in the space c_0 two norms

$$\|x\|_i = \sup_j |x(j)| + \Big(\sum_{j=1}^{\infty} |2^{-j}x(j+i)|^2\Big)^{\frac{1}{2}} \text{ for } x = (x(j)); i=0,1.$$

For $i=0,1$, let X_i be the space c_0 equipped with the norm $\|\cdot\|_i$. For $n=1,2,\ldots$, let $T_n : X_0 \to X_1$ be the map defined by

$$(x(1),x(2),\ldots) \to (x(n),x(1),\ldots,x(n-1),x(n+1),\ldots).$$

Then each T_n is an isomorphism of X_0 onto X_1 and $\lim_n \|T_n\|\|T_n^{-1}\| = 1$. Hence $d(X_0,X_1) = 1$. On the other hand, the norm $\|\cdot\|_0$ is strictly convex (for the definition see section 1 after 1.10) while $\|\cdot\|_1$ is not. Therefore X_0 is not isometrically isomorphic to X_1.

Let us mention that $d(c,c_0) = 3$, which is related to a question in [B], Rem. XI, §6, p. 151. Interesting generalizations of this fact are due to Cambern [1] and Gordon [1]; see also 10.19 and the comment after it.

From the compactness argument it follows that, for arbitrary Banach spaces X,Y of the same finite dimension, there exists a $d(X,Y)$-isomorphism of X onto Y.

The following important estimation is due to John [1]:

2.2. *If X is an n-dimensional Banach space, then $d(X,l_n^2) \leq \sqrt{n}$.*

Since $d(l_n^\infty, l_n^2) = \sqrt{n}$ (cf. 2.3), the estimation above is the best possible. The exact rate of growth of the sequence (d_n), where $d_n = \sup\{d(X,Y) : \dim X = \dim Y = n\}$, is unknown. From 2.2 and the "triangle inequality" $d(X,Z) \leq d(X,Y)\cdot d(Y,Z)$ it follows that $\sqrt{n} \leq d_n \leq n$ for $n=1,2,\ldots$

The computation of the Banach-Mazur distance between given isomorphic Banach spaces is rather difficult. Gurariĭ, Kadec and Macaev [1],[2] have found that

2.3. *If either $1 \leq p < q \leq 2$ or $2 \leq p < q \leq \infty$, then*

$$d(l_n^p, l_n^q) = n^{1/p-1/q} \quad (n=1,2,\ldots);$$

if $1 \leq p < 2 < q \leq \infty$. *then*

$$(\sqrt{2}-1)d(l_n^p, l_n^q) \leq \max(n^{1/p-1/2}, n^{1/2-1/q}) \leq \sqrt{2}d(l_n^p, l_n^q) \quad (n=1,2,\ldots).$$

For generalizations of 2.3 to the case of spaces with symmetric bases and some matrix spaces see Gurariĭ, Kadec and Macaev [2],[3], Garling and Gordon [1].

Estimations of the Banach-Mazur distance are related to the computation of so-called "projection constants". Let $a \geq 1$ and let X be a Banach space. A subspace Y of X is *a-complemented* in X if there exists a linear projection $P\colon X \underset{\text{onto}}{\to} Y$ with $\|P\| \leq a$. The infimum of the numbers a such that Y is a-complemented in X will be denoted $p(Y,X)$. For any Banach space E we let

$$p(E) = \sup p(i(E),X),$$

where the supremum is extended over all Banach spaces X and all isometrically isomorphic embeddings $i\colon E \to X$. The number $p(X)$ is called the *projection constant* of the Banach space E.

In general, if $\dim E = \infty$, then $p(E) = \infty$. No characterization of the class of Banach spaces E with $p(E) < \infty$ is known (cf. section 11). The projection constant of a Banach space E is closely related to extending linear operators with values in E.

2.4. *Let E be a Banach space. If $p(E) < \infty$, then, for every triple (X,Y,T) consisting of a Banach space X, its subspace Y and a continuous linear operator $T\colon Y \to E$ and for every $\varepsilon > 0$, there exists a linear operator $\tilde{T}\colon X \to E$ such that*

(*) $$\tilde{T} \text{ extends } T \text{ and } \|\tilde{T}\| \leq C\cdot\|T\|$$

with $C = p(E) + \varepsilon$. Conversely, if for every triple (X,Y,T) there is a $\tilde{T}$ satisfying (*), *then $p(E) \leq C$. We have $p(E) = \infty$ if and only if there exists a triple (X,Y,T) such that T admits no extension to a bounded linear operator defined on the whole of X.*

Using the theorem of John 2.2, Kadec and Snobar [1] have shown that

2.5. *If* $\dim X = n$, *then* $p(X) \leq \sqrt{n}$ $(n=1,2,\ldots)$.

The estimation 2.5 gives the best rate of growth. We find that (Grünbaum [1], Rutowitz [1], Daugavet [1])

2.6. $$p(l_n^2) = \pi^{-\frac{1}{2}} n\Gamma\left(\frac{n}{2}\right)/\Gamma\left(\frac{n+1}{2}\right) \sim \sqrt{2n/\pi} \quad (n=2,3,\ldots).$$

Rutovitz [1] and Garling and Gordon [1] estimated projection constants of the spaces l_n^p.

2.7. *If* $2 \leq p \leq \infty$, *then* $p(l_n^p) = n^{1/p}\alpha_p(n)$, *where* $1/\sqrt{2} < \alpha_p(n) \leq \alpha_\infty(n) = 1$ $(n = 1,2,\ldots)$. *If* $1 \leq p \leq 2$, *then* $p(l_n^p) = n^{\frac{1}{2}}\alpha_p(n)$, *where* $1 \geq \alpha_p(n) \geq \left(\sinh\frac{\pi}{2}\right)^{-1}$ $(n=1,2,\ldots)$.

Remark. Theorem 2.7 concerns real spaces l_n^p. However, in the complex case, the rate of growth is the same.

For generalizations of 2.7 to spaces with symmetric bases see Garling and Gordon [1] and the references therein.

By 2.7 we have in particular $p(l_n^\infty) = 1$ for $n=1,2,\ldots$; the last property isometrically characterizes the spaces l_n^∞ in the class of finite-dimensional Banach spaces (see Nachbin [1] and 10.15).

It is easy to show that $p(X) \leq d(X, l_n^\infty)$ for every n-dimensional Banach space X. It is not known whether the quantities $p(X)$ and $d(X, l_n^\infty)$ are of the same rate of growth, i.e. whether there exists a constant $K > 0$ independent of n and such that $d(X, l_n^\infty) \leq Kp(X)$ for every n-dimensional Banach space X. Also, the numbers

$$c_n = \sup \{p(X): \dim X = n\} \text{ for } n = 2,3,\ldots$$

have not been computed. Some results concerning the last problem are given in Gordon [2].

The Banach-Mazur distance and projection constants are connected with other isometric invariants of finite-dimensional Banach spaces. The asymptotic behaviour of these invariants in some classes of finite-dimensional Banach spaces with the dimensions growing to infinity gives rise to isomorphic invariants of infinite-dimensional Banach spaces. These problems have many points in common with the theory of Banach ideals. The interested reader is referred to Grothendieck [5],[6], Lindenstrauss and Pelczyński [1], Pietsch [1] with references, Gordon [2],[3],[4], Garling and Gordon [1], Gordon and Lewis [1], Gordon, Lewis and Retherford [1],[2], Snobar [1], Milman and Wolfson [1], Figiel, Lindenstrauss and Milman [1].

§3. Local representability of Banach spaces.

The following concept, introduced by Grothendieck [6] and James [10], originates from the Banach-Mazur distance.

Let $a \geq 1$. A Banach space X is *locally a-representable* in a Banach space Y, if for every $b > a$ every finite-dimensional subspace of X is b-isomorphic to a subspace of Y. If X is locally a-representable in Y and Y is locally a-representable in X, we say that X is *locally a-isomorphic* to Y. The space X is said to be *locally representable* in Y (*locally isometric* to Y) if X is locally 1-representable in Y (locally 1-isomorphic to Y).

First, we shall discuss the problem of finding Banach spaces which are locally representable in the spaces l^p $(1 \leq p < \infty)$ and c_0. We know (Grothendieck [5], Joichi [1], cf. also 9.7) that

3.1. *A Banach space X is locally a-representable in l^2 if and only if X is a-isomorphic to l^2.*

Theorem 3.1 can be generalized to the case of l^p with $1 \leq p < \infty$ (Bretagnolle, Dacunha-Castelle and Krivine [1], Bretagnolle and Dacunha-Castelle [1], Dacunha-Castelle and Krivine [1], Lindenstrauss and Pelczyński [1]) as follows:

3.2. *Let $1 \leq p < \infty$ and let $a \geq 1$. A Banach space X is locally a-representable in l^p if and only if X is a-isomorphic to a subspace of a space $L^p(\mu)$ (in particular to a subspace of L^p when X is separable).*

Thus, by the results of Schoenberg [1],[2], the local representability of a Banach space X in some l^p for $1 \leq p \leq 2$ can be characterized by the fact that the norm of X is negative definite. For $2n < p \leq 2n+2$ $(n=1,2,\ldots)$ more sophisticated conditions have been found by Krivine [1].

The last theorem is also valid for $p = \infty$. In fact, we have

3.3. (i) *For every cardinal $n \geq \aleph_0$, there is a compact Hausdorff space K such that the topological weight of the space $C(K)$ is n and every Banach space whose topological weight is $\leq n$ is isometrically isomorphic to a subspace of the space $C(K)$.*

(ii) *Every Banach space is locally representable in the space c_0.*

Statement (i) generalizes the classical Banach-Mazur theorem ([B], Chap. XI, Theorem 9), which says that every separable Banach space is isometrically isomorphic to a subspace of C. The proof of (i) is almost the same as that of Theorem 9 but, instead of using the fact that every compact metric space is a continuous image of the Cantor set, it employs the theorem of Esenin-Volpin [1] (which was proved under the continuum hypothesis), stating that for every cardinal $n \geq \aleph_0$ there is a compact Hausdorff space K of the topological weight n such that every compact Hausdorff space of topological weight $\leq n$ is a continuous image of K.

Statement (ii) follows from the fact that every centrally symmetric k-dimensional polyhedron with $2n$ vertices is affinely equivalent to the intersection of the cube $[-1,1]^n$ (the unit ball of the space l_n^∞) with a k-dimensional subspace of l_n^∞ for $k=1,2,\ldots;$ $n \geq k$ (Klee [2]).

Next consider the problem: Given $p \in [1,\infty]$, characterize Banach spaces in which l^p is locally representable. We present answers for $p=1,2,\infty$. (The case of arbitrary p, due to Krivine [2] (cf. also Maurey and Pisier [3], Rosenthal [9]) is much more difficult.) The following beautiful result is due to Dvoretzky [1]:

3.4. *The space l^2 is locally representable in every infinite-dimensional Banach space.*

This result is a simple consequence of the following fact concerning convex bodies:

3.5. (*Dvoretzky's theorem on almost spherical sections*). *For every $\varepsilon > 0$ and for every positive integer k, there exists a positive integer $N = N(k,\varepsilon)$ such that every bounded convex body (= convex set with non-empty interior) B in the real or complex space l_N^2 which is symmetric with respect to the origin admits an intersection with a k-dimensional subspace Y which approximates up to ε a Euclidean k-ball, i.e.*

$$\sup\{\|x\|: x \in Y \cap K\}/\inf\{\|x\|: x \in Y\setminus K\} < 1 + \varepsilon.$$

The proof of the real version of 3.5 is due to Dvoretzky [2] (previously it was announced in Dvoretzky [1]). Some completions and simplifications can be found in Figiel [2]. An essentially simpler proof, based on a certain isoperimetric theorem of P. Levy, has been given by Milman [2], cf. also Figiel, Lindenstrauss and Milman [1]. The proof of Figiel [5] based on an idea of Szankowski [1] is short and elegant.

Banach spaces with unconditional bases (for the definition see §7) have the following property (Tzafriri [1]):

3.6. *If X is an infinite-dimensional Banach space with an unconditional basis, then there exists a constant M, a sequence of projections $P_n\colon X \to X$ with $\|P_n\| \leq M$ for $n=1,2,\ldots$ and a $p \in \{1,2,\infty\}$ such that $\sup_n d(P_n(X), l_n^p) \leq M$.*

The proof of 3.6 is based on the Brunel-Sucheston [1] technique of constructing sub-symmetric bases, which employs a certain combinatorial theorem of Ramsey [1]. A similar argument yields also the following weaker version of Dvoretzky's theorem: *For every infinite-dimensional Banach space X there is an $a \geq 1$ such that l^2 is a-representable in X.*

Characterizations of Banach spaces in which c_0, equivalently l^∞, is locally represented are connected with the theory of random series. Recall that a measurable real function f on a probabilistic space (Ω,μ) is called a *standard Gaussian random variable* if

$$\mu\{\omega \in \Omega\colon f(\omega) < t\} = \frac{1}{\sqrt{2\pi}} \int_{-\infty}^{t} e^{-s^2/2}\,ds.$$

The Rademacher functions $(r_j)_{1 \leq j < \infty}$ are defined on the interval $[0,1]$ by the formula

$$r_j(t) = \operatorname{sgn} \sin 2^j \pi t,\quad j=1,2,\ldots$$

We have

3.7. *For every Banach space X the following statements are equivalent:*

(i) *The space c_0 is not locally a-representable in X for any $a \geqq 1$.*

(ii) *The space c_0 is not locally representable in X.*

(iii) *The space c_0 is not locally representable in the product space $(X \times X \times \dots)_{l^p}$ for any $p \in [1,\infty)$.*

(iv) *There are a $q \in [2,\infty)$ and a constant $C > 0$ such that*

$$\Big(\sum_{j=1}^{n} \|x_j\|^q\Big)^{1/q} \leq C\int_0^1 \Big\|\sum_{j=1}^{n} r_j(t)x_j\Big\| dt$$

for arbitrary $x_1,\dots,x_n \in X$ and $n=1,2,\dots$

(v) *For every sequence (x_n) of elements of X and for every sequence of independent standard Gaussian random variables, the series $\sum_n f_n(\omega)x_n$ converges almost everywhere iff so does the series $\sum_n r_n(t)x_n$.*

The equivalence between (i) and (ii) has been proved by Giesy [1]. The other implications in 3.7 are due to Maurey and Pisier [2]. Other equivalent conditions, stated in terms of factorizations of compact linear operators, can be found in Figiel [3].

The next theorem characterizes Banach spaces in which the space is not locally representable.

3.8. *For every Banach space X the following statements are equivalent:*

(i) *The space l^1 is not locally a-representable in X for any $a \geq 1$.*

(ii) *The space l^1 is not locally representable in X.*

(iii) *The space l^1 is not locally representable in the product space $(X \times X \times \dots)_{l^p}$ for any $p \in (1,\infty)$.*

(iv) *There are a $q \in (1,\infty)$ and a constant $C > 0$ such that*

$$\int_0^1 \Big\|\sum_{i=1}^{n} r_i(t)x_i\Big\| dt \leq C\Big(\sum_{i=1}^{n} \|x_i\|^q\Big)^{1/q}$$

for arbitrary $x_1,\dots,x_n \in X$ and $n=1,2,\dots$

(v) *There are a $q \in (1,\infty)$ and a constant $C > 0$ such that*

$$\operatorname*{ess\,inf}_{0\leqq t\leqq 1} \Big\|\sum_{i=1}^{n} r_i(t)x_i\Big\| \leq C\Big(\sum_{i=1}^{n} \|x_i\|^q\Big)^{1/q}$$

for arbitrary $x_1,\dots,x_n \in X$ and $n=1,2,\dots$

The equivalence between (i) and (ii) has been proved by Giesy [1]. The other implications in 3.8 are due to Pisier [1].

Let us notice in connection with 3.7 and 3.8 that if a Banach space X has a subspace isomorphic either to l^1 or to c_0, then, for every $a \geq 1$, there is a subspace of X which is a-isomorphic to l^1 or c_0, respectively (James [9]). It is not known whether the spaces l^p with $1 < p < \infty$ have an analogous property.

Obviously, if a Banach space X has a subspace isometrically isomorphic to a space l^p or c_0, then the space l^p or c_0, respectively, is locally a-representable in X for some $a \geqq 1$. Converse implications are, in general, false. The spaces l^p for $1 \leqq p < \infty$, $p \neq 2$, and c_0 do not contain any subspace isomorphic to l^2 (cf. 12.) in contrast to Dvoretzky's theorem 3.5. Even more "pathological" in this respect is the example due to Tzirelson [1]. Below we present a modified version of this example given by Figiel and Johnson [2].

EXAMPLE. Let E_0 be the space of all scalar sequences having at most finitely many non-zero coordinates and let $(\|\cdot\|_n)$ be the sequence of norms on E_0 defined by

$$\|x\|_0 = \sup_k |x(k)|$$

$$\|x\|_{n+1} = \max\left(\|x\|_n, \frac{1}{2}\sum_{j=1}^{m}\left\|\sum_{i=\nu(j-1)+1}^{\nu(j)} x(i)e_i\right\|_n\right),$$

where $e_i = (0,0,\ldots,\underset{\text{the } i\text{th place}}{1},0,\ldots)$, and the supremum is extended over all increasing finite sequences of indices $\nu(0) < \nu(1) < \ldots < \nu(m)$ such that $\nu(0) \geqq m$. Let

$$\|x\| = \lim_n \|x\|_n \text{ for } x \in E_0.$$

It is easy to show that the limit above exists. Let E be the completion of E_0 in the norm $\|\cdot\|$. Then

3.9. *E is a separable Banach space with an unconditional basis which does not contain isomorphically any space l^p $(1 \leqq p \leqq \infty)$ or c_0.*

Concluding this section, we shall state a theorem of general nature indicating the difference between the local and the global structure of Banach spaces.

3.10. (*The Principle of Local Reflexivity*). *Every Banach space is locally isometric to its second dual.*

This fact is a consequence of the following result. (For simplicity we identify the Banach space X with its canonical image $\varkappa(X)$ in X^{**}.)

3.11. *Let X be a Banach space, let E and G be finite-dimensional subspaces of X^{**} and X^*, respectively, and let $0 < \varepsilon < 1$. Assume that there is a projection P of X^{**} onto E with $\|P\| \leqq M$. Then there are a continuous linear operator T: $E \to X$ and a projection P_0 of X onto $T(E)$ such that*

(a) $T(e) = e$ *for* $e \in E \cap X$.

(b) $f(Te) = e(f)$ *for* $e \in E$ *and* $f \in G$.

(c) $\|T\| \cdot \|T^{-1}\| \leqq 1 + \varepsilon$.

(d) $\|P_0\| \leqq M(1 + \varepsilon)$.

Moreover, if $P = Q^*$ where Q is a projection of X^* into X^*, then the projection P_0 can be chosen so as to satisfy (d) *and the additional condition*

(e) $P_0^{**}(x^{**}) = P(x^{**})$ *whenever* $P(x^{**}) \in X$.

Theorem 3.10 and a part of 3.11 have been given by Lindenstrauss and Rosenthal [1]. Theorem 3.11 in the present formulation is due to Johnson, Rosenthal and Zippin [1]. For an alternative proof see Dean [1].

§4. The moduli of convexity and smoothness; super-reflexive Banach spaces. Unconditionally convergent series.

Intensive research efforts have been devoted to the invariants of the local structure of Banach spaces related to the geometrical properties of their unit spheres. In this section we shall discuss two invariants of this type: the modulus of convexity (Clarkson [1]) and the modulus of smoothness (Day [3]).

Let X be a Banach space; for $t > 0$, we set

$$\delta_X(t) = \inf\{1-\tfrac{1}{2}\|x+y\|: \|x\| = \|y\| = 1, \|x-y\| \geqq t\},$$

$$\rho_X(t) = \tfrac{1}{2} \sup \{\|x+y\| + \|x-y\| - 2 : \|x\| = 1, \|y\| = t\}.$$

The functions δ_X and ρ_X are called, respectively, the *modulus of convexity* and the *modulus of smoothness* of the Banach space X. The space X is said to be *uniformly convex* (resp. *uniformly smooth*) if $\delta_X(t) > 0$ for $t > 0$ (resp. $\lim_{t\to 0} \rho_X(t)/t = 0$).

The moduli of convexity and smoothness are in a sense dual to each other. We have (Lindenstrauss [8], cf. also Figiel [6]).

4.1. *For every Banach space* X, $\rho_{X^*}(t) = \sup_{0\le s\le 2} \big(ts/2 - \delta_X(s)\big)$.

The next result characterizes the class of Banach spaces for which one can define an equivalent uniformly convex (smooth) norm.

4.2. *For every Banach space X the following conditions are equivalent:*

(a) *X is isomorphic to a Banach space which is both uniformly convex and uniformly smooth.*

(b) *X is isomorphic to a uniformly smooth space.*

(c) *X is isomorphic to a uniformly convex space.*

(d) *Every Banach space which is locally a-representable in X, for some $a > 1$, is reflexive.*

(e) *Every Banach space locally representable in X is reflexive.*

(f) *The dual space X^* satisfies conditions* (a)-(e).

A Banach space satisfying the equivalent conditions of 4.2 is said to be *super-reflexive*.

Theorem 4.2 is a product of combined efforts of R. C. James [10], [11] and Enflo [2]. The implication: "(b) and (c) ⇒ (a)" has been proved by Asplund [2]. For the characterizations of super-reflexivity in terms of "goodesics" on the unit spheres see James and Schaffer [1], and in terms of basic sequences, see V. I. Gurariĭ and N. I. Gurariĭ [1] and James [12]

If X is a super-reflexive Banach space, then by (e) neither l^1 nor c_0 is locally representable in X. Therefore the product

$$(l_1^1 \times l_2^1 \times l_3^1 \times \dots)_{l^2}$$

is an example of a reflexive Banach space which is not super-reflexive. A much more sophisticated example is due to James [13], who proved that

4.3. *There exists a reflexive Banach space RJ which is not super-reflexive but is such that l^1 is not locally representable in RJ*

Clarkson [1] has shown that, for $1 < p < \infty$, the spaces L^p and l^p are uniformly convex. The exact values of $\delta_X(t)$ for $X = L^p, l^p$ have been computed by Hanner [1] and Kadec [5]. Their results together with 4.1 yield the following asymptotic formulae:

4.4. *If X is either L^p or l^p with $1 < p < \infty$, then*

$$\delta_X(t) = a_p t^k + o(t^k), \quad \rho_X(t) = b_p t^m + o(t^m),$$

with $k = \max(2,p)$, $m = \min(2,p)$, *where a_p and b_p are suitable positive constants depending only on p. Moreover, if Y is a uniformly convex (resp. uniformly smooth) Banach space which is isomorphic to L^p or l^p, then, for small positive t, we have $\delta_Y(t) \leqq \delta_{l^p}(t)$ (resp. $\rho_Y(t) \geqq \rho_{l^p}(t)$).*

Orlicz spaces (i.e. the spaces (o) and (O) in the terminology of [B], p. 138) admit equivalent uniformly convex norms iff they are reflexive (see Milnes [1]).

The moduli of convexity and smoothness are connected with the properties of unconditionally convergent series in the space X. Let us notice that the property: "*the series* $\sum_n \epsilon_n x_n$ *of elements of a Banach space* X *is convergent for every sequence of signs* (ϵ_n)" is equivalent to the unconditional convergence of the series in the sense of Orlicz [3], cf. [B], Rem. IX, §4.

We have

4.5. *If* $\sum_n \epsilon_n x_n$ *with* x_n*'s in a uniformly convex Banach space* X *is convergent for every sequence of signs* (ϵ_n), *then* $\sum_{n=1}^{\infty} \delta_X(\|x_n\|) < \infty$.

If $\sum_{n=1}^{\infty} \epsilon_n x_n$ *with* x_n*'s in a uniformly smooth Banach space* X *is divergent for every sequence of signs* (ϵ_n), *then* $\sum_{n=1}^{\infty} \rho_X(\|x_n\|) = \infty$.

The first statement of 4.5 is due to Kadec [5], the second to Lindenstrauss [8].

Combining 4.4 with 4.5, we obtain (Orlicz [1],[2])

4.6. *Let* $1 < p < \infty$. *If* $\sum_n f_n$ *is an unconditionally convergent series in the space* L^p (*or more generally, in* $L^p(\mu)$), *then* $\sum_{n=1}^{\infty} \|f_n\|^{c(p)} < \infty$, *where* $c(p) = \max(p,2)$.

The last fact is also valid for the space L^1, which is non-reflexive, and hence is not uniformly convex. We have (Orlicz [1])

4.7. *If in the space* L^1 *the series* $\sum_n f_n$ *is unconditionally convergent, then* $\sum_{n=1}^{\infty} \|f_n\|^2 < \infty$.

The exponents $c(p)$ in 4.6 and 2 in 4.7 are the best possible. This can easily be checked directly for $p > 2$; for $1 \leq p \leq 2$ it follows from the crucial theorem on unconditionally convergent series due to Dvoretzky and Rogers [1] (cf. also Figiel, Lindenstrauss and Milman [1]).

4.8. *Let* (a_n) *be a sequence of positive numbers such that* $\sum_{n=1}^{\infty} a_n^2 < \infty$. *Then in every infinite-dimensional Banach space* X *there exists an unconditionally convergent series* $\sum_n x_n$ *such that* $\|x_n\| = a_n$ *for* $n=1,2,\ldots$ *In particular, in every infinite-dimensional Banach space there is an unconditionally convergent series* $\sum_n x_n$ *such that* $\sum_{n=1}^{\infty} \|x_n\| = \infty$.

Combining 4.8 with 4.5, we get

4.9. *For every Banach space* X *there exist positive constants* a *and* b *such that* $\delta_X(t) \leq at^2$ *and* $\rho_X(t) \geq bt^2$ *for small* $t > 0$.

Concluding our discussion, we shall state another theorem on unconditionally convergent series, which generalizes the theorem of Orlicz [1] (mentioned in [B], Rem. IX, §4, p. 149).

4.10. *For every Banach space* X *the following statements are equivalent:*

(a) *For every series* $\sum_n x_n$ *of elements of* X, *if* $\sum_{n=1}^{\infty} |x^*(x_n)| < \infty$ *for every* $x^* \in X^*$, *then the series* $\sum_n x_n$ *is unconditionally convergent.*

(b) *For every series* $\sum_n x_n$ *of elements of* X *the condition*

$$\sup_n \left\| \sum_{k=1}^{n} r_k(t) x_k \right\| < \infty$$

almost everywhere on $[0,1]$ *implies the unconditional convergence of the series* $\sum_n x_n$. (*Here* r_n *denotes the* n*-th Rademacher function for*

$n=1,2,\ldots$)

(c) *No subspace of X is isomorphic to* c_0.

The equivalence of conditions (a) and (c) is proved in Bessaga and Pelczyński [3]. The equivalence of (b) and (c) is due to Kwapień [2].

There is ample literature concerning the moduli of convexity and smoothness and other related invariations of Banach spaces. In addition to the references already given in the text, the reader may consult books by Day [1], Chapt. VII, §2, Lindenstrauss and Tzafriri [1],[2], the surveys by Milman [1], Zizler [1], Cudia [2], Lindenstrauss [4],[6] and papers by Asplund [1], Bonic and Frampton [1], Cudia [1], Day [4], Day, James and Swaminathan [1], Figiel [1], Figiel and Pisier [1], V. I. Gurarii [2],[3],[4], Henkin [1], Lovaglia [1], Nordlander [1],[2].

The theory of unconditionally convergent series is related to the theory of absolutely summing operators, originated by Grothendieck, and radonifying operators in the sense of L. Schwartz, which is a branch of measure theory in infinite-dimensional linear spaces. The interested reader is referred to the following books and papers: Grothendieck [1],[2], Pietsch [1],[2],[3], Persson and Pietsch [1], Lindenstrauss and Pelczyński [1], Maurey [1], Kwapień [1], L. Schwartz [1],[2].

For further information see "Added in proof".

CHAPTER III

The approximation property and bases

There are many instances in operator theory where it is convenient to represent a given linear operator as a limit of a sequence of operators with already known properties. The best investigated classes of operators are finite rank operators and compact operators, therefore it is natural to ask whether every continuous linear operator can be approximated by linear operators from these classes. Such a question was raised in [B], Rem. VI, §1, p. 146. Banach, Mazur and Schauder have already observed that the approximation problem is related to the problem of existence of a basis, and to some questions on the approximation of continuous functions (cf. Scottish Book [1], problem 157). A detailed study by Grothendieck [4] published in the middle fifties explained the fundamental role of the approximation problem in the structure theory of Banach spaces, and that this problem arises in various contexts (for instance, if one attempts to determine the trace of a nuclear operator). Substantial progress was made in 1972 by Enflo [3], who constructed the first example of a Banach space which does not have the approximation property.

§5. The approximation property.

We begin with some notation. By an *operator* we shall mean a continuous linear operator. For arbitrary Banach spaces X and Y, we denote

$\boldsymbol{B}(X,Y)$ = the space of all operators from X into Y,
$\boldsymbol{K}(X,Y)$ = the space of all compact operators from X into Y,
$\boldsymbol{F}(X,Y)$ = the space of all finite rank operators from X into Y.

For any $T \in \boldsymbol{B}(X,Y)$, we let $|||T||| = \sup\{\|Tx\| : \|x\| \leq 1\}$, the operator norm of T.

Definition. A Banach space Y *has the* ap(= *the approximation property*) if every compact operator with range in Y is the limit, in the operator norm, of a sequence of finite rank operators, i.e. for every Banach space X and for every $K \in \boldsymbol{K}(X,Y)$, there exist $F_n \in \boldsymbol{F}(X,Y)$ $(n=1,2,\ldots)$ such that $\lim_n |||F_n - K||| = 0$.

The approximation property can easily be expressed in intrinsic terms of Y. We have (cf. Grothendieck [4] and Schaefer [1], Chap. III, §9).

5.1. *For every Banach space Y the following statements are equivalent:*

(i) *Y has the* ap.

(ii) *given a compact subset C of Y, there exists a finite rank operator $F \in \boldsymbol{F}(Y,Y)$ such that $\|Fy - y\| < 1$ for all $y \in C$.*

The celebrated result of Enflo [3] on the existence of a Banach space which fails the ap has been improved by Davie [1],[2], Figiel [4] and Szankowski [2] as follows:

5.2. *For every* $p \in [1,\infty]$, $p \neq 2$, *there exists a subspace* E_p *of the space* l^p *which does not have the approximation property. Moreover,* $E_\infty \subset c_0$.

Davie's proof is short and elegant. It uses some properties of random series. Figiel's proof seems to be the most elementary. For other proofs of Enflo's theorem and related theorems see Figiel and Pelczyński [1] and Kwapień [4]. Kwapień's result seems to be interesting also from the point of view of harmonic analysis. He has shown that

5.3. *For each p with* $2 < p < \infty$, *there exist increasing sequences* (n_k) *and* (m_k) *of positive integers such that the closed linear subspace of* L^p *spanned by the functions* $f_k(t) = e^{in_k 2\pi t} + e^{im_k 2\pi t}$ $(k=1,2,\ldots)$ *fails the approximation property.*

It is interesting to compare 5.2 with the observation by W. B. Johnson [3] that there is a Banach space which is not isomorphic to a Hilbert space but such that every subspace of the space has ap.

Starting from one example of a Banach space which does not have the ap, one can construct further examples by passing to the dual space and taking products, because the approximation property is preserved under these operations. We have

5.4. *Any complemented subspace of a Banach space having the* ap *has the* ap.

5.5. *Let* (E_i) *be a sequence of Banach spaces each having the* ap. *Then the product* $(E_1 \times E_2 \times \ldots)_{l^p}$ *has the* ap *for* $1 \leq p < \infty$.

5.6. (Grothendieck [4]). *If* X^* *has the* ap, *then so does* X.

The last result is an easy consequence of the improved Local Reflexivity Principle 3.11.

It is interesting to note that the converse of 5.6 is false. Namely, from 1.8 it follows that

5.7. (Lindenstrauss [5]). *There exists a Banach space which has the* ap *(even has a basis) but whose dual does not have the* ap.

W. B. Johnson [1] gave a simple construction of such a space. Let (B_n) be a sequence of finite-dimensional Banach spaces such that, for every $\varepsilon > 0$ and for every finite-dimensional Banach space B, there exists an index n_0 such that $d(B, B_{n_0}) < 1 + \varepsilon$. Let us set

$$BJ = (B_1 \times B_2 + \ldots)_{l^1}.$$

Then the space BJ has the following universality property:

5.8. *The conjugate of any separable Banach space is isomorphic to a complemented subspace of the space* $(BJ)^*$.

The space E_p of 5.2, being separable and reflexive for $1 < p < \infty$, is a conjugate of a separable Banach space. Hence, by 5.4 and 5.8, $(BJ)^*$ does not have the ap. On the other hand, it follows from 5.5 and the fact that every finite-dimensional Banach space has the ap that the space BJ has the approximation property.

The next two results do not directly concern the general theory of Banach spaces; however, they are closely related to theorem 5.2.

5.9. *There exists a continuous real function f defined on the square* $[0,1] \times [0,1]$ *which cannot be uniformly approximated by functions of the form*

$$g(s,t) = \sum_{j=1}^{n} a_j f(s,t_j) f(s_j,t)$$

where $a_1,\ldots,a_n$ *are arbitrary real numbers,* $s_1,\ldots,s_n,t_1,\ldots,t_n$, *belong to the interval* $[0,1]$, *and* $n=1,2,\ldots$

5.10. *We have*

(a) *For every real* β *with* $\frac{2}{3} < \beta \leq 1$ *there exists a real matrix* $A = (a_{ij})_{i,j=1}^{\infty}$ *such that*

(+) $$A^2 = 0, \text{ i.e. } \sum_{j=1}^{\infty} a_{ij}a_{jk} = 0 \text{ for } i,k=1,2,\ldots,$$

(++) $$\sum_{j=1}^{\infty} \sup_i |a_{ij}|^{\beta} < \infty,$$

(+++) $$\sum_{i=1}^{\infty} a_{ii} \neq 0.$$

(b) *If a matrix* $A = (a_{ij})$ *satisfies* (+) *and* (++) *with* $\beta = \frac{2}{3}$, *then* $\sum_{i=1}^{\infty} a_{ii} = 0$.

Grothendieck [1] has proved that 5.9 and 5.10 (a) for $\beta = 1$ are equivalent to the existence of a Banach space not having the ap. (The implication "5.9 $\Rightarrow$ 5.2 for $p = \infty$" was already known to Mazur around the year 1936.) 5.10 (a) for $2/3 < \beta < 1$ was observed by Davie [3]. 5.10 (b) is due to Grothendieck [4].

Finally note that there are uniform algebras (Milne [1]) and Banach lattices (Szankowski [3]) which fail to have ap.

§6. The bounded approximation property.

In general, a proof that a particular Banach space has the approximation property shows that the space in question already has a stronger property. Several properties of that type are discussed by Lindenstrauss [1], Johnson, Rosenthal and Zippin [1], Grothendieck [4] and Pełczyński and Rosenthal [1]. Here we shall only discuss the bounded approximation property, and in the next section the existence of a Schauder basis.

Definition. A Banach space Y is said to *have the* bap (= *the bounded approximation property*) if there exists a constant $a \geq 1$ such that, for every $\varepsilon > 0$ and for every compact set $C \subset Y$, there exists an $F \in \mathcal{F}(X,X)$ such that

(*) $$\|Fx - x\| < \varepsilon \text{ for } x \in C \text{ and } \|\|F\|\| \leq a.$$

More precisely, we then say that Y has the bap with a constant a.

It is not difficult to show that

6.1. *A separable Banach space* Y *has the* bap *if and only if there exists a sequence* (F_n) *of finite rank operators such that*

$$\lim_n \|F_n y - y\| = 0 \text{ for all } y \in Y.$$

From 5.1 we immediately get

6.2. *If a Banach space has the* bap, *then it has the* ap.

Figiel and Johnson [1] have shown that the converse of 6.2 is not true.

6.3. *There exists a Banach space FJ which has the* ap *but fails the* bap.

The idea of the proof of 6.3 is the following. Let X be a Banach space with the bap and such that X^* does not have the ap. For instance let $X = BJ$ of 5.8. Next we make use of the following lemma:

6.4. *Let* Y *be a Banach space and let* $a \geq 1$. *If every Banach space isomorphic to* Y *has the* bap *with the constant* a, *then* Y^* *has the* bap.

It follows from 6.4 that, for every positive integer n, there exists a Banach space X_n isomorphic to X and such that X_n does not have the bap with any constant a less than n. We put

$$FJ = (X_1 \times X_2 \times \ldots)_{l^2}.$$

Clearly, every isomorphic image of a space having the ap has the ap. Thus each X_n has the ap. Hence, by 5.5, the space FJ has the ap. On the other hand, FJ fails the bap. This follows from the fact that if a Banach space Y has the bap with a constant a and if Z is a subspace of Y which is the range of a projection of norm ≤ 1, then Z has the bap with a constant $\leq a$.

The space FJ also has the following interesting property:

6.5. *There is no sequence* (K_n) *of compact linear operators such that* $\lim_n \|K_n x - x\| = 0$ *for all* $x \in FJ$.

Indeed, the existence of such a sequence combined with the fact that FJ has the ap would imply the existence of a sequence (F_n) of finite rank operators such that $|||F_n - K_n||| \leq 2^{-n}$ for $n=1,2,\ldots$ Hence we would have $\lim_n \|F_n x - x\| = 0$ for all $x \in X$, which, by 6.1, would contradict the fact that the space FJ does not have the bap.

The result 6.5 answers in the negative a question raised in [B], Rem. VI, §1, p. 146.

Freda Alexander [1] has observed that, for $p > 2$, there exists a subspace X_p of the space L_p such that $F(X_p, X_p)$ is not dense (in the norm topology) in $\boldsymbol{K}(X_p, X_p)$.

Example 6.3 of Figiel and Johnson contrasts with the following deep result (Grothendieck [4], cf. Lindenstrauss-Tzafriri [1] for a simple proof).

6.6. *If a Banach space X is either reflexive or separable and conjugate to a Banach space and if X has the* ap, *then X has the* bap.

Next observe that the improved Local Reflexivity Principle 3.11 yields an analogue of 5.6.

6.7. (Grothendieck [4]). *If X is a Banach space such that X* has the* bap *with a constant a, then X has the* bap *with a constant* $\leq a$.

We conclude this section with a result which gives a characterization of the bounded approximation property in an entirely different language.

Let S be a closed subset of a compact metric space T and let E and X be closed linear subspaces of the spaces $C(S)$ and $C(T)$, respectively. The pair $(E.X)$ is said to *have the bounded extension property*, if, given $\varepsilon > 0$, every function $f \in E$ has a bounded family of extensions

$$\Phi(f,\varepsilon) = \{f_{\varepsilon,W}: W \supset S,\ W \text{ is open in } T\} \subset X$$

such that $|f_{\varepsilon,W}(t)| \leq \varepsilon$ whenever $t \in T \backslash W$.

6.8. *For every separable Banach space Y the following conditions are equivalent:*

(i) *Y has the* bap,

(ii) *for every closed subset of a compact metric space T, for every isometrically isomorphic embedding* $i: Y \to C(S)$ *and for every closed linear subspace X of the space C(T) such that the pair* $(i(Y), X)$ *has the bounded extension property, there exists a bounded linear operator* $L: i(Y) \to X$ *such that* $(Lf)(s) = f(s)$ *for* $s \in S$ *and* $f \in i(Y)$.

The proof of the implication (i) ⇒ (ii) is due to Ryll-Nardzewski, cf. Pelczyński and Wojtaszczyk [1] and Michael and Pelczyński [1]. The implication (ii) ⇒ (i) has been established by Davie [2].

§7. Bases and their relation to the approximation property.

The bounded approximation property is closely connected with the property of the existence of a basis in the space. Recall that a sequence (e_n) of elements of a Banach space X constitutes a *basis* for X if, for every $x \in X$, there exists a unique sequence of scalars $(f_n(x))$ such that

$$x = \sum_{n=1}^{\infty} f_n(x) e_n.$$

The map $x \to f_n(x)$ is a continuous linear functional on X called the *n-th coefficient functional* of the basis (e_n) ([B], Chap. VII, §3). Let us set

$$S_n(x) = \sum_{m=1}^{n} f_m(x) e_m \text{ for } x \in X;\ n=1,2,\ldots$$

Clearly (S_n) is a sequence of finite rank projections with the property: $\lim_n \|S_n(x) - x\| = 0$ for $x \in X$. Thus, by 6.1, we get

7.1. *If a Banach space X has a basis, then X is separable and has the bounded approximation property.*

Hence every example of a separable Banach space which fails the bap provides an example of a separable Banach space which does not have any basis. No example of a Banach space which has the bap and does not have any basis is known.

On the other hand, we have also a "positive" result relating the bap and the existence of a basis.

7.2. *A separable Banach space has the* bap *if and only if it is isomorphic to a complemented subspace of a Banach space with a basis.*

This has been established by Johnson, Rosenthal and Zippin [1] and Pełczyński [6].

Let us mention some theorems related to 7.2.

7.3. (Lindenstrauss [5], Johnson [1]). *Let X be a separable conjugate (resp. separable reflexive) Banach space. Then X has the* bap *if and only if X is isomorphic to a complemented subspace of a separable conjugate (resp. reflexive) space with a basis.*

Note that, by 6.6, one can replace in 7.3 the "bap" by the "ap".

7.4. *There exists a Banach space UB, unique up to an isomorphism, with a basis (e_n) with the coefficient functionals (f_n) such that:*

(a) *every separable Banach space with the* bap *is isomorphic to a complemented subspace of UB;*

(b) *for every basis (y_k) of a Banach space Y, there exist an increasing sequence (m_k) of indices, an isomorphic embedding $T: Y \to UB$ and a projection $P: UB \to T(Y)$ such that $Ty_k = \|y_k\| e_{n_k}$ for $k=1,2,\ldots$ and $P(x) = \sum_{k=1}^{\infty} f_{n_k}(x) e_{n_k}$ for $x \in UB$.*

Part (b) has been proved by Pelczyński [8]. (a) follows from (b) via 7.2. Schechtman [2] gave a simple proof of 7.3 (b). Johnson and Szankowski [1], completing 7.3 (a), have shown that if E is a Banach space such that every separable Banach space with ap is isomorphic to a complemented subspace of E, then E is not separable.

A still open question is "the finite-dimensional basis problem". For a basis (e_n) with the coefficient functionals (f_n), we put

$$K(e_n) = \sup_m \sup_{\|x\|\le 1} \left\| \sum_{n=1}^{m} f_n(x) e_n \right\|.$$

Next, if X is a Banach space with a basis, we let $K(X) = \inf K(e_n)$ where the infimum is taken over all bases for X. Finally, we define

$$K^{(n)} = \sup \{K(X)\colon \dim X = n\}.$$

The finite-dimensional basis problem is the following: is it true that $\lim_n K^{(n)} = \infty$.

It is easy to show that $K^{(2)} = 1$ and it is known that $K^{(n)} > 1$ for $n > 2$ (Bohnenblust [2]). It follows from John's theorem 1.1 that $K^{(n)} \leqq n^{\frac{1}{2}}$. Enflo [4] has proved that there exists a Banach space X isomorphic to the Hilbert space l^2 and such that $K(X) > 1$. Using 7.2 it is easy to show that Johnson's space BJ of 5.8 has a basis. Thus, by 6.4, we infer that, for each n, there exists a Banach space X_n (isomorphic to BJ) with a basis and such that $K(X_n) \geqq n$.

In the same way as for the ap and bap we have

7.5 (Johnson, Rosenthal and Zippin [1]). *If X^* has a basis, then so does X. Conversely, if X has a basis, X^* is separable and has the* ap, *then X^* has a basis.*

On the other hand, it follows from Lindenstrauss [5] that there exists a Banach space Z with a basis such that Z^* is separable and fails the ap, and hence Z^* does not have any basis.

For the most common Banach spaces bases have been constructed. We mention here two results of this nature.

7.6 (Johnson, Rosenthal and Zippin [1]). *If X is a separable Banach space such that either X or X^* is isomorphic to a complemented subspace of a space E which is either C or L^p* $(1 \leqq p < \infty)$, *then X has a basis.*

Let Ω be a compact finite-dimensional differentiable manifold with or without boundary. Denote by $C^k(\Omega)$ the Banach space of all real functions on Ω which have all continuous partial derivatives of order $\leqq k$.

7.7. *The space $C^k(\Omega)$ has a basis.*

In particular, for $\Omega = [0,1] \times [0,1]$ and $k = 1$, we obtain a positive answer to the question ([B], Rem. VII, §3, p. 147) whether the space $C^1([0,1] \times [0,1])$ has a basis.

The proof of 7.7 is reduced to the case of concrete manifolds by the following result of Mityagin [3]:

7.8. *For a fixed pair (k,n) of natural numbers, if Ω_1 and Ω_2 are n-dimensional differentiable manifolds with or without boundary, then the spaces $C^k(\Omega_1)$ and $C^k(\Omega_2)$ are isomorphic.*

Now 7.7 follows from Ciesielski [1], Ciesielski and Domsta [1], and independently from Schonefeld [1],[2], where explicit constructions of bases in $C^k(\Omega)$ are given, for Ω being either the n-cube $[0,1]^n$ or the n-torus T^n $(n,k=1,2,\ldots)$.

Bockariev [1] answering a question of [B], Rem. VII, §3, p. 147, has shown that the Disc Algebra = the space of [B], Example 10, p. 7 has a basis.

The theorem of Banach stating that

7.9. *Every infinite-dimensional Banach space contains an infinite-dimensional subspace with a basis;*

and announced in [B], Rem. VII, §3, p. 147, has been improved and modified in several papers (cf. Bessaga and Pelczyński [3],[4], Day [5], Gelbaum [1], Davis and Johnson [2], Johnson and Rosenthal [1], Kadec and Pelczyński [2], Milman [1], Pelczyński [7]). In particular, it has been shown that

7.10. (Pelczyński [7]). *Every non-reflexive Banach space contains a non-reflexive subspace with a basis.*

7.11 (Johnson and Rosenthal [1]). *Every infinite Banach space which is the conjugate of a separable Banach space contains an infinite-dimensional subspace which has a basis and which is a conjugate space.*

7.12 (Johnson and Rosenthal [1]). *Every separable infinite-dimensional Banach space admits an infinite-dimensional quotient with a basis.*

The separability assumption in 7.12 is related to the open question whether every Banach space has a separable infinite-dimensional quotient.

There is a huge literature concerning the classification of bases and their generalizations, and also concerning the properties of special bases. The reader may consult the books by Day [1], Lindenstrauss and Tzafriri [1], Singer [1] and the surveys by Milman [1] and McArthur [1], where bases in Banach spaces are discussed, the book by Rolewicz [2] and the surveys by Dieudonné [2],[3], Mityagin [1],[2] and McArthur [1], where bases in general linear topological spaces are treated.

Concluding this section, we add that the question raised in [B], Rem. VII, §1, p. 147 has been answered by Ovsepian and Pełczyński [1]. We have (cf. Pełczyński [9])

7.13. *Every separable Banach space X admits a biorthogonal system (x_n, f_n) such that $\|x_n\| = 1$ for $n=1,2,\ldots$, $\lim_n \|f_n\| = 1$, and* (a) *if $f \in X^*$ and $f(x_n) = 0$ for all n, then $f = 0$, and* (b) *if $x \in X$ and $f_n(x) = 0$ for all n, then $x = 0$. Moreover, given $c > 1$ the biorthogonal sequence can be chosen so that $\sup_n \|f_n\| < c$.*

It is unknown whether the "Moreover" part of 7.13 is true for $c = 1$.

§8. Unconditional bases.

A basis (e_n) for a Banach space X is *unconditional* if

$$\sum_{n=1}^{\infty} |f_n(x)x^*(e_n)| < \infty \text{ for all } x \in X;\ x^* \in X^*,$$

where (f_n) is the sequence of coefficient functionals of the basis (e_n).

The existence of an unconditional basis in the space is a very strong property. It determines on the space the Boolean algebra of projections (P_σ), where, for any subset σ of positive integers, the projection $P_\sigma \in B(X,X)$ is defined by

$$P_\sigma(x) = \sum_{n \in \sigma} f_n(x) e_n,$$

and, in the real case, it determines also the lattice structure on X induced by the partial ordering: $x < y$ iff $f_n(x) \leq f_n(y)$ for $n=1,2,\ldots$

Several results on unconditional bases can be generalised to an arbitrary Boolean algebra of projections, and Banach lattices. The reader is referred to Dunford and Schwartz [1] and Part III, Lindenstrauss and Tzafriri.

To illustrate the consequences of the existence of an unconditional basis in a Banach space, we state an already classical result due to R. C. James [1].

8.1. *A Banach space with an unconditional basis is reflexive if and only if none of its subspaces is isomorphic either to c_0 or to l^1.*

From 8.1, 1.5 and 1.6 it immediately follows that the spaces J and DJ defined in §1 have no unconditional bases. In fact, these spaces cannot be isomorphically embedded into any Banach space with an unconditional basis. Therefore the universal space C ([B], Chap. XI, §8) has no unconditional basis.

The existence of unconditional bases in sequence spaces like l^p ($1 \leq p < \infty$), c_0 and in separable Orlicz sequence spaces (= the space (o) in the notation of [B], Rem. Introduction, §7, p. 138) is trivial. The next result of Paley [2] and Marcinkiewicz [1] is much more difficult.

8.2. *The Haar system is an unconditional basis in the spaces* L^p *for* $1 < p < \infty$.

For a relatively simple proof of this theorem see Burkholder [1].

The Paley-Marcinkiewicz theorem can be generalised to symmetric function spaces. A *symmetric function space* is a Banach space E consisting of equivalence classes of Lebesgue measurable functions on [0,1] such that

(a) $L^\infty \subset E \subset L^1$,

(b) if $f_1 \in E, f_2$ is a measurable function on [0,1] such that if $|f_2|$ is equidistributed with $|f_1|$, then $f_2 \in E$, and $\|f_2\|_E = \|f_1\|_E$.

The following result is due to Olevskiĭ [1], cf. Lindenstrauss and Pelczyński [2] for a proof.

8.3. *A symmetric function space* E *has an unconditional basis if and only if the Haar system is an unconditional basis for* E.

Combining 8.2 with the interpolation theorem of Semenov [1], we get

8.4. *Let* E *be a symmetric function space and let* $g_E(t) = \|\chi_{[0,t]}\|_E$ *where* $\chi_{[0,t]}$ *denotes the characteristic function of the interval* $[0,t]$. *If* $1 < \liminf_{t\to 0} g_E(2t)/g_E(t) \leq \limsup_{t\to 0} g_E(2t)/g_E(t) < 2$, *then the Haar system is an unconditional basis for* E.

A corollary to this theorem is the following result, established earlier in a different way by Gaposhkin [1]:

8.5. *An Orlicz function space (= the space* (O) *in the notation of* [B], *p.* 138) *has an unconditional basis if and only if it is reflexive.*

An important class of unconditional bases is that of symmetric bases. A basis (e_n) for X with the sequence of coefficient functionals (f_n) is called *symmetric* if for every $x \in X$ and for every permutation $p(\cdot)$ of the indices, the series $\sum_{n=1}^{\infty} f_n(x)e_{p(n)}$ converges.

The next result is due to Lindenstrauss [9].

8.6. *Let* (y_k) *be an unconditional basis in a Banach space* Y. *Then there exist a symmetric basis* (x_n) *in a Banach space* X *and an isomorphic embedding* $T: Y \to X$ *whose values on the vectors* y_k *are*

$$Ty_k = c_k \cdot \sum_{n_k < n \leq n_{k+1}} x_n \quad \text{for } k = 1,2,\ldots,$$

for some scalars c_k *and indices* $1 \leq n_1 < n_2 < \ldots$

For every symmetric basis (e_n) with the coefficient functionals f_n $(n=1,2,\ldots)$ and for every increasing sequence of indices (n_k), the operator $P: X \to X$ defined by

$$P(x) = \sum_{k=1}^{\infty} \left((n_{k+1} - n_k)!\right)^{-1} \cdot \sum_{p\in\Pi_k} \sum_{j=n_k+1}^{n_{k+1}} f_{p(j)}(x)e_j,$$

where Π_k denotes the set of all permutations of the indices

$n_k+1,\ldots,n_{k+1}$, is a bounded projection onto the subspace of X spanned by the blocks $\sum_{j=n_k+1}^{n_{k+1}} e_j$ $(k=1,2,\ldots)$. Hence, by 8.6, we have

8.7 (Lindenstrauss [9]). *Every Banach space with an unconditional basis is isomorphic to a complemented subspace of a Banach space with a symmetric basis.*

It is not known whether the converse of 8.7 is true or, equivalently, whether every complemented subspace of a Banach space with an unconditional basis has an unconditional basis. The question is open even for complemented subspaces of L^p $(1 < p < \infty;\ p \neq 2)$.

The next result is similar to 7.4.

8.8. *There exists, a unique up to an isomorphism, Banach space US, with a symmetric basis such that every Banach space with an unconditional basis is isomorphic to a complemented subspace of US. Moreover, the space US has an unconditional but not symmetric basis* (e_n) *with the following property:*

(*) *for every unconditional basis* (y_k) *in any Banach space Y, there exist an isomorphic embedding* $T\colon Y \to US$ *and an increasing sequence of indices* (n_k) *such that* $Ty_k = \|y_k\| e_{n_k}$ *for* $k=1,2,\ldots$

The existence of an unconditional basis with property (*) has been established by Pełczyński [8], see also Zippin [2] for an alternative simpler proof. Combining (*) with 8.7 one gets the first statement of 8.8.

In contrast to 7.5, we have

8.9. *There exists a Banach space X which does not have any unconditional basis, but its conjugate X* does.*

An example of such a space is $C(\omega^\omega)$, the space of all scalar-valued continuous functions on the compact Hausdorff space of all ordinals $\leqq \omega^\omega$, whose conjugate is l^1 (cf. Bessaga and Pełczyński [2], p. 62 and Lindenstrauss and Pełczyński [1], p. 297). The existence of a Banach lattice without ap (Szankowski [3]) yields that $(US)^*$ fails to have ap. (However, if X^* is separable and X has an unconditional basis, then X^* also has an unconditional basis!)

We do not know whether every infinite-dimensional Banach space contains an infinite-dimensional subspace with an unconditional basis (compare with 7.9).

We shall end this section with the discussion of the "unconditional finite-dimensional basis problem", which has been solved by Y. Gordon and D. Lewis. For an unconditional basis (e_n) with the coefficient functionals (f_n), we let

$$K_u(e_n) = \sup \{\sum_n |f_n(x)x^*(e_n)| : \|x\| \leqq 1,\ \|x^*\| \leqq 1\}.$$

Next, if X is a Banach space with an unconditional basis, we set $K_u(X) = \inf K_u(e_n)$, where the infimum is taken over all unconditional bases for X. Finally, we define

$$K_u^{(n)} = \sup \{K_u(X) : \dim X = n\}.$$

Let $B_n = B(l_n^2, l_n^2)$, the n^2 dimensional Banach space of all linear operators from the n-dimensional Euclidean space into itself.

Gordon and Lewis [1] have proved that

8.10. *There exists a* $C > 0$ *such that* $C\sqrt{n} \leqq K_u(B_n) \leqq \sqrt{n}$, *for* $n=1,2,\ldots$

In fact, they have obtained a slightly stronger result:

8.11. *If Y is a Banach space with an unconditional basis and Y contains a subspace isometrically isomorphic to* B_n, *then for every projection P of Y onto this subspace, we have*

$$\|P\| \cdot K_u(Y) \geqq C\sqrt{n},$$

where $C > 0$ *is a universal constant independent of* n.

The exact rate of growth of the sequence $(K_u^{(n)})$ has recently been found by Figiel, Kwapień and Pelczyński [1] who proved that $K_u^{(n)} \geqq C\sqrt{n}$. It follows from John's Theorem 2.2 that $K_u^{(n)} \leqq \sqrt{n}$.

CHAPTER IV

§9. Characterizations of Hilbert spaces in the class of Banach spaces.

The problems concerning isometric and isomorphic characterizations of Hilbert spaces in the class of Banach spaces, posed in [B], p. 151-2, have stimulated the research activity of numerous mathematicians. Isomorphic characterizations of Hilbert spaces have proved to be much more difficult than the isometric characterizations.

We say that a property (P) *isometrically (isomorphically) characterizes Hilbert spaces in the class of Banach spaces* if the following statement is true: "A Banach space X has property (P) iff X is isometrically isomorphic (is isomorphic) to a Hilbert space". By a *Hilbert space* we mean any Banach space H (separable, non-separable, or finite-dimensional) whose norm is given by $\|x\| = (x,x)^{\frac{1}{2}}$, where $(\cdot,\cdot)\colon H \times H \to K$ is an inner product and K is the field of scalars (real or complex numbers).

We shall first discuss isometric characterizations of Hilbert spaces. Results in this field are extensively presented in Day's book [1], Chap. VII, §3. Therefore here we shall restrict ourselves to discussing the most important facts and giving supplementary information.

The basic isometric characterization of Hilbert spaces is due to Jordan and von Neumann [1].

9.1. *A Banach space X is isometrically isomorphic to a Hilbert space iff it satisfies the parallelogram identity:*

$$\|x + y\|^2 + \|x - y\|^2 = 2(\|x\|^2 + \|y\|^2) \text{ for all } x,y \in X.$$

As an immediate corollary of 9.1 we get

9.2. *A Banach space X is isometrically isomorphic to a Hilbert space if and only if every two-dimensional subspace of X is isometric to a Hilbert space.*

An analogous characterization but with 2-dimensional subspaces replaced by 3-dimensional ones was earlier discovered by Fréchet [1]. In the thirties Aronszajn [1] found other isometric characterizations of a Hilbert space, which, as 9.2, are of a two-dimensional character, i.e. are stated in terms of properties of a pair of vectors in the space.

A characterization of an essentially 3-dimensional character was given by Kakutani [1] (see also Phillips [1]) in the case of real spaces, and by Bohnenblust [1] in the complex case. It states that

9.3. *For a Banach space X with* $\dim X \geq 3$ *the following statements are equivalent:*

(i) *X is isometrically isomorphic to a Hilbert space,*

(ii) *every 2-dimensional subspace of X is the range of a projection of norm* 1.

(iii) *every subspace of* X *is the range of a projection of norm* 1.

Here and in the sequel, by "dim" we mean the algebraic dimension with respect to the corresponding field of scalars.

Assume that H is a Hilbert space with $2 < \dim H \leq \infty$ and $2 \leq k < \dim H$. Obviously all k-dimensional subspaces of H are isometrically isomorphic to each other. The question ([B], Rem. XII, p. 152, properties (4) and (5)) whether the property above characterizes Hilbert spaces has been solved only partially, i.e. under certain dimensional restrictions. Let us say that a real (resp. complex) Banach space X *has the property* H^k, for $k=2,3,\ldots$, if $\dim X \geq k$ and all subspaces of X of real (resp. complex) dimension k are isometrically isomorphic to each other.

9.4. *The following two tables give the dimensional restrictions on Banach spaces* X *under which the property* H^k *implies that* X *is isometrically isomorphic to a Hilbert space.*

The real case

k even	$k+1 \leq \dim X \leq \infty$
k odd	$k+2 \leq \dim X \leq \infty$

The complex case

k even	$k+1 \leq \dim X \leq \infty$
k odd	$2k \leq \dim X \leq \infty$

The real case of $k = 2$, $\dim X < \infty$, was solved by Auerbach, Mazur and Ulam [1]. The case of $\dim X = \infty$ is a straightforward consequence of Dvoretzky's [2] theorem on almost spherical sections (see 3.5). This was observed in Dvoretzky [1]. The remaining statements are due to Gromov [1]. The simplest unsolved case is $k = 3$, $\dim X = 4$.

We shall mention two more isometric characterizations of Hilbert space.

9.5 (Foias [1], von Neumann [1]). *A complex Banach space* X *is isometrically isomorphic to a Hilbert space if and only if, for every linear operator* $T\colon X \to X$ *and for every polynomial* P *with complex coefficients, the inequality* $\|P(T)\| \leq \|T\| \cdot \sup_{|z|=1} |P(z)|$ *holds.*

9.6 (Auerbach [1], von Neumann [2]). *A finite-dimensional Banach space* X *is isometrically isomorphic to a Hilbert space if and only if the group of linear isometries of* X *acts transitively on the unit sphere of* X, *i.e. for every pair of points* $x, y \in X$ *such that* $\|x\| = \|y\| = 1$, *there is a linear isometry* $T\colon X \xrightarrow[\text{onto}]{} X$ *such that* $T(x) = y$.

Remark. Let $1 \leq p < \infty$ and let μ be an arbitrary non-sigma-finite non-atomic measure. Then the group of linear isometries of the space $L^p(\mu)$ acts transitively on the unit sphere of the space. Therefore the assumption of 9.6 that X is finite-dimensional is essential. The question whether there exists a separable Banach space other than a Hilbert space whose group of linear isometries acts transitively on the unit sphere remains open (cf. [B], Rem. XI, §5, p. 151).

Now we shall discuss various isomorphic characterizations of a Hilbert space. The simplest among them reflects the fact that all subspaces of a fixed dimension of a Hilbert space are isometric, and hence are "equi-isomorphic". More precisely, we have

9.7. *For every Banach space* X *the following statements are equivalent:*

(1) *X is isomorphic to a Hilbert space,*

(2) $$\sup_n \sup_{E \in \mathcal{U}_n(X)} d(E, l_n^2) < \infty$$

(3) $$\sup_n \sup_{E\in\mathcal{U}^n(X)} d(E,l^2_n) < \infty,$$

where $\mathcal{U}_n(X)$ (*resp.* $\mathcal{U}^n(X)$) *denotes the family of all n-dimensional subspaces (resp. quotient spaces) of the space X.*

From the theorem of Dvoretzky, it follows that conditions (2) and (3) can be replaced, respectively, by

(2′) $$\sup_n \sup_{E,F\in\mathcal{U}_n(X)} d(E,F) < \infty,$$

(3′) $$\sup_n \sup_{E,F\in\mathcal{U}^n(X)} d(E,F) < \infty.$$

Theorem 9.7 is implicitly contained in Grothendieck [5]. The equivalence between (1) and (2) was explicitly stated by Joichi [1], cf. here 3.1. In connection with 9.7 note that the following question is still unanswered: "If X is a Banach space and all infinite-dimensional subspaces of X are isomorphic to each other, is X then isomorphic to a Hilbert space?" ([B], Rem. XII, p. 153).

The following elegant result of Lindenstrauss and Tzafriri [3] (cf. also Kadec and Mitjagin [1]) is an isomorphic analogue of theorem 9.3.

9.8. *A Banach space X is isomorphic to a Hilbert space if and only if:*

(*) *each subspace of X is complemented.*

This theorem shows that property (7) discussed in [B] on p. 152-3 is a feature of Banach spaces isomorphic to a Hilbert space only.

The proof of 9.8 starts with an observation of Davis, Dean and Singer [1] that condition (*) implies

$$\infty > \sup_n P_n(X) = \sup_{E\in\mathcal{U}_n(X)} \inf\{\|P\|: P \text{ is a projection of } X \text{ onto } E\}.$$

Next, by an ingenious use of Dvoretzky's Theorem 3.4, it is shown that $\sup_n P_n(X) < \infty$ implies condition (2) of 9.7.

Historical remark. Theorem 9.8 states that every Banach space which is not isomorphic to any Hilbert space has a non-complemented subspace. The construction of such subspaces in concrete Banach spaces was relatively difficult. Banach and Mazur [1] showed that every isometrical isomorph of l^1 in the space C is not complemented. Murray [1] constructed non-complemented subspaces in the spaces L^p. For a large class of Banach spaces with a symmetric basis an elegant construction of non-complemented subspaces was given by Sobczyk [2].

Combining 9.8 with earlier results of Grothendieck [4], we obtain

9.9. *The only, up to an isomorphism, locally convex complete linear metric spaces with property* (*) *are the Hilbert spaces, the space s of all scalar sequences, and the product* $s\times H$, *where H is an infinite-dimensional Hilbert space.*

In the same way as 9.8 one can prove (cf. Lindenstrauss and Tzafriri [3])

9.10. *A Banach space X is isomorphic to a Hilbert space if and only if, for every subspace Y of X and for every compact linear operator* $T\colon Y\to Y$, *there exists a linear operator* $\tilde{T}\colon X\to Y$ *which extends T.*

An interesting characterization of a Hilbert space is due to Grothendieck [5] (cf. also Lindenstrauss and Pełczyński [1]).

9.11. *A Banach space X is isomorphic to a Hilbert space if and only if*

(**) *there is a constant K such that, for every scalar matrix* $(a_{ij})_{i,j=1}^{n}$ $(n=1,2,\ldots)$ *and every* $x_1,\ldots,x_n \in X$ *of norm* 1, $x_1^*,\ldots,x_n^* \in X^*$ *of norm* 1, *there are scalars* $s_1,\ldots,s_n,t_1,\ldots,t_n$ *each of absolute value* ≤ 1 *such that*

$$\Big|\sum_{i,j} a_{ij}x_i^*(x_j)\Big| \leq K\Big|\sum_{i,j} a_{ij}s_i t_j\Big|.$$

In contrast to the previous characterizations, it is not easy to show that Hilbert spaces have property (**). Interesting proofs of this fact were recently given by Maurey [1], Maurey and Pisier [1], Krivine [3].

Closely related to 9.12 is the following characterization (cf. Grothendieck [5], Lindenstrauss and Pelczyński [1]).

9.12. *A separable Banach space X is isomorphic to a Hilbert space iff X and X* are isomorphic to subspaces of the space* L^1 *iff X and X* are isomorphic to quotient spaces of C.*

In the above theorem the assumption of separability of X can be dropped if one replaces the spaces L^1 and C by "sufficiently big" $\mathcal{L}_1$ and $\mathcal{L}_\infty$ spaces. (For the definition see section 10.)

Let us notice that every separable Hilbert space is isometrically isomorphic to a subspace of L^1 (cf. e.g. Lindenstrauss and Pelczyński [1]). We do not know whether 9.12 admits an isometrical version, i.e. whether every infinite dimensional Banach space X such that X and X^* are isometrically isomorphic to subspaces of L^1 is isometrically isomorphic to a Hilbert space. For partial results see Bolker [1]. For dim $X < \infty$ the answer is negative (R. Schneider [1]).

From the parallelogram identity one obtains by induction, for $n=2,3,\ldots$ and for arbitrary elements of a Hilbert space,

$$2^{-n}\sum_{\varepsilon}\|\varepsilon_1 x_1 + \varepsilon_2 x_2 + \ldots + \varepsilon_n x_n\|^2 = \sum_{j=1}^{n}\|x_j\|^2,$$

where $\sum_{\varepsilon}$ denotes the sum extended over all sequences $(\varepsilon_1,\ldots,\varepsilon_n)$ of ±1's. The following isomorphic characterization of Hilbert spaces, due to Kwapień [1], is related to the above identity.

9.13. *A Banach space X is isomorphic to a Hilbert space if and only if there exists a constant A such that*

$$A^{-1}\sum_{j=1}^{n}\|x_j\|^2 \leq \sum_{\varepsilon}\Big\|\sum_{j=1}^{n}\varepsilon_j x_j\Big\|^2 \leq A\sum_{j=1}^{n}\|x_j\|^2$$

for arbitrary $x_1,\ldots,x_n \in X$ *and for* $n=2,3,\ldots$

From 9.13 Kwapień [1] has derived another isomorphic characterization of Hilbert spaces. In order to state it, we shall need some additional notation. Let $L_0^2(R,X)$ denote the normed linear space consisting of simple functions with values in the Banach space X and with supports of finite Lebesgue measure in R. We define $|f| = (\int_{-\infty}^{+\infty}\|f(t)\|^2 dt)^{\frac{1}{2}}$ for $f \in L_0^2(R,X)$. By $L^2(R,X)$ we denote the completion of $L_0^2(R,X)$ in the norm $|\cdot|$. The Fourier transformation $F\colon L_0^2(R,X) \to L^2(R,X)$ is defined by the classical formula

$$F(f)(t) = (2\pi)^{-\frac{1}{2}}\int_{-\infty}^{+\infty} e^{-ist}f(s)\,ds.$$

Under this notation we have

9.14. *For every complex Banach space X the following statements are equivalent*

(i) *X is isomorphic to a Hilbert space.*

(ii) *There is a constant $A > 0$ such that*

$$\sum_{j=-n}^{n} \|x_j\|^2 \leq A \int_0^{2\pi} \Big\| \sum_{j=-n}^{n} e^{jit} x_j \Big\|^2 dt$$

for arbitrary $x_{-n}, \ldots, x_0, \ldots, x_n \in X$ and for $n=1,2,\ldots$

(iii) *There exists a constant $A > 0$ such that*

$$\int_0^{2\pi} \Big\| \sum_{j=-n}^{n} e^{ijt} x_j \Big\|^2 dt \leq A \sum_{j=-n}^{n} \|x_j\|^2$$

for arbitrary $x_{-n}, \ldots, x_0, \ldots, x_n \in X$ and for $n=1,2,\ldots$

(iv) *The Fourier transformation $F\colon L_0^2(R,X) \to L^2(R,X)$ is a bounded linear operator.*

Using 9.13 Figiel and Pisier [1] have proved that

9.15. *A Banach space X is isomorphic to a Hilbert space if and only if there exist a constant $A > 0$ and Banach spaces X_1 and X_2 isomorphic to X such that X_1 is uniformly convex, X_2 is uniformly smooth and the moduli of convexity and smoothness satisfy the inequalities $\delta_{X_1}(t) \geq At^2$, $\rho_{X_2}(t) \leq At^2$ for small $t > 0$.*

Meškov [1] improving a result of Sundaresan [1] has shown that

9.16. *A real Banach space X is isomorphic to a Hilbert space if and only if X and X^* have equivalent norms which are twice differentiable everywhere except the origins of X and X^*.*

An operator $T\colon X \to Y$ is *nuclear* if there are $x_j^* \in X^*$, $y_j \in Y$ $(j=1,2,\ldots)$ with $\sum_{j=1}^{\infty} \|x_j^*\| \|y_j\| < \infty$ and $Tx = \sum_{j=1}^{\infty} x_j^*(x) y_j$ for $x \in X$. P. Ørno observed (cf. Johnson, König, Maurey and Retherford [1]).

9.17. *A Banach space X is isomorphic to a Hilbert space iff every nuclear $T\colon X \to X$ has summable eigenvalues.*

Enflo [1] gave a non-linear characterization of Hilbert spaces.

9.18. *A Banach space X is isomorphic to a Hilbert space if and only if X is uniformly homeomorphic to a Hilbert space H, i.e. there is a homeomorphism $h\colon X \xrightarrow{\text{onto}} H$ such that h and h^{-1} are uniformly continuous functions in the metrics induced by the norms of X and H.*

CHAPTER V

Classical Banach spaces

The spaces $L^p(\mu)$ and $C(K)$ are distinguished among Banach spaces by their regular properties. However, most of those properties, of both isomorphic and isometric character, extend to some wider classes of spaces, which can easily be defined in terms of finite-dimensional structure, i.e. by requiring certain properties of finite-dimensional subspaces of a given space.

Definition. (Lindenstrauss and Pełczyński [1]). Let $1 \leq p \leq \infty$ and let $\lambda > 1$. A Banach space X is an $\mathcal{L}_{p,\lambda}$ *space* if, for every finite-dimensional subspace $E \subset X$, there is a finite-dimensional subspace $F \subset X$ such that $F \supset E$ and $d(F, l_k^p) < \lambda$, where $k = \dim F$. The space X is an $\mathcal{L}_p$ *space* provided that it is an $\mathcal{L}_{p,\lambda}$ space for some $\lambda \in (1,\infty)$.

The class $\mathcal{L}_p = \bigcup_{\lambda>1} \mathcal{L}_{p,\lambda}$ is the required class of spaces which have most of the isomorphic properties of the spaces $L^p(\mu)$ and $C(K)$ (for $p = \infty$). From the point of view of the isometric theory the natural class is the subclass of $\mathcal{L}_p$ consisting of all those spaces X which are $\mathcal{L}_{p,\lambda}$ for every $\lambda > 1$, i.e. the class $\bigcap_{\lambda>1} \mathcal{L}_{p,\lambda}$.

§10. The isometric theory of classical Banach spaces.

First, we shall discuss the case $1 \leq p < \infty$, which is simpler than that of $p = \infty$. We have

10.1. *Let $1 \leq p < \infty$. A Banach space X is isometrically isomorphic to an $L^p(\mu)$ space if and only if X is an $\mathcal{L}_{p,\lambda}$ space for every $\lambda > 1$.*

Recall that a projection $P\colon X \to X$ is said to be *contractive* if $\|P\| \leq 1$.

10.2. *If P is a contractive projection in a space $L^p(\mu)$, then $Y = P(L^p(\mu))$ is an $\mathcal{L}_{p,\lambda}$ space for every $\lambda > 1$.*

The proofs of 10.1 and 10.2 are due to the combined effort of many mathematicians (for the history see Lacey [1]). They are based in an essential way on the following theorem on the representation of Banach lattices, which (in a less general form) has been discovered by Kakutani and Bohnenblust.

Recall, that if x is a vector in a Banach lattice, then $|x|$ is defined to be $\max(x,0) + \max(-x,0)$.

10.3. *Let $1 \leq p \leq \infty$. A Banach lattice X is lattice-isometrically isomorphic to a Banach lattice $L^p(\mu)$ if and only if $(\|x\|^p + \|y\|^p)^{1/p} = \|x + y\|$ whenever* $\min(|x|,|y|) = 0$, *for $x, y \in X$. (If $p = \infty$, then by $(\|x\|^p + \|y\|^p)^{1/p}$ we mean* $\max(\|x\|,\|y\|)$).

We also have (Ando [1])

10.4. *If X is a Banach lattice with $\dim X \geq 3$, then X is lattice-isometrically isomorphic to a lattice $L^p(\mu)$ if and only if every proper sublattice of X is the image of a positive contractive projection.*

In particular, if $1 \leq p < \infty$, then every separable subspace of $L^p(\mu)$ is contained in a subspace of the space which is isomorphic to a

space $L^p(\nu)$ *and which is the image of a contractive projection.*

For $1 < p < \infty$ the spaces $L^p(\mu)$ are reflexive (and even uniformly convex and uniformly smooth). We have

10.5. $(L^p(\mu))^* = L^{p^*}(\mu)$, *with* $p^* = p/(p-1)$. *The equality means here the canonical isomorphism given by* $f \to \int \cdot f d\mu$ *for* $f \in L^{p^*}(\mu)$.

This is a generalization of the classical theorem of Riesz [1] (cf. [B], p. 37).

Theorem 10.5 remains valid for $p = 1$ ($p^* = \infty$) in the case of sigma-finite measures. For arbitrary measures we have only the following fact (see e.g. Pelczyński [2]):

10.6. *For every measure* μ *there exists a measure* ν *(which in general is defined on another sigma-field of sets) such that the spaces* $L^1(\mu)$ *and* $L^1(\nu)$ *are isomorphic and such that the map* $f \to \int \cdot f d\nu$ *is an isometrical isomorphism of* $L^\infty(\nu)$ *onto* $(L^1(\nu))^*$.

The following theorem is due to Grothendieck [2]:

10.7. *If* X^* *is isometrically isomorphic to a space* $C(K)$, *then* X *is isometrically isomorphic to a space* $L^1(\nu)$.

The isometric classification of spaces $L^p(\nu)$ reduces to the Boolean classification of measure algebras (S, Σ, μ). The latter is relatively simple in the case of sigma-finite measures. We have

10.8. *If* μ *is a sigma-finite measure, then the space* $L^p(\mu)$ *is isometrically isomorphic to a finite or infinite product*

$$\left(l^p(A) \times L^p(\lambda^{n_1}) \times L^p(\lambda^{n_2}) \times \ldots\right)_p$$

where A *is the set of atoms of the measure* μ *and* $n_1, n_2, \ldots$ *is a sequence of distinct cardinals and* λ^n *denotes the measure which is the product of* n *copies of the measure* λ *defined on the field of all subsets of the two-point set* $\{0,1\}$ *such that* $\lambda(\{0\}) = \lambda(\{1\}) = \frac{1}{2}$.

Theorem 10.8 is a consequence of a profound result of Maharam [1] stating that every homogeneous measure algebra is isomorphic to a measure algebra of the measure λ^n for some cardinal n.

From 10.8 and the remark after 10.4 it easily follows that every separable space $L^p(\mu)$ is isometrically isomorphic to the image of a contractive projection in the space L^p (for $1 \leq p < \infty$).

Now we shall discuss the case $p = \infty$.

Definition. A Banach space X is called a *Lindenstrauss space* if its dual X^* is isometrically isomorphic to a space $L^1(\mu)$.

The classical theorem of Riesz on the representation of linear functionals on $C(K)$ (for the proof see, for instance, Dunford and Schwartz [1] and Semadeni [2]) combined with theorem 10.3 shows that all the spaces $C(K)$ are Lindenstrauss spaces. It is particularly interesting to note that the class of Lindenstrauss spaces is essentially wider than the class of spaces $C(K)$, for instance c_0 is a Lindenstrauss space which is not isometrically isomorphic to any space $C(K)$. Also, if S is a Choquet simplex (for the definition see Alfsen [1]), then the space Af(S) of all affine scalar functions on S is a Lindenstrauss space; so is the space in 11.15. Now we state several results.

10.9. *For every Banach space* X *the following statements are equivalent:*

(1) X *is an* $\mathcal{L}_{\infty,\lambda}$ *space for every* $\lambda > 1$,

(2) X *is a Lindenstrauss space,*

(3) *the second dual* X^{**} *is isometrically isomorphic to a space* $C(K)$.

10.10. *A Lindenstrauss space X is isometrically isomorphic to a space C(K) if and only if the unit ball of X has at least one extreme point and the set of extreme points of X* is w*-closed.*
Every space $L^\infty(\mu)$ is isometrically isomorphic to a space C(K).

The following is an analogue of 10.2:

10.11. *If P is a contractive projection in a Lindenstrauss space X, then P(X) is a Lindenstrauss space.*

It should be noted that not all Lindenstrauss spaces are images of spaces C(K) under contractive projections (cf. Lazar and Lindenstrauss [1] for details). However, we have

10.12 (Lazar and Lindenstrauss [1]). *Every separable Lindenstrauss space is isometrically isomorphic to the image of a contractive projection in a space* Af(*S*).

Grothendieck [4] has observed that in the class of Banach spaces Lindenstrauss spaces can be characterized by some properties of the extension of linear operators, and spaces $L^1(\mu)$ can be characterized by properties of lifting linear operators. We have

10.13. *For every Banach space X the following statements are equivalent:*

(a1) *X is a Lindenstrauss space.*

(a2) *For arbitrary Banach spaces E,F, an isometrically isomorphic embedding $j: F \to E$, a compact linear operator $T: F \to X$ and $\varepsilon > 0$, there exists a compact linear operator $\tilde{T}: E \to X$ which extends T (i.e. $T = \tilde{T}j$) and is such that $\|\tilde{T}\| \leq (1+\varepsilon)\|T\|$.*

(a3) *For arbitrary Banach spaces Y,Z, an isometrically isomorphic embedding $j: X \to Y$ and a compact linear operator $T: X \to Z$ there exists a compact linear operator $\tilde{T}: Y \to Z$ such that $T = \tilde{T}j$ and $\|\tilde{T}\| = \|T\|$.*

10.14. *For every Banach space X the following statements are equivalent:*

(a*1) *X is isometrically isomorphic to a space $L^1(\mu)$.*

(a*2) *For an arbitrary Banach space E, its quotient space F, a compact linear operator $T: X \to F$ and $\varepsilon > 0$ there exists a compact linear operator $\tilde{T}: X \to E$ with $\|\tilde{T}\| \leq (1+\varepsilon)\|T\|$ which lifts T, i.e. $T = \phi\tilde{T}$, where ϕ is the quotient map of E onto F.*

(a*3) *For arbitrary Banach spaces Y,Z, a linear operator $\phi: Y \underset{\text{onto}}{\to} X$ and a compact linear operator $T: Z \to X$ there exists a compact linear operator $\tilde{T}: Z \to Y$ such that $\|\tilde{T}\| = \|T\|$ and $T = \phi\tilde{T}$.*

Other interesting characterizations can be found in Lindenstrauss [1],[2].

Omitting in (a2),(a3) (resp. in (a*2),(a*3)) the requirement that the linear operators T and $\tilde{T}$ should be compact, we obtain characterizations of important classes of injective (resp. projective) Banach spaces. They are narrow subclasses of Lindenstrauss spaces (resp. of spaces $L^1(\mu)$); see the theorems below.

Recall that a compact Hausdorff space *K* is said to be *extremally disconnected* if the closure of every open set in *K* is open.

10.15 (Nachbin-Goodner-Kelley). *For every Banach space X the following statements are equivalent:*

(b1) *X is isometrically isomorphic to a space C(K) with K extremally disconnected.*

(b2) *For arbitrary Banach spaces E,F, an isometrically isomorphic embedding $j: E \to F$, and a linear operator $T: E \to X$, there exists a* *linear operator* $\tilde{T}$ *such that* $T = \tilde{T}j$ *and* $\|T\| = \|\tilde{T}\|$.

(b3) *X satisfies* (a2) *with "compact linear operator" replaced by "linear operator".*

(b4) *X satisfies* (a3) *with "compact linear operator" replaced by "linear operator".*

10.16. *For every Banach space X the following statements are equivalent:*

(b*1) *X is isometrically isomorphic to a space* $l^1(S)$.

(b*2) *For an arbitrary Banach space E, its quotient space F and a linear operator* $T: X \to F$ *there exists a linear operator* $\tilde{T}: X \to E$ *such that* $\|\tilde{T}\| = \|T\|$ *and* $T = \phi\tilde{T}$ *where* $\phi: E \to F$ *is the quotient map.*

(b*3) *X satisfies* (a*2) *with "compact linear operator" replaced by "linear operator".*

(b*4) *X satisfies* (a*3) *with "compact linear operator" replaced by "linear operator".*

The isometrical classification of the spaces $C(K)$ reduces to the topological classification of compact Hausdorff spaces. For compact metric spaces this fact has been established by Banach (see [B], Chap. IX, Theorem 3). The general result is due to M. H. Stone [1] and S. Eilenberg [1]. It is as follows:

10.17. *Compact Hausdorff spaces* K_1 *and* K_2 *are homeomorphic if and only if the spaces* $C(K_1)$ *and* $C(K_2)$ *are isometrically isomorphic.*

D. Amir [1] and M. Cambern [1] have strengthened this result as follows: *If there is an isomorphism T of* $C(K_1)$ *onto* $C(K_2)$ *such that* $\|T\|\cdot\|T^{-1}\| < 2$, *then* K_1 *and* K_2 *are homeomorphic.* The constant 2 is the best possible; there are compact metric spaces K_1 and K_2 such that $d\big(C(K_1), C(K_2)\big) = 2$ (H. B. Cohen [1]). However, if K_1 and K_2 are countable compacta, then $d\big(C(K_1), C(K_2)\big) \geqq 3$ (Y. Gordon [1]).

An isometric classification of Lindenstrauss spaces is not known. Many interesting partial results can be found in Lindenstrauss and Wulbert [1] and Lazar and Lindenstrauss [1]. Let us note that the space c_0 is minimal among Lindenstrauss spaces in the following sense.

10.18 (Zippin [1]). *Every infinite-dimensional Lindenstrauss space X contains a subspace V which is isometrically isomorphic to the space* c_0. *Moreover, if X is separable, then the subspace V can be chosen so as to be the image of a contractive projection in the space X.*

The class of separable Lindenstrauss spaces admits a maximal member. More precisely:

10.19 (Pelczyński and Wojtaszczyk [1]). *There exists a separable Lindenstrauss space* Γ *with the property that for every separable Lindenstrauss space X and for every* $\varepsilon > 0$ *there is an isometrically isomorphic embedding* $T: X \to \Gamma$ *with* $\|x\| \leqq \|Tx\| \leqq (1+\varepsilon)\|x\|$ *for* $x \in X$ *and such that* $T(X)$ *is the image of a contractive projection from X.*

Wojtaszczyk [1] has shown that the space Γ with the above properties can be constructed in such a way that it is a Gurariĭ space of the universal arrangement (cf. Gurariĭ [1]), i.e. it has the following property:

(*) *For every pair* $F \supset E$ *of finite-dimensional Banach spaces, for every isometrically isomorphic embedding* $T: E \to \Gamma$ *and for every* $\varepsilon > 0$, *there is an extension* $T: F \to \Gamma$ *such that* $\|e\| \leqq \|Te\| \leqq (1+\varepsilon)\|e\|$ *for* $e \in E$.

Gurariĭ [1] has shown that every Banach space satisfying condition (*) is a Lindenstrauss space and that the Gurariĭ space is unique up to an almost-isometry, i.e. if Γ_1 and Γ_2 are Gurariĭ spaces, then $d(\Gamma_1, \Gamma_2) = 1$. Luski [1] proved that the Gurariĭ space is isometrically unique.

The reader interested in the topics of this section is referred to the monograph by Lacey [1], which contains, among other things, proofs of the majority of the results stated here both for the real and for the complex scalars. Many results and an extensive bibliography on $C(K)$ spaces can be found in Semadeni's book [2]. For the connections of Lindenstrauss spaces with Choquet simplexes see Alfsen [1]. Further information can be found in the following surveys: Bernau-Lacey [1], Edwards [1], Lindenstrauss [2],[4], Proceedings of Conference in Swansea [1], and in the papers: Effros [1], [2],[3], Lazar [1],[2],[3], Lindenstrauss and Tzafriri [2].

§11. The isomorphic theory of $\mathcal{L}_p$ spaces.

The isomorphic theory of $\mathcal{L}_p$ spaces is, in general, much more complicated than the metric theory of $L^p(\mu)$ spaces and Lindenstrauss spaces. The theory is still far from being completed. Many problems remain open. The only case in which the situation is clear is that of $p = 2$. From 9.7 it immediately follows

11.1. *A Banach space X is an $\mathcal{L}_2$ space if and only if it is isomorphic to a Hilbert space.*

The basic theorem of the general theory of $\mathcal{L}_p$ spaces is the following result, due to Lindenstrauss and Rosenthal [1]. (Recall that $p^* = p/(p-1)$ for $1 < p < \infty$; $p^* = 1$ for $p = \infty$; $p^* = \infty$ for $p = 1$.)

11.2. *Let $1 \leq p \leq \infty$ and $p \neq 2$. For every Banach space X which is not isomorphic to a Hilbert space the following statements are equivalent:*

(1) *X is an $\mathcal{L}_p$ space.*

(2) *There is a constant $c > 1$ such that, for every finite-dimensional subspace E of X, there are a finite-dimensional space l_n^p, a linear operator $T: l_n^p \to X$ and a projection P of X onto $T(l_n^p)$ such that $\|y\| \leq \|Ty\| \leq c\|y\|$ for $y \in l_n^p$, $T(l_n^p) \supset E$, $\|P\| \leq c$.*

(3) *X^* is isomorphic to a complemented subspace of a space $L^{p^*}(\mu)$.*

(4) *X^* is an $\mathcal{L}_{p^*}$ space.*

This yields the following corollary:

11.3. *We have*

(a) *Let $1 < p < \infty$ and let X be a Banach space which is not isomorphic to any Hilbert space. Then X is an $\mathcal{L}_p$ space if and only if X is isomorphic to a complemented subspace of a space $L^p(\mu)$.*

(b) *Every $\mathcal{L}_1$ space (resp. $\mathcal{L}_\infty$ space) is isomorphic to a subspace of an $L^1(\mu)$ space (resp. $L^\infty(\mu)$).*

(c) *If X is an $\mathcal{L}_1$ space (resp. an $\mathcal{L}_\infty$ space), then X^{**} is isomorphic to a complemented subspace of a space $L^1(\mu)$ (resp. $L^\infty(\mu)$).*

A Hilbert space can be isomorphically embedded as a complemented subspace of an $L^p(\mu)$ space for $1 < p < \infty$. (The subspace of L^p spanned by the Rademacher system $\{\operatorname{sgn} \sin 2^n \pi t : n = 0, 1, \ldots\}$ is such an example.) On the other hand, by Grothendieck [3], no complemented subspace of a space $L^1(\mu)$ is isomorphic to an infinite-dimensional Hilbert space. This is the reason why the assumption that X is not isomorphic to any Hilbert space does not appear in (b) and (c).

The paper Lindenstrauss and Rosenthal [1] contains many interesting characterizations of $\mathcal{L}_p$ spaces. Here we shall quote the following analogues of 10.13 and 10.14. Recall that a Banach space G is said to be *injective* if for every pair of Banach spaces $Z \supset Y$ and for every linear operator $T: Y \to G$, there is a linear operator $\tilde{T}: Z \to G$ which extends T.

11.4. *For every Banach space X the following statements are equivalent:*

(1) *X is an $\mathcal{L}_1$ space.*

(2) *For all Banach spaces Z and Y and any surjective linear operator $\Phi: Z \to Y$, every compact linear operator $T: X \to Y$ has a compact lifting $\tilde{T}: X \to Z$ (i.e. $T = \Phi\tilde{T}$).*

(3) *For all Banach spaces Z and Y and any surjective linear operator $\Phi: Z \to X$, every compact linear operator $T: Y \to X$ has a compact lifting $\tilde{T}: Y \to Z$.*

(4) *X^* is an injective Banach space.*

The reader interested in characterizations of $\mathcal{L}_p$ spaces in terms of Boolean algebras of projections (due to Lindenstrauss, Zippin and Tzafriri) is referred to Lindenstrauss and Tzafriri [2]. Other characterizations, in the language of operator ideals, can be found in Retherford and Stegall [1], Lewis and Stegall [1], in the surveys by Retherford [1] and Gordon, Lewis and Retherford [1] and in the monograph by Pietsch [1].

Now we shall discuss the problem of isomorphic classification of the spaces $\mathcal{L}_p$. If $1 < p < \infty$, then by 11.3, the problem reduces to that of isomorphic classification of complemented subspaces of spaces $L^p(\mu)$; also in the general case it is closely related to the latter problem. The latter problem is completely answered only for $l^p(S)$ spaces for $1 \leq p < \infty$. We have (Pelczyński [3], Köthe [2], Rosenthal [2]).

11.5. *Let $1 \leq p < \infty$. If X is a complemented subspace of a space $l^p(S)$ (resp. of $c_0(S)$), then X is isomorphic to a space $l^p(T)$ (resp. $c_0(T)$).*

To classify all separable $\mathcal{L}_p$ spaces for $1 < p < \infty$ one has to describe all complemented subspaces of L^p. This programme is far from being completed. Lindenstrauss and Pelczyński [1] have observed that $L^p, l^p, l^p \times l^2$ and $E_p = (l^2 \times l^2 \times \ldots)_{l^p}$ are isomorphically distinct $\mathcal{L}_p$ spaces for $1 < p < \infty$, $p \neq 2$. Next Rosenthal [3],[4] has discovered less trivial examples of $\mathcal{L}_p$ spaces.

Let $\infty > p > 2$. Let X_p be the space of scalar sequences $x = (x(n))$ such that

$$\|x\| = \max\left(\left(\sum_{n=1}^{\infty} |x(n)|^p\right)^{1/p}, \left(\sum_{n=1}^{\infty} |x(n)|^2/\log(n+1)\right)^{\frac{1}{2}}\right) < \infty.$$

Let $B_p = (B_{p,1} \times B_{p,2} \times \ldots)_{l^p}$, where $B_{p,n}$ is the space of all square summable scalar sequences equipped with the norm

$$\|x\|_{B_{p,n}} = \max\left(n^{1/p-1/2}\left(\sum_{j=1}^{\infty} |x(j)|^2\right)^{\frac{1}{2}}, \left(\sum_{j=1}^{\infty} |x(j)|^p\right)^{1/p}\right).$$

For $1 < p < 2$ we put $X_p = (X_{p^*})^*$ and $B_p = (B_{p^*})^*$.

11.6. (Rosenthal). *Let $1 < p < \infty$, $p \neq 2$. The spaces X_p, B_p, $(X_p \times X_p \times \ldots)_{l^p}, X_p \times E_p$ and $X_p \times B_p$ are isomorphically distinct $\mathcal{L}_p$ spaces each different from $L^p, l^p, l^p \times l^2, E_p$.*

Taking "L_p-tensor powers" of X_p Schechtman [1] proved

11.7. *There exists infinitely many mutually non-isomorphic infinite-dimensional separable* $\mathcal{L}_p$ *spaces* $(1 < p < \infty,\ p \neq 2)$.

Johnson and Odell [1] have proved

11.8. *If* $1 < p < \infty$, *then every infinite-dimensional separable* $\mathcal{L}_p$ *space which does not contain* l^2 *is isomorphic to* l^p.

11.8 yields the following earlier result of Johnson and Zippin[1].

11.9. *Let X be an infinite-dimensional* $\mathcal{L}_p$ *space with* $1 < p < \infty$. *If X is either a subspace or a quotient of* l^p, *then X is isomorphic to* l^p.

The above fact is also valid for the space c_0.
Now let us pass to $p = 1$. The problem of isomorphic classification of complemented subspaces of spaces $L^1(\mu)$ is a very particular case of that of isomorphic classification of $\mathcal{L}_1$ spaces. Even in the separable case neither of these problems is satisfactorily solved.
In contrast to 11.9 we have

11.10. *Among subspaces of* l_1 *there are infinitely many isomorphically distinct infinite-dimensional* $\mathcal{L}_1$ *spaces.*

This has been established by Lindenstrauss [7]. His construction of the required subspaces $X_1, X_2, \dots$ of l^1 is inductive and based on the fact that every separable Banach space is a linear image of l^1. $X_1 = \ker h_1$, where h_1 is a linear operator of l^1 onto L^1, and $X_{n+1} = \ker h_n$, where h_n is a linear operator of l^1 onto X_n for $n=2,3,\dots$
We do not know whether the set of all isomorphic types of separable $\mathcal{L}_p$ spaces is countable $(1 \leq p < \infty,\ p \neq 2)$.
In contrast to 11.10 the following conjecture is probable.

CONJECTURE. *Every infinite-dimensional complemented subspace of* L^1 *is isomorphic either to* l^1 *or to* L^1.

What we know is:

11.11 (Lewis and Stegall [1]). *If X is an infinite-dimensional complemented subspace of* L^1 *and X is isomorphic to a subspace of a separable dual space (in particular, to a subspace of* l^1*), then X is isomorphic to* l^1.

This implies that:

(a) *The space* L^1 *is not isomorphic to any subspace of a separable dual Banach space* (Gelfand [1], Pełczyński [2]).

(b) *The space* l^1 *is the only (up to isomorphisms) separable infinite-dimensional* $\mathcal{L}_1$ *space which is isomorphic to a dual space.*

The proof of (b) follows from 11.11, 11.3 (c) and the observation that every dual Banach space is complemented in its second dual.
In the non-separable case it is not known whether every dual $\mathcal{L}_1$ space is isomorphic to a space $L^1(\mu)$. Also it is not known which $L^1(\mu)$ spaces are isomorphic to dual spaces. For sigma-finite measures μ, $L^1(\mu)$ is isomorphic to a dual space iff μ is purely atomic (Pełczyński [2], Rosenthal [5]).
Now we shall discuss the situation for $p = \infty$. It seems to be the most complicated because of new phenomena which appear both in the separable and in the non-separable case. First, in contrast to the case of $1 \leq p < \infty$ (where there were only two isomorphic types of infinite-dimensional separable $L^p(\mu)$ spaces, namely L^p and l^p), there are infinitely many isomorphically different separable infinite-dimensional spaces $C(K)$. The complete isomorphic classification of such spaces is given in the next two theorems.

11.12 (Milutin [1]). *If K is an uncountable compact metric space, then the space $C(K)$ is isomorphic to the space C.*

For every countable compact space K, let $\alpha(K)$ denote the first ordinal α such that the αth derived set of K is empty.

11.13 (Bessaga and Pelczyński [2]). *Let K_1 and K_2 be countable infinite compact spaces such that $\alpha(K_1) \leq \alpha(K_2)$. Then the spaces $C(K_1)$ and $C(K_2)$ are isomorphic if and only if there is a positive integer n such that $\alpha(K_1) \leq \alpha(K_2) \leq \alpha(K_1)^n$.*

The theorem of Milutin 11.12 answers positively the question of Banach (cf. [B], p. 112).

It is easy to show that if K is a countable infinite compact space then the Banach space $(C(K))^*$ is isomorphic to l^1. Hence, by 11.13, there are uncountably many isomorphically different Banach spaces whose duals are isometrically isomorphic. This answers another question in [B], Rem. XI, §9.

The problem of describing all isomorphic types of complemented subspaces of separable spaces $C(K)$ is open. The answer is known for c being isomorphic to c_0 (cf. 11.5) and $C(\omega^\omega)$ (Alspach [1]). This problem can be reduced to that of isomorphic classification of complemented subspaces of the space C. It is very likely that

CONJECTURE. *Every complemented subspace of C is isomorphic either to C or to $C(K)$ for some countable compact metric space K.*

The following result of Rosenthal [6] strongly supports this conjecture.

11.14. *If X is a complemented subspace of C such that X^* is non-separable, then X is isomorphic to C.*

The class of isomorphic types of Lindenstrauss spaces is essentially bigger than that of complemented subspaces of $C(K)$. We have

11.15 (Benyamini and Lindenstrauss [1]). *There exists a Banach space BL with $(BL)^*$ isometrically isomorphic to l^1 and such that BL is not isomorphic to any complemented subspace of any space $C(K)$.*

From the construction of Benyamini and Lindenstrauss [1] it easily follows that, in fact, there are uncountably many isomorphically different spaces with the above property. Combining 11.15 and 10.19, we conclude that the Gurariĭ space Γ is also an example of a Lindenstrauss space which is not isomorphic to any complemented subspace of any $C(K)$.

Bourgain [1] gave a striking example of an infinite dimensional separable $\mathcal{L}_\infty$ space which does not have subspaces isomorphic to c_0; hence, by 10.18, it is not isomorphic to any Lindenstrauss space. Let us note that the results of Pelczyński [3] and Kadec and Pelczyński [1] imply

11.16. *If $1 \leq p < \infty$, then every infinite-dimensional $\mathcal{L}_p$ space has a complemented subspace isomorphic to l^p. Every infinite-dimensional complemented subspace of a space $C(K)$ contains isomorphically the space c_0.*

Our last result on separable $\mathcal{L}_\infty$ spaces is the following characterisation of c_0.

11.17. *Every Banach space E isomorphic to c_0 has the following property:*

(*S*) *If F is a separable Banach space containing isometrically E, then E is complemented in F.*

Conversely, if an infinite-dimensional separable Banach space E has property (S), then E is isomorphic to c_0.

The first part of 11.17 is due to Sobczyk [1] (cf. Veech [1] for a simple proof). The second part is due to Zippin [3]. A particular case of Zippin's result, assuming that E is isomorphic to a $C(K)$ space, was earlier obtained by Amir [2].

Now we shall be concerned with the problem of isomorphic classification of non-separable spaces $C(K)$. The multitude of different non-separable spaces $C(K)$ and the variety of their isomorphical invariants is so rich that there is almost no hope of obtaining any complete description of the isomorphic types of non-separable spaces $C(K)$, even for K's of cardinality continuum. The results which have been obtained concern special classes of spaces $C(K)$ and their complemented subspaces. Among general conjectures the following seems to be very probable.

CONJECTURE. *Every* $C(K)$ *space is isomorphic to a space* $C(K_0)$ *for some compact totally disconnected Hausdorff space* K_0.

The following result is due to Ditor [1].

11.18. *For every compact Hausdorff space* K, *there exist a totally disconnected compact Hausdorff space* K_0, *a continuous surjection* $\phi: K \to K_0$ *and a contractive positive projection* $P: C(K_0) \underset{\text{onto}}{\to} \phi^0[C(K)]$, *where* $\phi^0: C(K) \to C(K_0)$ *is the isometric embedding defined by* $\phi^0(f) = f \circ \phi$ *for* $f \in C(K)$. *Hence* $C(K)$ *is isometric to a complemented subspace of* $C(K_0)$.

An analogous result for compact metric spaces was earlier established by Milutin [1], cf. Pełczyński [4].

The theorem of Milutin 11.12 can be generalised only to special classes of non-metrizable compact spaces. Recall that the *topological weight* of a topological space K is the smallest cardinal n such that there exists a base of open subsets of K of cardinality n. We have (Pełczyński [4])

11.19. *Let* K *be a compact Hausdorff space whose topological weight is an infinite cardinal* n. *If* K *is either a topological group or a product of a family of metric spaces, then* $C(K)$ *is isomorphic to* $C([0,1]^n)$.

In particular, for every compact space K satisfying the assumptions of 11.19, the space $C(K)$ is isomorphic to its Cartesian square. This property is not shared by arbitrary infinite compact Hausdorff spaces. We have (Semadeni [1])

11.20. *Let* ω_1 *be the first uncountable ordinal and let* $[\omega_1]$ *be the space of all ordinals which are* $\leq \omega_1$ *with the natural topology determined by the order. Then the space* $C([\omega_1])$ *is not isomorphic to its Cartesian square.*

Numerous mathematicians have studied injective spaces (whose definition was given before 11.4). Theorem 10.15 of Nachbin, Goodner and Kelley suggests the following

CONJECTURE. *Every injective Banach space is isomorphic to a space* $C(K)$ *for some extremally disconnected compact Hausdorff space* K.

It is easy to see that: (1) every complemented subspace of an injective space is injective, (2) every space $l^\infty(S)$ is injective, (3) a Banach space is injective if and only if it is complemented in every Banach space containing it isometrically, (4) every Banach space X is isometrically isomorphic to a subspace of the space $l^\infty(S)$, where S is the unit sphere of X^*. From the above remarks it follows that

11.21. *A Banach space* X *is injective if and only if it is isometrically isomorphic to a complemented subspace of a space* $l^\infty(S)$.

Lindenstrauss [3] has shown (cf. 11.5):

11.22. *Every infinite-dimensional complemented subspace of* l^∞ ($= l^\infty(S)$ *for a countable infinite* S) *is isomorphic to* l^∞.

As a corollary from this theorem we get the following earlier result of Grothendieck [3].

11.23. *Every separable injective Banach space is finite-dimensional.*

Theorem 11.22 cannot be generalized to the spaces $l^\infty(S)$ with uncountable S. In fact, we have

11.24 (Akilov [1]). *For every measure* μ *the space* $L^\infty(\mu)$ *is injective.*

11.25 (Pelczyński [3],[5], Rosenthal [5]). *Let* μ *be a sigma-finite measure. Then the space* $L^\infty(\mu)$ *is isomorphic to* $l^\infty(S)$ *if and only if the measure* μ *is separable (i.e. the space* $L^1(\mu)$ *is separable).*

Theorem 11.24 is closely related to the following

11.26. (a) *An* $\mathcal{L}_\infty$ *space isomorphic to a dual space is injective.*

(b) *An injective bidual space is isomorphic to an* $L^\infty(\mu)$.

11.26 (a) follows from 11.4 (4) because by Dixmier [1] every dual Banach space is complemented in its second dual. 11.26 (b) is due to Haydon [1].

Applying deep results of Solovay and Gaifman concerning complete Boolean algebras, Rosenthal [5] has shown that

11.27. *There exists an injective Banach space which is not isomorphic to any dual Banach space.*

Let us mention that Isbell and Semadeni [1] have proved that

11.28. *There exists a compact Hausdorff space* K *which is not extremally disconnected and is such that* $C(K)$ *is injective.*

Concluding this section, let us notice that the "dual problem" to the last conjecture is completely solved. Namely (cf. 10.16) we have

11.29 (Köthe [2]). *For every Banach space* X *the following statements are equivalent:*

(1) X *is projective, i.e. for every pair* E,F *of Banach spaces, for every linear surjection* $h\colon F \to E$ *and for every linear operator* $T\colon X \to E$, *there exists a linear operator* $\tilde{T}\colon X \to F$ *which lifts* T, *i.e.* $h\tilde{T} = T$.

(2) X *is isomorphic to a space* $l^1(S)$.

The reader interested in the problems discussed in this section is referred to Lindenstrauss and Tzafriri [1],[2], Semadeni [2], Bade [1], Pelczyński [4] and Ditor [1], Lindenstrauss [2],[4], Rosenthal [9], and to the references in the above mentioned books and papers, see also "Added in proof".

§12. The isomorphic structure of the spaces $L^p(\mu)$.

The starting point for the discussion of this section is [B], Chap. XII. We shall discuss the following question:

I. Given $1 \leq p_1 < p_2 < \infty$. What are the Banach spaces E which are simultaneously isomorphic to a subspace of L^{p_1} and to a subspace of L^{p_2}?

One can ask more generally:

II. Which Banach spaces X are isomorphic to subspaces of a given space $L^p(\mu)$?

One of the basic results in this direction is theorem 3.2 of this survey, which can be restated as follows:

12.1 *A Banach space E is (isometric) isomorphic to a subspace of a space $L^p(\mu)$ iff E is locally (isometrically) isomorphically representable in l^p.*

We shall restrict our discussion to the case where $1 \leq p < \infty$ and E is a separable Banach space. Since every separable subspace of the space $L^p(\mu)$ is isometrically isomorphic to a subspace of L^p, in the sequel we shall study isomorphic properties of the spaces L^p. It turns out that the case $2 < p < \infty$ is much simpler than that of $1 \leq p < 2$. The following concepts will be useful in our discussion.

Definition. Let $1 \leq p < \infty$. We shall say that a subspace E of the space L^p is a *standard image of l^p* if there exist isomorphisms $T: l^p \underset{\text{onto}}{\to} E$ and $U: L^p \underset{\text{onto}}{\to} L^p$ such that, for $n \neq m$ $(n,m=1,2,\ldots)$, the intersections of the supports of the functions $UT(e_n)$ and $UT(e_m)$ have measure zero. Here e_n (for $n=1,2,\ldots$) denotes the nth unit vector in the space l^p.

A subspace E of the space L^p will be called *stable* if it is closed in the topology of the convergence in measure, i.e. for every sequence (f_n) of elements of E, the condition

$\lim_n \int_0^1 |f_n(t)|/(1+|f_n(t)|)\,dt = 0$ implies $\lim_n \|f_n\|_p = 0$.

It is easy to see that

12.2. (a) *Every sequence of functions in L^p which have pair-wise disjoint supports spans a standard image of l^p.*

(b) *Every standard image of l^p is complemented in L^p.*

Much deeper, especially for $1 \leq p < 2$, is the next result, which shows that the property of subspaces of L^p of being stable does not depend on the location of the subspace in the space.

12.3. *Let $1 \leq p < \infty$ and $p \neq 2$. Then, for every infinite-dimensional subspace E of the space L^p, the following statements are equivalent:*

(1) *E is stable.*

(2) *No subspace of E is a standard image of l^p.*

(3) *No subspace of E is isomorphic to l^p.*

Moreover, if $p > 1$, conditions (1)-(3) are equivalent to those stated below:

(4) *There exists a $q \in [1,p)$ and a constant C_q such that*

(*) $$\|f\|_p \leq \|f\|_q \leq C_q\|f\|_p \text{ for } f \in E.$$

(5) *For every $q \in [1,p)$ there is a C_q such that* (*) *holds.*

The last theorem, for $p > 2$, is due to Kadec and Pelczyński [1], and for $1 \leq p < 2$, is due to Rosenthal [7]. The following result of Kadec and Pelczyński [1] is an immediate corollary of 12.3.

12.4. *Let E be an infinite-dimensional subspace of a space L^p with $2 < p < \infty$. Then E is stable if and only if E is isomorphic to a Hilbert space.*

Suppose that $2 < p < \infty$ and E is a subspace of L^p which is isomorphic to a Hilbert space. Then, by 12.4 and by the condition 12.3 (5) with $q = 2$, the orthogonal (with respect to the L^2 inner product) projection of L^p onto E is continuous as an operator from L^p into L^p.

Hence, by 12.3 (2) and 12.2 (b), we get

12.5. *Let* $2 < p < \infty$ *and let* E *be a subspace of* L^p. *Then:*

(a) *if* E *is isomorphic to a Hilbert space, then* E *is complemented in* L^p;

(b) *if* E *is not isomorphic to any Hilbert space, then* E *contains a complemented subspace isomorphic to* l^p.

The next result is due to Johnson and Odell [1].

12.6. *Suppose that* E *is a subspace of a space* L^p *with* $2 < p < \infty$. *Then* E *is isomorphic to a subspace of the space* l^p *if and only if no subspace of* E *is isomorphic to a Hilbert space.*

The assumption of 12.6 that $p > 2$ is essential. For each p with $1 \leq p < 2$, there is a subspace E of L^p such that E is not isomorphic to any subspace of l^p and no infinite dimensional subspace of E is stable (Johnson and Odell [1]).

Now we shall discuss the situation for $1 \leq p < 2$. In this case there are many isomorphically different stable subspaces of the space L^p. The crucial fact is the following theorem, which goes back to P. Levy [1]; however, it was stated in the Banach space language much later (by Kadec [4] for l^q, and by Bretagnolle, Dacunha-Castelle and Krivine [1] and Lindenstrauss and Pelczyński [1] in the general case).

12.7. *If* $1 \leq p < q \leq 2$; *then the space* L^p *contains a subspace* E_q *isometrically isomorphic to* L^q.

The proof of 12.7 employs a probabilistic technique. Its idea is the following:

1. For every q with $1 < q \leq 2$, there exists a random variable (= measurable function) $\xi_q\colon R \to R$ which has the characteristic function

$$\hat{\xi}_q(s) = \int_R \exp(\xi_q(t)\cdot is)dt = \exp(-|s|^q)$$

and is such that, for each $p < q, \xi_q \in L^p(R)$. By $L^p(R^n)$ we denote here the space $L^p(\lambda)$, where λ is the n-dimensional Lebesgue measure for R^n.

2. Let $\xi_{q1},\ldots,\xi_{qn}$ be independent random variables each of the same distribution as ξ_q, for instance let $\xi_{qj} \in L^p(R^n)$ be defined by $\xi_{qj}(t_1,t_2,\ldots,t_n) = \xi_q(t_j)$. Assume that $c_1,\ldots,c_n$ are real numbers such that $\sum_{j=1}^n |c_j|^q = 1$, and let $\eta = \sum_{j=1}^n c_j\xi_{qj}$. Since the random variables $\xi_{q1},\ldots,\xi_{qn}$ are independent and have the same distribution and hence the same characteristic functions as ξ_q, we have

$$\hat{\eta}(s) = \sum_{j=1}^n c_j\hat{\xi}_{qj}(s) = \sum_{j=1}^n \exp(-|sc_j|^q)$$

$$= \exp(-|s|^q\cdot \sum_{j=1}^n |c_j|^q) = \exp(-|s|^q) = \hat{\xi}_q(s).$$

Hence η has the same distribution as ξ_q and therefore

$$(*) \qquad \Big\| \sum_{j=1}^n c_j\xi_{pj} \Big\|_p = \|\eta\|_p = \|\xi_q\|_p \quad \text{if} \quad \sum_{j=1}^n |c_j|^q = 1,$$

for every p with $1 \leq p < q$.

3. By (*), the linear operator $T\colon l_n^p \to L^p(R^n)$ defined by $T(c_1,\ldots,c_n) = \|\xi_q\|_p^{-1}\cdot\sum_{j=1}^n c_j\xi_{qj}$ is an isometric embedding. Hence L^q is locally representable in l^p. Applying 12.1 we complete the proof.

By Banach [B], p. 124, Theorem 10, and the fact that the space l^1 is not reflexive, it follows that if $1 \leq p < q < 2$, then l^p is not isomorphic to any subspace of L^q. Hence, by 12.3, the subspaces E_q of 12.7 are stable.

Theorem 12.7 can be generalized as follows (Maurey [1]):

12.8. *Let* $1 < p \leq q < 2$. *Then, for every measure* μ, *there exists a measure* ν *such that the space* $L^q(\mu)$ *is isometrically isomorphic to a subspace of the space* $L^p(\nu)$.

Rosenthal [7] has discovered another property of stable subspaces of L^p, which can be called the extrapolation property.

12.9. *If* $1 \leq p < \infty$, $p \neq 2$, *and* E *is a stable subspace of the space* L^p, *then there exist an isomorphism* U *of* L^p *onto itself and an* $\varepsilon > 0$ *such that* $U(E)$ *is a closed stable subspace of the space* $L^{p+\varepsilon}$, i.e. *there is a* $C > 0$ *such that* $\|f\|_p \leq \|f\|_{p+\varepsilon} \leq C\|f\|_p$ *for every* $f \in E$.

Combining 12.9 with the result of Kadec and Pelczyński [1] showing that

12.10. *Every non-reflexive subspace of* L^1 *contains a standard image of* l^1, we obtain the following:

12.11 (Rosenthal [7]). *Every reflexive subspace of the space* L^1 *is stable, hence isomorphic to a subspace of a space* L^p *for some* $p > 1$.

The results of Chap. XII of [B] and Orlicz [2], Satz 2 combined with 12.3, 12.4 and 12.7 yield an answer to question (I) stated at the beginning of this section and to the question in [B] on p. 124. We have

12.12. *Let* E *be an infinite-dimensional Banach space and let* $1 \leq p < q < \infty$. E *is isomorphic to a subspace of* L^p *and to a subspace of* L^q *if and only if* E *is isomorphic to a subspace of* $L^{\min(q,2)}$. *In particular, if* $q \leq 2$, *then* $\dim_l L^p \geq \dim_l L^q \geq \dim_l l^q$, *and if* $p \neq 2 < q$, *then* $\dim_l L^p$ *is incomparable with* $\dim_l L^q$ *and with* $\dim_l l^q$.

The fact that, for $2 < p < q$, the linear dimensions of L^p and l^q are incomparable has been established first by Paley [1]. The incomparability of $\dim_l L^p$ and $\dim_l L^q$ for $q > 2 > p$ is due to Orlicz [2]. For $1 < p < \infty$, $p \neq 2$, there exist the subspaces of the space L^p which are isomorphic to l^p but are not standard images of l^p. This is a consequence of the following theorem of Rosenthal [3],[8], and Bennett, Dor, Goodman, Johnson and Newman [1].

12.13. *If either* $1 < p < \infty$, $p \neq 2$, *then there exists a non-complemented subspace of* l^p *which is isomorphic to the whole space.*

It is not known whether every subspace of l^1 which is isomorphic to l^1 is complemented in the whole space.

By 12.7 and the fact that, for $p \neq q$ no subspace of l^p is isomorphic to l^q, it follows that the assumption $p > 2$ in 12.5 (b) is indispensable. The following result is related to 12.5 (a):

12.14. (a) *Let* $1 < p < 2$ *and let* E *be an infinite-dimensional subspace of the space* L^p. *If* E *is isomorphic to the Hilbert space, then* E *contains an infinite-dimensional subspace which is complemented in* L^p.

(b) *If* $1 \leq p < \infty$, $p \neq 2$, *then there exists a non-complemented subspace of* L^p *which is isomorphic to a Hilbert space.*

Part (a) is due to Pelczyński and Rosenthal [1], and part (b) - to Rosenthal [8] for $1 \leq p \leq 4/3$ and to Bennett, Dor, Goodman, Johnson and Newman for all p with $1 \leq p < 2$.

In connection with the table in [B], p. 154 (property (15)) let us observe (cf. Pelczyński [3] and 5.2) that

12.15. *If* $1 \leq p < \infty$, $p \neq 2$, *then there exists an infinite-dimensional closed linear subspace of* l^p *which is not isomorphic to the whole space.*

The following theorem of Johnson and Zippin [1] gives a description of subspaces with the approximation property of the spaces l^p.

12.16. *If* E *is a subspace of a space* l^p *with* $1 < p < \infty$, *and* E *has the approximation property, then* E *is isomorphic to a complemented subspace of a product space* $(G_1 \times G_2 \times \ldots)_{l^p}$, *where* G_n*'s are finite-dimensional subspaces of the space* l^p.

CHAPTER VI

§13. The topological structure of linear metric spaces.

The content of [B], Rem. XI, §4 was a catalyst for intensive investigations of the topological structure of linear metric spaces and their subsets. These investigations have led to the following theorem.

13.1. ANDERSON-KADEC THEOREM. *Every infinite-dimensional, separable, locally convex complete linear metric space is homeomorphic to the Hilbert space* l^2.

This result fully answers one of the questions raised in [B], Rem. XI, §4, p. 151 and disproves the statement that the space s is not homeomorphic to any Banach space ([B], Rem. IV, §1, p. 143). Theorem 13.1 is a product of combined efforts of Kadec [11],[12], Anderson [1] and Bessaga and Pelczyński [5],[6]. For alternative or modified proofs see Bessaga and Pelczyński [7] and Anderson and Bing [1]. Earlier partial results can be found in papers by Mazur [1], Kadec [6],[7],[8],[9],[10], Kadec and Levin [1], Klee [1], Bessaga [1].

In the proofs of 13.1 and other results on homeomorphisms of linear metric spaces three techniques are employed:

A. *Kadec's coordinate approach.* The homeomorphism between spaces X and Y is established by setting into correspondence the points $x \in X$ and $y \in Y$ which have the same "coordinates". The "coordinates" are defined in metric terms with respect to suitably chosen uniformly convex norms (see the text after 1.9 for the definition) of the spaces.

B. *The decomposition method,* which consists in representing the spaces in question as infinite products, and performing on the products suitable "algebraic computations" originated by Borsuk [1] (cf. [B], Chap. XI, §7, Theorems 6-8). For the purpose of stating some results, we recall the definition of topological factors. Let X and Y be topological spaces. Y is said to be a *factor of* X (written $Y|X$) if there is a space W such that X is homeomorphic to $Y \times W$. A typical result obtained with the use of the decomposition method is the following criterion, due to Bessaga and Pelczyński [5],[6]:

13.2. *Let X and H be a Banach space and an infinite-dimensional Hilbert space, respectively, both of the same topological weight. Then $H|X$ implies that X is homeomorphic to H.*

Many applications of 13.2 depend on the following result of Bartle and Graves [1] (see also Michael [1],[2],[3] for a simple proof and generalizations).

13.3. *Let X be a Banach space. If Y is either a closed linear subspace or a quotient space of X, then $Y|X$.*

Notice that both 13.2 and 13.3 are valid under the assumption that

X is merely a locally convex complete linear metric space.

Also the next result due to Toruńczyk [3],[4],[5], and some of its generalizations give rise to applications of the decomposition method.

13.4. *If X is a Banach space and A is an absolute retract for metric spaces which can be topologically embedded as a closed subset of X, then $A|(X \times X \times \ldots)_{l^2}$. If H is an infinite-dimensional Hilbert space and A is a complete absolute retract for metric spaces and the topological weight of A is less than or equal to that of H, then $A|H$.*

C. *The absorption technique*, which gives an abstract framework for establishing homeomorphisms between certain pairs (X,E) and (Y,F) consisting of metric spaces and their subsets, when X and Y are already known to be homeomorphic. (The pairs (X,E) and (Y,F) are said to be *homeomorphic*, in symbols $(X,E) \sim (Y,F)$, if there is a homeomorphism h of X onto Y which carries E onto F, and hence carries $X \setminus E$ onto $X \setminus F$). A particular model designed for identifying concrete spaces homeomorphic to R^∞ can briefly be described as follows. Consider the Hilbert cube $Q = [-1,1]^\infty$ and its pseudo-interior $P = (-1,1)^\infty$, which is obviously homeomorphic to R^∞. It turns out that every subset $A \subset Q$ which is such that $(Q,A) \sim (Q,Q\setminus P)$ can be characterized by certain property involving extensions and approximations of maps and related to Anderson's [2] theory of Z-sets, called cap (for compact absorption property). Hence, in order to show that a metric space E is homeomorphic to R^∞ it is enough to represent E as a subset of a space X homeomorphic to Q so that the complement $X \setminus E$ has cap. For applying this technique it is convenient to have many models for the Hilbert cube. An important role in this respect is played by the following classical theorem, due to Keller [1],

13.5. *Every infinite-dimensional compact convex subset of the Hilbert space l^2 is homeomorphic to the Hilbert cube,*

and the remark of Klee [4]

13.6. *Every compact convex subset of any locally convex linear metric space is affinely embeddable into l^2.*

For more details concerning the model presented here and other models of the absorption technique see papers by Anderson [4], Bessaga and Pelczyński [8],[7],[9], Toruńczyk [2] and the book by Bessaga and Pelczyński [10], Chapters IV, V, VI, VIII. The most general axiomatic setting for "absorption" with miscellaneous applications is presented by Toruńczyk [2] and Geoghegan and Summerhill [1].

During the years 1966-1977 several authors attempted to extend the Kadec-Anderson theorem to Banach spaces of an arbitrary topological weight; for the information see Bessaga and Pelczyński [1], Chap. VII, and also Toruńczyk [5], Terry [1]. The final solution has been obtained only recently by Toruńczyk [6] who proved

13.7. *Let X be a complete metric space which is an absolute retract for metric spaces and let $\aleph = wX$, the density character of X. Then X is homeomorphic to the Hilbert space $l_2(\aleph)$ if and only if the following two conditions are satisfied:*

(a) *$X \times l_2$ is homeomorphic to X,*

(b) *every closed subset A of X with $wA < \aleph$ is a Z-set, i.e. for every compact $K \subset X$ the identity embedding of K into X is the uniform limit of a sequence of continuous maps of K into $X \setminus A$.*

In particular,

13.8. *Every locally convex complete metric linear space is homeo-*

morphic to a Hilbert space.

Detailed proofs and other characterizations of Hilbert spaces and Hilbert space manifolds can be found in Toruńczyk [6].

It is natural to ask if in the Anderson-Kadec Theorem 13.1 the assumption of local convexity is essential. This problem is open and only very special non-locally convex spaces are known to be homeomorphic to l_2. For instance (Bessaga and Pelczyński [9]):

13.9. *The space* S ([B], Introduction, §7, p. 6) *is homeomorphic to* l^2. *More generally, if* X *is a separable complete metric space which has at least two different points, then the space* M_X *of all Borel measurable maps* $f: [0,1] \to X$ *with the topology of convergence in (the Lebesgue) measure is homeomorphic to* l^2.

More examples are presented in Bessaga and Pelczyński [10], Chap. VI.

It is known that a non-complete normed linear space cannot be homeomorphic to any Banach space. This easily follows from the theorem of Mazur and Sternbach [1] that every G_δ linear subspace of a Banach space must be closed. There are at least $\aleph_1$ topologically different separable normed linear spaces which can be distinguished by their absolute Borel types (Klee [5], and Mazur - unpublished). Henderson and Pelczyński have proved that even among sigma-compact normed linear spaces there are at least $\aleph_1$ topologically different (cf. Bessaga and Pelczyński [10], Chapter VIII, §5).

It is not known whether every normed linear space is homeomorphic to an inner product space.

Using suitable absorption models, one can prove (Bessaga and Pelczyński [8] and [10], Chap. VIII, §5, Toruńczyk [2])

13.10. *If* X *is an infinite-dimensional normed linear space which is a countable union of its finite-dimensional compact subsets, then* X *is homeomorphic to the subspace* ΣR *of* R^∞ *consisting of all sequences having at most finitely many non-zero coordinates. If* X *is a sigma-compact normed linear space containing an infinite-dimensional compact convex subset, then* X *is homeomorphic to the pseudo-boundary* $Q\backslash P$ *of the Hilbert cube.*

For more details on topological classification of non-complete linear metric spaces the reader is referred to Bessaga and Pelczyński [10], Chap. VIII and the references therein.

Another interesting problem is to find which subsets of a given infinite-dimensional Banach space are homeomorphic to the whole space. The situation is completely different from that in the finite-dimensional case. For instance, we have

13.11. *Let* X *be an infinite-dimensional Banach space. Then the following kinds of subsets* X *are homeomorphic to the whole space:*

(i) *spheres,*

(ii) *arbitrary closed convex bodies (= closed convex sets with non empty interior), in particular: closed balls, closed half-spaces, strips between two half-spaces and so on,*

(iii) *the sets* $X\backslash A$, *where* A *is sigma-compact.*

This result for the space l^2 and several other special spaces has been obtained by Klee [3],[6]. The general case can be reduced to that of l^2 by factoring from X a separable space, homeomorphic to l^2, and by applying some additional constructions, cf. Bessaga and Pelczyński [10], Chap. VI.

The investigations of topological structure of linear metric spaces resulted in active development of the theory of infinite-dimensional manifolds. If E is a linear metric space, then by a *topological man-*

ifold modelled on E (briefly: an E*-manifold*) we mean a metrizable topological space M which has an open cover by sets homeomorphic to open subsets of E. In the same manner one defines manifolds modelled on the Hilbert cube.

A fundamental theorem on topological classification of manifolds with a fixed model E, an infinite-dimensional linear metric space satisfying certain conditions, is due to Henderson (see Henderson [1],[2] and Henderson and Schori [1]). For simplicity we state this theorem in the case of Hilbert spaces.

13.12. *Let H be an infinite-dimensional Hilbert space. Then every connected H-manifold is homeomorphic to an open subset of H. H-manifolds M_1 and M_2 are homeomorphic if and only if they are of the same homotopy type*, i.e. *there are continuous maps $f\colon M_1 \to M_2$ and $g\colon M_2 \to M_1$ such that the compositions gf and fg are homotopic to the identities* id_{M_1} *and* id_{M_2}, *respectively.*

For analogous results on infinite-dimensional differential manifolds, see Burghelea and Kuiper [1], Eells and Elworthy [1], Elworthy [1], Moulis [1].

The systematic theory of manifolds modelled on the Hilbert cube has been developed by Chapman [2],[3],[4],[5] and is closely related to the simple homotopy theory of polyhedra (Chapman [5],[6], cf. Appendix to Cohen [1]) and has some points in common with Borsuk's shape theory (Chapman [1]). Chapman [7] is an excellent source of information.

We conclude this section with some comments concerning the classification of Banach spaces with respect to uniform homeomorphisms. Banach spaces X and Y are *uniformly homeomorphic* if there exists a homeomorphism $f\colon X \xrightarrow[\text{onto}]{} Y$ such that both f and f^{-1} are uniformly continuous.

There are non isomorphic but uniformly homeomorphic Banach spaces (Aharoni and Lindenstrauss [1]). However, Enflo [1] has proved that a Banach space which is uniformly homeomorphic to a Hilbert space is already isomorphic to the Hilbert space (cf. 9.13 here).

Combining the results of Lindenstrauss [10] and Enflo [5] we get

13.13. *If $1 \leq p < q \leq \infty$, then, for arbitrary measures μ and ν, the spaces $L^p(\mu)$ and $L^q(\nu)$ are not uniformly homeomorphic, except the case where* $\dim L^p(\mu) = \dim L^q(\nu) < \infty$.

To state the next result (due to Lindenstrauss [10]) we recall that a closed subspace S of a metric space M is said to be a *uniform retract of* M if there is a uniformly continuous map $r\colon M \to S$ such that $r(x) = x$ for $x \in S$.

13.14. *If a linear subspace Y of a Banach space X is a uniform retract of X and $\varkappa(Y)$ is complemented in Y^{**}, then Y is complemented in X.*

Observe that if Y is reflexive or, more generally, conjugate to a Banach space, then $\varkappa(Y)$ is complemented in Y^{**} (cf. Dixmier [1]).

On the other hand, we have (see Lindenstrauss [10])

13.15. *Let K be a compact metric space. Then every isometric image of $C(K)$ in an arbitrary metric space M is a uniform retract of M.*

Combining 13.14 and 13.15 with the result of Grothendieck [3] (cf. Pelczyński [3]) that no separable infinite-dimensional conjugate Banach space is complemented in a $C(K)$, we get

13.16. *If K is an infinite compact metric space, then the space $C(K)$ is not uniformly homeomorphic to any conjugate Banach space.*

Enflo [6] has shown that

13.17. *No subset of a Hilbert space is uniformly homeomorphic to the space C.*

In "Added in proof" we present Aharoni's and Ribe's contributions to the classification of Banach spaces with respect to uniform homeomorphisms.

Uniform homeomorphisms of locally convex complete metric spaces have been studied by Mankiewicz [1],[2], cf. also Bessaga [1], §11. In particular, Mankiewicz [2] has proved that

13.18. *If X is one of the spaces $l^2, s, l^2 \times s$ and Y is a locally convex linear metric space which is uniformly homeomorphic to X, then Y is isomorphic to X.*

From 13.18 it immediately follows that s is not uniformly homeomorphic to l_2 (a more general fact is proved in Bessaga [1], p. 282).

§14. Added in proof.

Ad §2. The following basic fact in the isomorphic theory of Banach spaces, due to H. P. Rosenthal, is related to the discussion in §9 Chap. IX and to Example 2 in §3 of this survey.

14.1. *Let (x_n) be a bounded sequence in a Banach space. Then (x_n) contains a subsequence equivalent to the standard vector basis of l^1 iff (x_n) has a subsequence whose no subsequence is a weak Cauchy sequence.*

For the proof (for real Banach spaces) see Rosenthal [11]; Dor [1] has adjusted Rosenthal's proof to cover the complex spaces. For related but more delicate results the reader is referred to the excellent survey by Rosenthal [12] and to the papers: Odell and Rosenthal [1] and Bourgain, Fremlin and Talagrand [1].

For further information on *WCG* spaces and renorming problems the reader is referred to the lecture notes by Diestel [1] and to the book by Diestel and Uhl [1].

Ad §3. Theorems 13.7 and 13.8 generalize to the case of arbitrary $p \in (1,\infty)$. We have

14.2 (Krivine [2]). *Let $1 < p < \infty$. Then l^p is locally representable in a Banach space X iff l^p is locally a-representable in X for some $a \geq 1$.*

For an alternative proof of 14.2 see Rosenthal [10].

Using 14.2, Maurey and Pisier [3] have established

14.3. *Let X be a Banach space, let p_X (resp. q_X) be the supremum (resp. infimum) of $p \in [1,\infty]$ such that there is a positive $C = C(q,X) < \infty$ with the property that, for every finite sequence (x_j) of elements*

$$\int_0^1 \Big\| \sum_j r_j(t) x_j \Big\| dt \leq C\Big(\sum_j \|x_j\|^q\Big)^{1/q}$$

$$\left(\text{resp. } \int_0^1 \Big\| \sum_j r_j(t) x_j \Big\| dt \geq C\Big(\sum_j \|x_j\|^q\Big)^{1/q}\right),$$

where (r_j) are the Rademacher functions.

Then l^{p_X} and l^{q_X} are locally representable in X.

Observe that $1 \leq p_X \leq 2$ and $\infty \geq q_X \geq 2$. (The right-hand side inequalities follow from Dvoretsky's Theorem.) In the limit case $p_X = 1$ (resp. $q_X = \infty$) Theorem 14.3 yields 13.8 equivalence (i) and (iv) (resp. 13.7).

Entirely different criterion of local representability of l^1 was discovered by Milman and Wolfson [1].

14.4. *Let X be an infinite-dimensional Banach space with the property that there is a $C<\infty$ such that for every $n=1,2,\dots$ there is an n-dimensional subspace, say E_n, of X with $d(E_n, l_n^2) \leq C\sqrt{n}$. Then l^1 is locally representable in X.*

Ad §4. R. C. James [14] improved 4.3 by constructing a non-reflexive Banach space of type 2, i.e. satisfying 13.8 (iv) with $q=2$.

The reader interested in the subject discussed in §4 is referred to the books and notes: Lindenstrauss and Tzafriri [1], volume II, Maurey and Schwartz [1] (various exposés by Maurey, Maurey and Pisier, and Pisier), Diestel [1], and to the papers: Figiel [6],[7], [8], and Pisier [2].

Ad §5. 14.5 (Szankowski [4]). *The space of all bounded linear operators from l^2 into itself fails to have the approximation property.*

Ad §8. The following result, due to Maurey and Rosenthal [1], is related to the question whether every infinite-dimensional Banach space contains an infinite-dimensional subspace with an unconditional basis.

14.6. *There exists a Banach space which contains a weakly convergent to zero sequence of vectors of norm one such that no infinite subsequence of the sequence forms an unconditional basis for the subspace which it spans.*

Ad §9. The paper by Enflo, Lindenstrauss and Pisier [1], contains an example of a Banach space X which is not isomorphic to a Hilbert space but which has a subspace, say Y, such that both Y and X/Y are isometrically isomorphic to l^2 (cf. also Kalton and Peck [1]).

Ad §§10 and 11. We recommend to the reader the surveys: Rosenthal [9],[12]. The reader might also consult the book by Diestel and Uhl [1].

Most of the recent works on $C(K)$ spaces concern non-separable $C(K)$ spaces. The reader is referred to Alspach and Benyamini [1], Argyros and Negropontis [1], Benyamini [2], Dashiell [1], Dashiell and Lindenstrauss [1], Ditor and Haydon [1], Etcheberry [1], Hagler [1],[2], Haydon [1],[2],[3],[4], Gulko and Oskin [1], Kislyakov [1], Talagrand [1], Wolfe [1]. The separable $C(K)$ spaces are studied in the papers: Alspach [1], Benyamini [1], Billard [1], Zippin [1].

Ad §12. The reader interested in the subject should consult the seminar notes by Maurey and Schwartz [1] and the memoir by Johnson, Maurey, Schechtman and Tzafriri [1]. The reader is also referred to the survey by Rosenthal [9] and to the papers: Alspach, Enflo and Odell [1], Enflo and Rosenthal [1], Enflo and Starbird [1], Gamlen and Gaudet [1], Stegall [1],[2].

Ad §13. The following result of Ribe [1] shows that, despite the example of Aharoni and Lindenstrauss [1] mentioned in §13, the classification of Banach spaces with respect to uniform homeomorphisms is "close" to linear topological classification.

14.7. *If Banach spaces X and Y are uniformly homeomorphic, then there is an $a \geq 1$ such that X is locally a-representable in Y and Y is locally a-representable in X.*

It is known, however (Enflo oral communication), that the spaces L^1 and l^1, which are obviously locally representable each into the other, are not uniformly homeomorphic. On the other hand, isomorphically different Banach spaces might have the same "uniform dimension".

14.8 (Aharoni [1]). *There is a constant* K *so that for every separable metric space* (X,d) *there is a map* $T: X \to c_0$ *satisfying the condition* $d(x,y) \leq \|Tx - Ty\| \leq Kd(x,y)$ *for every* $x,y \in X$. *Hence every separable Banach space is uniformly homeomorphic to a bounded subset of* c_0.

14.9 (Aharoni [2]). *For* $1 \leq p \leq 2, 1 \leq q < \infty$, L^p *is uniformly homeomorphic to a subset of* l^q, *i.e. there is a subset* $Z \subset l^q$ *and a homeomorphism* $f: L^p \to Z$ *such that* f *and* f^{-1} *are uniformly continuous. Moreover,* L^p *is uniformly homeomorphic to a bounded subset of itself.*

Bibliography

G. P. Akilov

[1] *On the extension of linear operations*, Dokl. Akad. Nauk SSSR (1947), pp. 643-646 (Russian).

Freda E. Alexander

[1] *Compact and finite rank operators on subspaces of l_p*, Bull. London Math. Soc. 6 (1974), pp. 341-342.

E. Alfsen

[1] *Compact convex sets and boundary integrals*, Springer Verlag, Berlin 1971.

D. Amir

[1] *On isomorphisms of continuous function spaces*, Israel J. Math. 3 (1965), pp. 205-210.

[2] *Projections onto continuous function spaces*, Proc. Amer. Math. Soc. 15 (1964), pp. 396-402.

D. Amir and J. Lindenstrauss

[1] *The structure of weakly compact sets in Banach spaces*, Ann. of Math. 88 (1968), pp. 35-46.

R. D. Anderson

[1] *Hilbert space is homeomorphic to the countable infinite product of lines*, Bull. Amer. Math. Soc. 72 (1966), pp. 515-519.

[2] *On topological infinite deficiency*, Michigan Math. J. 14 (1967), pp. 365-383.

[3] *Homeomorphism on infinite-dimensional manifolds*, Proc. Inter. Math. Congress, Nice 1970, vol. 2, pp. 13-18.

[4] *On sigma-compact subsets of infinite-dimensional manifolds*, preprint.

T. Ando

[1] *Banachverbände und positive Projektionen*, Math. Z. 109 (1969), pp. 121-130.

N. Aronszajn

[1] *Caractérisation métrique de l'espace de Hilbert, des espaces vectoriels et de certains groups métriques*, Comptes Rendus Acad. Sci. Paris 201 (1935), pp. 811-813 and pp. 873-875.

E. Asplund

[1] *Fréchet differentiability of convex functions*, Acta Math. 121 (1968), pp. 31-47.

[2] *Averaged norms*, Israel J. Math. 5 (1967), pp. 227-233.

H. Auerbach

[1] *Sur les groups bornés de substitutions linéaires*, Comptes Rendus Acad. Sci. Paris 195 (1932), pp. 1367-1369.

H. Auerbach, S. Mazur et S. Ulam

[1] *Sur une propriété caractéristique de l'ellipsoïde*, Monatshefte für Mathematik und Physik 42 (1935), pp. 45-48.

W. G. Bade

[1] *The Banach space C(S)*, Lecture Notes 26, Aarhus University 1971.

S. Banach and S. Mazur

[1] *Zur Theorie der Linearen Dimension*, Studia Math. 4 (1933), pp. 100-112.

R. G. Bartle and L. M. Graves

[1] *Mappings between function spaces*, Trans. Amer. Math. Soc. 72 (1952), pp. 400-413.

Y. Benyamini and J. Lindenstrauss

[1] *A predual of* l_1 *which is not isomorphic to a C(K) space*, Israel J. Math. 13 (1972), pp. 246-254.

S. Bernau and H. E. Lacey

[1] *Characterisations and classifications of some classical Banach spaces*, Advances in Math. 12 (1974), pp. 367-401.

C. Bessaga

[1] *On topological classification of complete linear metric spaces*, Fund. Math. 55 (1965), pp. 251-288.

[2] *Topological equivalence of non-separable Banach spaces*, Symp. on Infinite-Dimensional Topology, Ann. of Math. Studies 69 (1972), pp. 3-14.

C. Bessaga and A. Pelczyński

[1] *Banach spaces non-isomorphic to their Cartesian squares I*, Bull. Acad. Polon. Sci. 8 (1960), pp. 77-80.

[2] *Banach spaces of continuous functions IV*, Studia Math. 19 (1960), pp. 53-62.

[3] *On bases and unconditional convergence of series in Banach spaces*, ibid. 17 (1958), pp. 151-164.

[4] *Properties of bases in* B_0 *spaces*, Prace Mat. 3 (1959), pp. 123-142 (Polish).

[5] *Some remarks on homeomorphisms of Banach spaces*, Bull. Acad. Polon. Sci. Sér. Sci. Math. Astronom. Phys. 8 (1960), pp. 757-760.

[6] *Some remarks on homeomorphisms of F-spaces*, ibid. 10 (1962), pp. 265-270.

[7] *A topological proof that every separable Banach space is homeomorphic to a countable product of lines*, ibid. 17 (1969), pp. 487-493.

[8] *The estimated extension theorem, homogeneous collections and skeletons, and their applications to the topological classification of linear metric spaces and convex sets*, Fund. Math. 69 (1970), pp. 153-190.

[9] *On spaces of measurable functions*, Studia Math. 44 (1972), pp. 597-615.

[10] *Selected topics in infinite-dimensional topology*, Monografie Matematyczne 58, PWN, Warszawa 1975.

C. Bessaga, A. Pelczyński and S. Rolewicz

[1] *On diametral approximative dimension and linear homogeneity of F-spaces*, Bull. Acad. Polon. Sci. 9 (1961), pp. 677-683.

E. Bishop and R. R. Phelps

[1] *A proof that every Banach space is subreflexive*, Bull. Amer. Math. Soc. 67 (1961), pp. 97-98.

[2] *The support functionals of a convex set*, Proceedings of Symposia in Pure Mathematics, vol. VII, Convexity, Amer. Math. Soc., Providence, Rhode Island 1963.

S. V. Bočkariev

[1] *Existence of bases in the space of analytic functions and some properties of the Franklin system*, Math. Sbornik 95 (137) (1974), pp. 3-18 (Russian).

F. Bohnenblust

[1] *A characterization of complex Hilbert spaces*, Portugal. Math. 3 (1942), pp. 103-109.

[2] *Subspaces of $l_{p,n}$ spaces*, Amer. J. Math. 63 (1941), pp. 64-72.

E. D. Bolker

[1] *A class of convex bodies*, Trans. Amer. Math. Soc. 145 (1969), pp. 323-345.

R. Bonic and J. Frampton

[1] *Smooth functions on Banach manifolds*, J. Math. Mech. 15 (1966), pp. 877-898.

K. Borsuk

[1] *Über Isomorphie der Funktionalräume*, Bull. Int. Acad. Pol. Sci. (1933), pp. 1-10.

N. Bourbaki

[1] *Eléments de mathématique, Livre V, Espaces vectoriels topologiques*, Hermann, Paris 1953.

[2] *Eléments d'histoire des mathématiques*, Hermann, Paris 1960.

J. Bretagnolle et D. Dacunha-Castelle

[1] *Application de l'étude de certaines formes linéaires aléatoires au plongement d'espaces de Banach dans des espaces L^p*, Ann. Ecole Normale Supérieure 2 (1969), pp. 437-480.

J. Bretagnolle, D. Dacunha-Castelle et J. L. Krivine

[1] *Lois stables et espaces L_p*, Ann. Inst. Henri Poincaré, Sér. B. 2 (1966), pp. 231-259.

A. Brunel and L. Sucheston

[1] *On B-convex Banach spaces*, Math. Systems Theory 7 (1973).

D. Burghelea and N. H. Kuiper

[1] *Hilbert manifolds*, Ann. of Math. 90 (1969), pp. 379-417.

D. L. Burkholder

[1] *Distribution function inequalites for martingales*, Annals of Probability 1 (1973), pp. 19-42.

M. Cambern

[1] *A generalised Banach-Stone theorem*, Proc. Amer. Math. Soc. 17 (1966), 396-400.

T. A. Chapman

[1] *On some application of infinite-dimensional manifolds to the theory of shape*, Fund. Math. 76 (1972), pp. 181-193.

[2] *On the structure of Hilbert cube manifolds*, Compositio Math. 24 (1972), pp. 329-353.

[3] *Contractible Hilbert cube manifolds*, Proc. Amer. Math. Soc. 35 (1972), pp. 254-258.

[4] *Compact Hilbert cube manifolds and the invariance of the Whitehead torsion*, Bull. Amer. Math. Soc. 79 (1973), pp. 52-56.

[5] *Classification of Hilbert cube manifolds and infinite simple homotopy types*, Topology.

[6] *Surgery and handle straightening in Hilbert cube manifolds*, Pacific J. Math. 45 (1973), pp. 59-79.

Z. Ciesielski

[1] *A construction of basis in* $C^1(I^2)$, Studia Math, 33 (1969), pp. 243-247.

Z. Ciesielski and J. Domsta

[1] *Construction of an orthonormal basis in* $C^m(I^d)$ *and* $W_p^m(I^d)$, Studia Math. 41 (1972), pp. 211-224.

J. A. Clarkson

[1] *Uniformly convex spaces*, Trans. Amer. Math. Soc. 40 (1936), pp. 396-414.

M. Cohen

[1] *A course in simple homotopy theory*, Springer Verlag, New York-Heidelberg-Berlin 1973.

D. F. Cudia

[1] *The geometry of Banach spaces. Smoothness*, Trans. Amer. Math. Soc. 110 (1964), pp. 284-314.

[2] *Rotundity*, Proc. Sympos. Pure Math. 7, Amer. Math. Soc., Providence, Rhode Island 1963.

D. Dacunha-Castelle et J. L. Krivine

[1] *Applications des ultraproduits à l'étude des espaces et des algèbres de Banach*, Studia Math. 41 (1972), pp. 315-334.

I. K. Daugavet

[1] *Some applications of the generalized Marcinkiewicz-Berman identity*, Vestnik Leningrad. Univ. 23 (1968), pp. 59-64 (Russian).

A. M. Davie

[1] *The approximation problem for Banach spaces*, Bull. London Math. Soc. 5 (1973), pp. 261-266.

[2] *Linear extension operators for spaces and algebras of functions*, American J. Math. 94 (1972), pp. 156-172.

[3] *The Banach approximation problem*, J. Approx. Theory 13 (1975), pp. 392-394.

W. J. Davis, D. W. Dean and I. Singer

[1] *Complemented subspaces and Λ-systems in Banach spaces*, Israel J. Math. 6 (1968), pp. 303-309.

W. J. Davis, T. Figiel, W. B. Johnson and A. Pelczyński

[1] *Factoring weakly compact operators*, J. Functional Analysis 17 (1974), pp. 311-327.

W. J. Davis and W. B. Johnson

[1] *A renorming of non reflexive Banach spaces*, Proc. Amer. Math. Soc. 37 (1973), pp. 486-488.

[2] *Basic sequences and norming subspaces in non-quasi-reflexive Banach spaces*, Israel J. Math. 14 (1973), pp. 353-367.

[3] *On the existence of fundamental and total bounded biorthogonal systems in Banach spaces*, Studia Math. 45 (1973), pp. 173-179.

M. M. Day

[1] *Normed linear spaces*, Ergebnisse d. Math., Springer Verlag 1958.

[2] *Strict convexity and smoothness of normed spaces*, Trans. Amer. Math. Soc. 78 (1955), pp. 516-528.

[3] *Uniform convexity in factor and conjugate spaces*, Ann. of Math. 45 (1944), pp. 375-385.

[4] *Reflexive Banach spaces not isomorphic to uniformly convex spaces*, Bull. Amer. Math. Soc. 47 (1941), pp. 313-317.

[5] *On the basis problem in normed spaces*, Proc. Amer. Math. Soc. 13 (1962), pp. 655-658.

M. M. Day, R. C. James and S. Swaminathan

[1] *Normed linear spaces that are uniformly convex in every direction*, Canad. J. Math. 23 (1971), pp. 1051-1059.

D. W. Dean

[1] *The equation $L(E,X^{**}) = L(E,X)^{**}$ and the principle of local reflexivity*, Proc. Amer. Math. Soc. 40 (1973), pp. 146-149.

J. Dieudonné

[1] *Complex structures on real Banach spaces*, Proc. Amer. Math. Soc. 3 (1952), pp. 162-164.

[2] *Recent developments in the theory of locally convex vector spaces*, Bull. Amer. Math. Soc. 59 (1953), pp. 495-512.

[3] *On biorthogonal systems*, Michigan J. Math. 2 (1954), pp. 7-20.

S. Ditor

[1] *On a lemma of Milutin concerning averaging operators in continuous function spaces*, Trans. Amer. Math. Soc. 149 (1970), 443-452.

J. Dixmier

[1] *Sur un théorème de Banach*, Duke Math. J. 15 (1948), pp. 1057-1071.

E. Dubinsky

[1] *Every separable Fréchet space contains a non stable dense subspace*, Studia Math. 40 (1971), pp. 77-79.

N. Dunford and J. T. Schwartz

[1] *Linear operators, I. General theory; II. Spectral theory; III. Spectral operators*, Interscience Publ., New York-London 1958; 1963; 1971.

A. Dvoretsky

[1] *A theorem on convex bodies and applications to Banach spaces*, Proc. Nat. Acad. Sci. USA 45 (1959), pp. 223-226.

[2] *Some results on convex bodies and Banach spaces*, Proc. Symp. Linear Spaces, Jerusalem (1961), pp. 123-160.

A. Dvoretsky and C. A. Rogers

[1] *Absolute and unconditional convergence in normed linear spaces*, Proc. Nat. Acad. Sci. USA 36 (1950), pp. 192-197.

W. F. Eberlein

[1] *Weak compactness in Banach spaces*, Proc. Nat. Acad. Sci. USA 33 (1947), pp. 51-53.

D. A. Edwards

[1] *Compact convex sets*, Proc. Int. Math. Congress Nice D, pp. 359-362.

J. Eells and K. D. Elworthy

[1] *Open embeddings of certain Banach manifolds*, Ann. of Math. 91 (1970), pp. 465-485.

E. G. Effros

[1] *On a class of real Banach spaces*, Israel J. Math. 9 (1971), pp. 430-458.

[2] *Structure in simplex spaces*, Acta Math. 117 (1967), pp. 103-121.

[3] *Structure in simplexes II*, J. Functional Analysis 1 (1967), pp. 379-391.

S. Eilenberg

[1] *Banach space methods in topology*, Ann. of Math. 43 (1942), pp. 568-579.

K. D. Elworthy

[1] *Embeddings isotopy and stability of Banach manifolds*, Compositio Math. 24 (1972), pp. 175-226.

P. Enflo

[1] *Uniform structures and square roots in topological groups I, II*, Israel J. Math. 8 (1970), pp. 230-252: pp. 253-272.

[2] *Banach spaces which can be given an equivalent uniformly convex norm*, ibid. 13 (1973), pp. 281-288.

[3] *A counterexample to the approximation problem in Banach spaces*, Acta Math. 130 (1973), pp. 309-317.

[4] *A Banach space with basis constant* > 1, Ark. Mat. 11 (1973), pp. 103-107.

[5] *On the non-existence of uniform homeomorphisms between* L_p*-spaces*, ibid. 8 (1971), pp. 103-105.

[6] *On a problem of Smirnov*, ibid. 8 (1971), pp. 107-109.

A. S. Esenin-Volpin

[1] *On the existence of the universal compactum of arbitrary weight*, Dokl. Akad. Nauk SSSR 68 (1949), pp. 649-653 (Russian).

T. Figiel

[1] *An example of infinite dimensional reflexive Banach space non-isomorphic to its Cartesian square*, Studia Math. 42 (1972), pp. 295-306.

[2] *Some remarks on Dvoretsky's theorem on almost spherical sections of convex bodies*, Colloq. Math. 24 (1972), pp. 241-252.

[3] *Factorization of compact operators and applications to the approximation problem*, Studia Math. 45 (1973), pp. 191-210.

[4] *Further counterexamples to the approximation problem*, preprint the Ohio State University (1973).

[5] *A short proof of Dvoretsky's theorem*, Compositio Math. 33 (1976), pp. 297-301.

T. Figiel and W. B. Johnson

[1] *The approximation property does not imply the bounded approximation property*, Proc. Amer. Math. Soc. 41 (1973), pp. 197-200.

[2] *A uniformly convex space which contains no* l_p, Compositio Math. 29 (1974), pp. 179-190.

T. Figiel and A. Pelczyński

[1] *On Enflo's method of construction of Banach spaces without the approximation property*, Uspehi Mat. Nauk 28 (1973), pp. 95-108 (Russian).

T. Figiel and G. Pisier

[1] *Séries aléatoires dans les espaces uniformément convexes ou uniformément lisses*, Comptes Rendus Acad. Sci. Paris 279 (1974), pp. 611-614.

C. Foiaş

[1] *Sur certains théorèmes de J. von Neumann concernant les ensembles spectraux*, Acta Sci. Math. (Szeged) 18 (1957), pp. 15-20.

M. Fréchet

[1] *Sur la définition axiomatique d'une classe d'espaces vectoriels distanciés applicables vectoriellement sur l'espaces de Hilbert*, Ann. of Math. 36 (1935), pp. 705-718.

V. F. Gaposhkin

[1] *On the existence of unconditional bases in Orlicz spaces*, Functional Anal. i Priloženn. 1 (1967), pp. 26-32 (Russian).

D. J. H. Garling and Y. Gordon

[1] *Relations between some constants associated with finite dimensional Banach spaces*, Israel J. Math. 9 (1971), pp. 346-361.

B. R. Gelbaum

[1] *Banach spaces and bases*, An. Acad. Brasil. Ci. 30 (1958), pp. 29-36.

I. M. Gelfand

[1] *Abstrakte Funktionen und lineare Operatoren*, Mat. Sb. 4 (1938), pp. 235-286.

R. Geoghegan and R. R. Summerhill

[1] *Pseudo-boundaries and pseudo-interiors in euclidean space*, Trans. Amer. Math. Soc. 194 (1974), pp. 141-165.

D. P. Giesy

[1] *A convexity condition in normed linear spaces*, Trans. Amer. Math. Soc. 125 (1966), pp. 114-146.

Y. Gordon

[1] *On the distance coefficient between isomorphic function spaces*, Israel J. Math. 8 (1970), pp. 391-397.

[2] *Asymmetry and projection constants of Banach spaces*, ibid. 14 (1973), pp. 50-62.

[3] *On the projection and Macphail constants of l_n^p spaces*, ibid. 6 (1968), pp. 295-302.

[4] *On p-absolutely summing constants of Banach spaces*, ibid. 7 (1969), pp. 151-163.

Y. Gordon and D. R. Lewis

[1] *Absolutely summing operators and local unconditional structures*, Acta Math. 133 (1974), pp. 27-48.

Y. Gordon, D. R. Lewis and J. R. Retherford

[1] *Banach ideals of operators with applications to the finite dimensional structure of Banach spaces*, Israel J. Math. 13 (1972), pp. 348-360.

[2] *Banach ideals of operators with applications*, J. Functional Analysis 14 (1973), pp. 85-129.

M. L. Gromov

[1] *On a geometric conjecture of Banach*, Izv. Akad. Nauk SSSR, Ser. Mat. 31 (1967), pp. 1105-1114 (Russian).

A. Grothendieck

[1] *Critères de compacité dans les espaces fonctionnels généraux*, Amer. J. Math. 74 (1952), pp. 168-186.

[2] *Une caractérisation vectorielle métrique des espaces L_1*, Canad. J. Math. 7 (1955), pp. 552-561.

[3] *Sur les applications linéaires faiblement compactes d'espaces du type $C(K)$*, ibid. 5 (1953), pp. 129-173.

[4] *Produits tensoriels topologiques et espaces nucléaires*, Mem. Amer. Math. Soc. 16 (1955).

[5] *Résumé de la théorie métrique des produits tensoriels topologiques*, Bol. Soc. Mat. São Paulo 8 (1956), pp. 1-79.

[6] *Sur certaines classes des suites dans les espaces de Banach et le théorème de Dvoretsky-Rogers*, ibid. 8 (1956), pp. 80-110.

B. Grünbaum

[1] *Projection constants*, Trans. Amer. Math. Soc. 95 (1960), pp. 451-465.

V. I. Gurariĭ

[1] *Space of universal disposition, isotopic spaces and the Mazur problem on rotations of Banach spaces*, Sibirsk. Mat. Ž. 7 (1966), pp. 1002-1013 (Russian).

[2] *On moduli of convexity and smoothness of Banach spaces*, Dokl. Akad. Nauk SSSR, 161 (1965), pp. 1105-1114 (Russian).

[3] *On dependence of certain geometric properties of Banach spaces on modulus of convexity*, Teor. Funkciĭ Funktional. Anal. i Priložen. 2 (1966), pp. 98-107 (Russian).

[4] *On differential properties moduli of convexity of Banach spaces*, Mat. Issled. 2.1 (1967), pp. 141-148 (Russian).

V. I. Gurariĭ and N. I. Gourariĭ

[1] *On bases in uniformly convex and uniformly smooth Banach spaces*, Izv. Akad. Nauk SSSR Ser. Mat. 35 (1971), pp. 210-215.

V. I. Gurariĭ, M. I. Kadec and V. I. Macaev

[1] *On Banach-Mazur distance between certain Minkowski spaces*, Bull. Acad. Polon. Sci. Sér. Sci. Math. Astronom. Phys. 13 (1965), pp. 719-722.

[2] *Distances between finite dimensional analogs of the L_p-spaces*, Mat. Sb. 70 (112) (1966), pp. 24-29 (Russian).

[3] *Dependence of certain properties of Minkowski spaces on asymmetry*, ibid. 71 (113) (1966), pp. 24-29 (Russian).

O. Hanner

[1] *On the uniform convexity of L^p and l^p*, Ark. Mat. 3 (1956), pp. 239-244.

D. W. Henderson

[1] *Stable classification of infinite-dimensional manifolds by homotopy type*, Invent. Math. 12 (1971), pp. 45-56.

[2] *Corrections and extensions of two papers about infinite-dimensional manifolds*, General Topol. and Appl. 1 (1971), pp. 321-327.

D. W. Henderson and R. M. Schori

[1] *Topological classification of infinite-dimensional manifolds by homotopy type*, Bull. Amer. Math. Soc. 76 (1970), pp. 121-124; cf. Henderson [2].

G. M. Henkin

[1] *On stability of unconditional bases in a uniformly convex space*, Uspehi Mat. Nauk 18 (1963), pp. 219-224 (Russian).

J. R. Isbell and Z. Semadeni

[1] *Projection constants and spaces of continuous functions*, Trans. Amer. Math. Soc. 107 (1963), pp. 38-48.

R. C. James

[1] *Bases and reflexivity of Banach spaces*, Ann. of Math. 52 (1950), pp. 518-527.

[2] *A non-reflexive Banach space isometric with its second conjugate space*, Proc. Nat. Acad. Sci. USA 37 (1957), pp. 174-177.

[3] *Characterisations of reflexivity*, Studia Math. 23 (1964), pp. 205-216.

[4] *Weakly compact sets*, Trans. Amer. Math. Soc. 17 (1964), pp. 129-140.

[5] *Weak compactness and reflexivity*, Israel J. Math. 2 (1964), pp. 101-119.

[6] *Reflexivity and the sup of linear functionals*, ibid. 13 (1972), pp. 289-300.

[7] *Separable conjugate spaces*, Pacific Math. J. 10 (1960), pp. 563-571.

[8] *A separable somewhat reflexive Banach space with nonseparable dual*, Bull. Amer. Math. Soc. 80 (1974), pp. 738-743.

[9] *Uniformly non square Banach spaces*, Ann. of Math. 80 (1964), pp. 542-550.

[10] *Super-reflexive Banach spaces*, Canad. J. Math. 24 (1972), pp. 896-904.
[11] *Some self dual properties of normed linear spaces*, Ann. Math. Studies 69, Princeton Univ. Press (1972), pp. 159-175.
[12] *Super-reflexive spaces with bases*, Pacific J. Math. 41 (1972), pp. 409-419.
[13] *The nonreflexive Banach space that is uniformly nonoctahedral*, Israel J. Math. 18 (1974), pp. 145-155.

R. C. James and J. J. Schaffer
[1] *Super-reflexivity and the girth of spheres*, Israel J. Math. 11 (1972), pp. 398-404.

F. John
[1] *Extremum problems with inequalities as subsidiary conditions*, R. Courant Anniversary Volume, Interscience, New York (1948), pp. 187-204.

W. B. Johnson
[1] *A complementably universal conjugate Banach space and its relation to the approximation problem*, Israel J. Math. 13 (1972), pp. 301-310.
[2] *Factoring compact operators*, ibid. 9 (1971), pp. 337-345

W. B. Johnson and J. Lindenstrauss
[1] *Some remarks on weakly compactly generated Banach spaces*, Israel J. Math. 17 (1974), pp. 219-230.

W. B. Johnson and E. Odell
[1] *Subspaces of L_p which embed into l_p*, Compositio Math. 28 (1974), pp. 37-51.

W. B. Johnson and H. P. Rosenthal
[1] *On ω^*-basic sequences and their applications to the study of Banach spaces*, Studia Math. 43 (1972), pp. 77-92.

W. B. Johnson, H. P. Rosenthal and M. Zippin
[1] *On bases finite dimensional decompositions and weaker structures in Banach spaces*, Israel J. Math. 9 (1971), pp. 488-506.

W. B. Johnson and M. Zippin
[1] *On subspaces and quotient of $(\Sigma G_n)l_p$ and $(\Sigma G_n)c_0$*, Israel J. Math. 13 (1972), pp. 311-316.

J. T. Joichi
[1] *Normed linear spaces equivalent to inner product spaces*, Proc. Amer. Soc. 17 (1966), pp. 423-426.

P. Jordan and J. von Neumann
[1] *On inner products in linear metric spaces*, Ann. of Math. 36 (1935), pp. 719-723.

M. I. Kadec
[1] *Spaces isomorphic to a locally uniformly convex space*, Izv. Vysš. Učebn. Zaved. Matematika 6 (13) (1959), pp. 51-57 (Russian).
[2] *Letter to the editor*, ibid. 6 (25) (1961), pp. 186-187 (Russian).
[3] *Conditions for differentiability of norm in Banach space*, Uspehi Mat. Nauk 20.3 (1965), pp. 183-188.
[4] *On linear dimension of the spaces L_p*, ibid. 13 (1958), pp. 95-98 (Russian).
[5] *Unconditional convergence of series in uniformly convex spaces*, ibid. 11 (1956), pp. 185-190 (Russian).
[6] *On homeomorphism of certain Banach spaces*, Dokl. Akad. Nauk SSSR 92 (1953), pp. 465-468 (Russian).

[7] *On the topological equivalence of uniformly convex spaces*, Uspehi Mat. Nauk 10 (1955), pp. 137-141 (Russian).

[8] *On strong and weak convergence*, Dokl. Akad. Nauk SSSR 122 (1958), pp. 13-16 (Russian).

[9] *On connection between weak and strong convergence*, Dopovidi Akad. Nauk Ukrain. RSR 9 (1959), pp. 465-468 (Ukranian).

[10] *On the topological equivalence of cones in Banach spaces*, Dokl. Akad. Nauk SSSR 162 (1965), pp. 1241-1244 (Russian).

[11] *On the topological equivalence of separable Banach spaces*, ibid. 167 (1966), pp. 23-25; English translation: Soviet Math. Dokl. 7 (1966), pp. 319-322.

[12] *A proof of the topological equivalence of all separable infinite-dimensional Banach spaces*, Funkcional. Anal. i Priložen. 1 (1967), pp. 53-62 (Russian).

M. I. Kadec and Ya. Levin

[1] *On a solution of Banach's problem concerning the topological equivalence of spaces of continuous functions*, Trudy Sem. Funkcional. Anal. Voronesh (1960) (Russian).

M. I. Kadec and B. S. Mityagin

[1] *Complemented subspaces in Banach spaces*, Uspehi Mat. Nauk 28 (1973), pp. 77-94.

M. I. Kadec and A. Pelczyński

[1] *Bases lacunary sequences and complemented subspaces in the spaces L_p*, Studia Math. 21 (1962), pp. 161-176.

[2] *Basic sequences, biorthogonal systems and norming sets in Banach and Fréchet spaces*, ibid. 25 (1965), pp. 297-323 (Russian).

M. I. Kadec and M. G. Snobar

[1] *On some functionals on Minkowski compactum*, Mat. Zametki 10 (1971), pp. 453-458 (Russian).

S. Kakutani

[1] *Some characterizations of Euclidean spaces*, Japan. J. Math. 16 (1939), pp. 93-97.

O. H. Keller

[1] *Die Homöomorphie der kompakten konvexen Mengen im Hilbertschen Raum*, Math. Ann. 105 (1931), pp. 748-758.

V. L. Klee

[1] *Mappings into normed linear spaces*, Fund. Math. 49 (1960), pp. 25-34.

[2] *Polyhedral sections of convex bodies*, Acta Math. 103 (1960), pp. 243-267.

[3] *Convex bodies and periodic homeomorphisms in Hilbert space*, Trans. Amer. Math. Soc. 74 (1953), pp. 10-40.

[4] *Some topological properties of convex sets*, ibid. 78 (1955), pp. 30-45.

[5] *On the Borelian and projective types of linear subspaces*, Math. Scand. 6 (1958), pp. 189-199.

[6] *Topological equivalence of a Banach space with its unit cell*, Bull. Amer. Math. Soc. 67 (1961), pp. 286-290.

G. Köthe

[1] *Topological vector spaces*, Springer Verlag, Berlin-Heidelberg-New York 1969.

[2] *Hebbare Lokalkonvexe Räume*, Math. Ann. 165 (1966), pp. 181-195.

J. L Krivine

[1] *Sous-espaces et cones convexes dans les espaces L^p*, Thèse, Paris 1967.

S. Kwapień

[1] *Isomorphic characterizations of inner product spaces by orthogonal series with vector valued coefficients*, Studia Math. 44 (1972), pp. 583-595.

[2] *On Banach spaces containing c_0. A supplement to the paper by J. Hoffman-Jorgensen "Sums of independent Banach space valued random variables"*, ibid. 52 (1974), pp. 187-188.

[3] *On a theorem of L. Schwartz and its applications to absolutely summing operators*, ibid. 38 (1970), pp. 193-201.

[4] *On Enflo's example of a Banach space without the approximation property*, Séminaire Goulaonic-Schwartz 1972-1973, Ecole Polytechnique, Paris.

H. E. Lacey

[1] *The isometric theory of classical Banach spaces*, Springer Verlag, Berlin-Heidelberg-New York 1974.

A. Lazar

[1] *Spaces of affine continuous functions on simplexes*, Trans. Amer. Math. Soc. 134 (1968), pp. 503-525.

[2] *The unit ball in conjugate L_1 space*, Duke Math. J. 36 (1972), pp. 1-8.

[3] *Polyhedral Banach spaces and extensions of compact operators*, Israel J. Math. 7 (1969), pp. 357-364.

A. Lazar and J. Lindenstrauss

[1] *Banach spaces whose duals are L_1-spaces and their representing matrices*, Acta Math. 126 (1971), pp. 165-195.

P. Levy

[1] *Théorie de l'addition de variables aléatoires*, Paris 1937.

D. R. Lewis and C. Stegall

[1] *Banach spaces whose duals are isomorphic to $l_1(\Gamma)$*, J. Functional Analysis 12 (1973), pp. 177-187.

J. Lindenstrauss

[1] *Extension of compact operators*, Mem. Amer. Math. Soc. 48 (1964), pp. 1-112.

[2] *The geometric theory of classical Banach spaces*, Proc. Int. Math. Congress Nice, pp. 365-373.

[3] *On complemented subspaces of m*, Israel J. Math. 5 (1967), pp. 153-156.

[4] *Some aspects of the theory of Banach spaces*, Advances in Math. 5 (1970), pp. 159-180.

[5] *On James's paper "Separable conjugate spaces"*, Israel J. Math. 9 (1971), pp. 279-284.

[6] *Weakly compact sets - their topological properties and the Banach spaces they generate*, Ann. Math. Studies 69, Princeton Univ. Press (1972), pp. 235-273.

[7] *A remark on L_1-spaces*, Israel J. Math. 8 (1970), pp. 80-82.

[8] *On the modulus of smoothness and divergent series in Banach spaces*, Michigan Math. J. 10 (1963), pp. 241-252.

[9] *A remark on symmetric bases*, Israel J. Math. 13 (1972), pp. 317-320.

[10] *On non-linear projections in Banach spaces*, Michigan Math. J. 11 (1964), pp. 263-287.

J. Lindenstrauss and A. Pelczyński

[1] *Absolutely summing operators in $\mathcal{L}_p$ spaces and their applications*, Studia Math. 29 (1968), pp. 275-326.

[2] *Contributions to the theory of the classical Banach spaces*, J. Functional Analysis 8 (1971), pp. 225-249.

J. Lindenstrauss and H. P. Rosenthal

[1] *The $\mathcal{L}_p$ spaces*, Israel J. Math. 7 (1969), pp. 325-349.

J. Lindenstrauss and C. Stegall

[1] *Examples of separable spaces which do not contain* l_1 *and whose duals are non-separable*, Studia Math. 54 (1975), pp. 81-103.

J. Lindenstrauss and L. Tzafriri

[1] *Classical Banach spaces: I. Sequence spaces; II. Function spaces*, Springer Verlag, Ergebnisse, Berlin-Heidelberg-New York 1977; 1979.

[2] *Classical Banach spaces*, Lecture Notes in Math. 333, Springer Verlag, Berlin 1973.

[3] *On complemented subspaces problem*, Israel J. Math. 9 (1971), pp. 263-269.

J. Lindenstrauss and D. E. Wulbert

[1] *On the classification of the Banach spaces whose duals are* L_1*-spaces*, J. Functional Analysis 4 (1969), pp. 332-349.

J. Lindenstrauss and M. Zippin

[1] *Banach spaces with sufficiently many Boolean algebras of projections*, J. Math. Anal. Appl. 25 (1969), pp. 309-320.

A. Lovaglia

[1] *Locally uniformly convex spaces*, Trans. Amer. Math. Soc. 78 (1955), pp. 225-238.

D. Maharam

[1] *On homogeneous measure algebras*, Proc. Nat. Acad. Sci. USA 28 (1942), pp. 108-111.

P. Mankiewicz

[1] *On the differentiability of Lipschitz mappings in Fréchet spaces*, Studia Math. 45 (1973), pp. 13-29.

[2] *On Fréchet spaces uniformly homeomorphic to the spaces* $H \times s$, Bull. Acad. Polon. Sci. Sér. Sci. Math. Astronom. Phys. 22 (1974), pp. 529-531.

J. Marcinkiewicz

[1] *Quelques théorèmes sur les séries orthogonales*, Ann. Soc. Polon. Math. 16 (1937), pp. 84-96.

B. Maurey

[1] *Théorèmes de factorisation pour les opérateurs lineaires à valeurs dans les espaces* L^p, Astérisque 11 (1974), pp. 1-163.

B. Maurey et G. Pisier

[1] *Un théorème d'extrapolation et ses conséquences*, Comptes Rendus Acad. Sci. Paris 277 (1973), pp. 39-42.

[2] *Caractérisation d'une classe d'espaces de Banach par des propriétés de séries aléatoires vectorielles*, ibid. 277 (1973), pp. 687-690.

S. Mazur

[1] *Une remarque sur l'homéomorphie des champs fonctionnels*, Studia Math. 1 (1929), pp. 83-85.

S. Mazur and L. Sternbach

[1] *Über die Borelschen Typen von linearen Mengen*, Studia Math. 4 (1933), pp. 48-54.

C. W. McArthur

[1] *Development in Schauder basis theory*, Bull. Amer. Math. Soc. 78 (1972), pp. 877-908.

E. Michael

[1] *Selected selection theorems*, Amer. Math. Monthly 58 (1956), pp. 233-238.

[2] *Continuous selections I*, Ann. of Math. 63 (1956), pp. 361-382.

[3] *Convex structures and continuous selections*, Canad. J. Math. 11 (1959), pp. 556-575.

E. Michael and A. Pelczyński

[1] *A linear extension theorem*, Illinois J. Math. 11 (1967), pp. 563-579.

H. Milne

[1] *Banach space properties of uniform algebras*, Bull. London Math. Soc. 4 (1972), pp. 323-327.

H. W. Milnes

[1] *Convexity of Orlicz spaces*, Pacific J. Math. 7 (1957), pp. 1451-1483.

V. D. Milman

[1] *Geometric theory of Banach spaces, I:* Uspehi Mat. Nauk 25.3 (1970), pp. 113-173; *II:* Uspehi Mat. Nauk 26.6 (1971), pp. 73-149 (Russian).

[2] *New proof of the theorem of A. Dvoretzky on intersections of convex bodies*, Funkcional. Anal. i Priložen. 5 (1971), pp. 28-37 (Russian).

A. A. Milutin

[1] *Isomorphisms of spaces of continuous functions on compacta of power continuum*, Teor. Funkciĭ, Funkcional. Anal. i Priložen. 2 (1966), pp. 150-156 (Russian).

B. S. Mityagin

[1] *Approximative dimension and bases in nuclear spaces*, Uspehi Mat. Nauk 14 (100) (1961), pp. 63-132 (Russian).

[2] *Fréchet spaces with the unique unconditional basis*, Studia Math. 38 (1970), pp. 23-34.

[3] *The homotopy structure of the linear group of a Banach space*, Uspehi Mat. Nauk 25 (1970), pp. 63-106 (Russian).

N. Moulis

[1] *Structures de Fredholm sur les variétés Hilbertiennes*, Lecture Notes in Mathematics 259, Springer Verlag, Berlin-Heidelberg-New York.

F. J. Murray

[1] *On complementary manifolds and projections in spaces L_p and l_p*, Trans. Amer. Math. Soc. 41 (1937), pp. 138-152.

L. Nachbin

[1] *A theorem of the Hahn-Banach type for linear transformations*, Trans. Amer. Math. Soc. 68 (1950), pp. 28-46.

J. von Neumann

[1] *Eine Spektraltheorie für allgemeine Operatoren eines unitären Raumes*, Math. Nachr. 4 (1951), pp. 258-281.

G. Nordlander

[1] *The modulus of convexity in normed linear spaces*, Ark. Mat. 4 (1960), pp. 15-17.

[2] *On sign-independent and almost sign-independent convergence in normed linear spaces*, ibid. 4 (1960), pp. 287-296.

A. M. Olevskiĭ

[1] *Fourier series and Lebesgue functions*, Summary of a lecture to the Moscow Math. Soc. Uspehi Mat. Nauk 22 (1967), pp. 236-239 (Russian).

W. Orlicz

[1] *Über unbedingte Konvergenz in Funktionenräumen I*, Studia Math. 4 (1933), pp. 33-37.

[2] *Über unbedingte Konvergenz in Funktionenräumen II*, ibid. 4 (1933), pp. 41-47.

[3] *Beiträge zur Theorie der Orthogonalentwicklungen II*, ibid. 1 (1929), pp. 243-255.

R. E. A. C. Paley

[1] *Some theorems on abstract spaces*, Bull. Amer. Math. Soc. 42 (1936), pp. 235-240.

[2] *A remarkable series of orthogonal functions I*, Proc. London Math. Soc. 34 (1932), pp. 247-268.

A. Pelczyński

[1] *A proof of Eberlein-Šmulian theorem by an application of basic sequences*, Bull. Acad. Polon. Sci. 12 (1964), pp. 543-548.

[2] *On Banach spaces containing $L_1(\mu)$*, Studia Math. 30 (1968), pp. 231-246.

[3] *Projections in certain Banach spaces*, ibid. 19 (1960), pp. 209-228.

[4] *Linear extensions, linear averagings, and their applications to linear topological classification of spaces of continuous functions*, Dissertationes Math. 58 (1968).

[5] *On the isomorphism of the spaces m and M*, Bull. Acad. Polon. Sci. 6 (1958), pp. 695-696.

[6] *Any separable Banach space with the bounded approximation property is a complemented subspace of a Banach space with a basis*, Studia Math. 40 (1971), pp. 239-242.

[7] *A note to the paper of I. Singer "Basic sequences and reflexivity of Banach spaces"*, ibid. 21 (1962), pp. 371-374.

[8] *Universal bases*, ibid. 32 (1969), pp. 247-268.

A. Pelczyński and H. P Rosenthal

[1] *Localization techniques in L^p spaces*, Studia Math. 52 (1975), pp. 263-289.

A. Pelczyński and P. Wojtaszczyk

[1] *Banach spaces with finite dimensional expansions of identity and universal bases of finite dimensional subspaces*, Studia Math. 40 (1971), pp. 91-108.

A. Persson and A. Pietsch

[1] *p-nukleare and p-integrale Abbildungen in Banachräumen*, Studia Math. 33 (1969), pp. 19-62.

R. S. Phillips

[1] *A characterization of Euclidean spaces*, Bull. Amer. Math. Soc. 46 (1940), pp. 930-933.

A. Pietsch

[1] *Operator Ideals*, Akademie-Verlag, Berlin 1978.

[2] *Nukleare lokal konvexe Räume*, Akademie Verlag, Berlin 1965.

[3] *Absolut p-summierende Abbildungen in normierten Räumen*, Studia Math. 28 (1967), pp. 333-353.

G. Pisier

[1] *Sur les espaces de Banach qui ne contiennent pas uniformément de l_n^1*, Comptes Rendus Acad. Sci. Paris 277 (1973), pp. 991-994.

V. Ptak

[1] *On a theorem of W. F. Eberlein*, Studia Math. 14 (1954), pp. 276-284.

F. D. Ramsey

[1] *On a problem of formal logic*, Proc. London Math. Soc. 30 (1929), pp. 338-384.

G. Restrepo

[1] *Differentiable norms in Banach spaces*, Bull. Amer. Math. Soc. 70 (1964), pp. 413-414.

J. R. Retherford

[1] *Operator characterisations of L_p spaces*, Israel J. Math. 13 (1972), pp. 337-347.

J. R. Retherford and C. Stegall

[1] *Fully nuclear and completely nuclear operators with applications to $\mathcal{L}_1$ and $\mathcal{L}_\infty$ spaces*, Trans. Amer. Math. Soc. 163 (1972), pp. 457-492.

F. Riesz

[1] *Untersuchungen über Systeme integrierbarer Funktionen*, Math. Ann. 69 (1910), pp. 449-497.

S. Rolewicz

[1] *An example of a normed space non-isomorphic to its product by the real line*, Studia Math. 40 (1971), pp. 71-75.

[2] *Metric linear spaces*, Monografie Matematyczne 56, PWN, Warszawa 1972.

H. P. Rosenthal

[1] *The heredity problem for weakly compactly generated Banach spaces*, Compositio Math. 28 (1974), pp. 83-111.

[2] *On relatively disjoint families of measures with some applications to Banach space theory*, Studia Math. 37 (1970), pp. 13-36.

[3] *On the subspaces of L_p $(p > 2)$ spanned by sequences of independent random variables*, Israel J. Math. 8 (1970), pp. 273-303.

[4] *On the span in L_p of sequence of independent random variables II*, Proc. 6th Berkeley Symp. on Prob. and Statis. vol. II. Probability theory (1972), pp. 149-167.

[5] *On injective Banach spaces and the spaces $L^\infty(\mu)$ for finite measures μ*, Acta Math. 124 (1970), pp. 205-248.

[6] *On factors of $C[0,1]$ with non-separable dual*, Israel J. Math. 13 (1972), pp. 361-378.

[7] *On subspaces of L_p*, Ann. of Math. 97 (1973), pp. 344-373.

[8] *Projections onto translation invariant subspaces of $L_p(G)$*, Memoirs of Amer. Math. Soc. 63 (1966).

D. Rutovitz

[1] *Some parameters associated with finite-dimensional Banach spaces*, J. London Math. Soc. 40 (1965), pp. 241-255.

H. H. Schaeffer

[1] *Topological vector spaces*, Springer Verlag, New York-Heidelberg-Berlin 1971.

I. J. Schoenberg

[1] *Metric spaces and positive definite functions*, Trans. Amer. Math. Soc. 44 (1938), p. 522-536.

[2] *Metric spaces and completely continuous functions*, Ann. of Math. 39 (1938), pp. 809-841.

S. Schonefeld

[1] *Schauder bases in spaces of differentiable functions*, Bull. Amer. Math. Soc. 75 (1969), pp. 586-590.

[2] *Schauder bases in the Banach space $C^k(T^q)$*, Trans. Amer. Math. Soc. 165 (1972), pp. 309-318.

L. Schwartz

[1] *Applications p-radonifiantes et théorème de dualité*, Studia Math. 38 (1970), pp. 203-213.

[2] *Applications radonifiantes*, Séminaire L. Schwartz, Ecole Polytechnique, Paris 1969-1970.

Z. Semadeni

[1] *Banach spaces non-isomorphic to their Cartesian squares II*, Bull. Acad. Polon. Sci. 8 (1960), pp. 81-84.

[2] *Banach spaces of continuous functions*, Vol. I, Monografie Matematyczne 55, PWN, Warszawa 1971.

E. M. Semenov

[1] *The new interpolation theorem*, Funkcional. Anal. i Priložen. 2 (1968), pp. 68-80.

W. Sierpiński

[1] *Cardinal and ordinal numbers*, Monografie Matematyczne 94, PWN, Warszawa 1958.

I. Singer

[1] *Bases in Banach spaces I*, Springer Verlag, Berlin-Heidelberg-New York 1970.

V. L. Šmulian

[1] *Über Lineare topologische Räume*, Mat. Sbornik 7 (1940), pp. 425-448.

M. G. Snobar

[1] *On p-absolutely summing constants*, Teor. Funkciĭ, Funkcional. Anal. i Priložen. 16 (1972), pp. 38-41.

A. Sobczyk

[1] *Projection of the space m on its subspace c_0*, Bull. Amer. Math. Soc. 47 (1941), pp. 938-947.

[2] *Projections in Minkowski and Banach spaces*, Duke Math. J. 8 (1941), pp. 78-106.

M. H. Stone

[1] *Application of the theory of boolean rings to general topology*, Trans. Amer. Math. Soc. 41 (1937), pp. 375-481.

K. Sundaresan

[1] *Smooth Banach spaces*, Bull. Amer. Math. Soc. 72 (1966), pp. 520-521.

A. Szankowski

[1] *On Dvoretzky's theorem on almost spherical sections of convex bodies*, Israel J. Math. 17 (1974), pp. 325-338.

W. E. Terry

[1] *Any infinite-dimensional Fréchet space homeomorphic with its countable product is topologically a Hilbert space*, Trans. Amer. Math. Soc. 196 (1974), pp. 93-104.

H. Toruńczyk

[1] *Skeletonized sets in complete metric spaces and homeomorphisms of the Hilbert cube*, Bull. Acad. Polon. Sci. Sér. Sci. Math. Astronom. Phys. 18 (1970), pp. 119-126.

[2] *Skeletons and absorbing sets in complete metric spaces*, preprint.

[3] *Compact absolute retracts as factors of the Hilbert spaces*, Fund. Math. 83 (1973), pp. 75-84.

[4] *Absolute retracts as factors of normed linear spaces*, ibid. 86 (1974), pp. 53-67.

[5] *On Cartesian factors and topological classification of linear metric spaces*, ibid. 88 (1975), pp. 71-86.

S. L. Troyanski

[1] *On locally convex and differentiable norms in certain non-separable Banach spaces*, Studia Math. 37 (1971), pp. 173-180.

[2] *On the topological equivalence of the spaces $c_0(\aleph)$ and $l(\aleph)$*, Bull. Acad. Polon. Sci. Sér. Sci. Math. Astronom. Phys. 15 (1967), pp. 389-396.

L. Tzafriri

[1] *On Banach spaces with unconditional bases*, Israel J. Math. 17 (1974), pp. 84-93.

B. S. Tzirelson

[1] *Not every Banach space contains l_p or c_0*, Funkcional. Anal. i Priložen. 8 (1974), pp. 57-60 (Russian).

J. H. M. Whitfield

[1] *Differentiable functions with bounded non empty support on Banach spaces*, Bull. Amer. Math. Soc. 72 (1966), pp. 145-146.

R. J. Whitley

[1] *An elementary proof of the Eberlein-Šmulian theorem*, Math. Ann. 172 (1967), pp. 116-118.

P. Wojtaszczyk

[1] *Some remarks on the Gurarii space*, Studia Math. 41 (1972), pp. 207-210.

M. Zippin

[1] *On some subspaces of Banach spaces whose duals are L_1 spaces*, Proc. Amer. Math. Soc. 23 (1969), pp. 378-385.

[2] *A remark on Pełczyński's paper "Universal bases"*, ibid. 26 (1970), pp. 294-300.

V. Zizler

[1] *On some rotundity and smoothness properties of Banach spaces*, Dissertationes Math. 87 (1971).

Additional Bibliography

I. Aharoni

[1] *Every separable metric space is Lipschitz equivalent to a subset of c_0*, Israel J. Math. 19 (1974), pp. 284-291.

[2] *Uniform embeddings of Banach spaces*, ibid. 27 (1977), pp. 174-179.

I. Aharoni and J. Lindenstrauss

[1] *Uniform equivalence between Banach spaces*, Bull. Amer. Math. Soc. 84 (1978), pp. 281-283.

D. Alspach

[1] *Quotients of $C[0,1]$ with separable dual*, Israel J. Math. 29 (1978), pp. 361-384.

D. Alspach and Y. Benyamini

[1] *Primariness of spaces of continuous functions on ordinals*, Israel J. Math. 27 (1977), pp. 64-92.

D. Alspach, P. Enflo and E. Odell

[1] *On the structure of separable* $\mathcal{L}_p$ *spaces* ($1 < p < \infty$), Studia Math. 60 (1977), pp. 79-90.

S. Argyros and S. Negropontis

[1] *Universal embeddings of* l^1_α *into* $C(X)$ *and* $L(\mu)$, to appear.

G. Bennett, L. E. Dor, V. Goodman, W. B. Johnson and C. M. Newman

[1] *On uncomplemented subspaces of* L^p, $1 < p < 2$, Israel J. Math. 26 (1977), pp. 178-187.

Y. Benyamini

[1] *An extension theorem for separable Banach spaces*, Israel J. Math. 29 (1978), pp. 24-30.

[2] *An M-space which is not isomorphic to a* $C(K)$ *space*, ibid. 28 (1977), pp. 98-102.

P. Billard

[1] *Sur la primarité des espaces* $C(\alpha)$, Studia Math. 62 (1978), pp. 143-162.

J. Bourgain

[1] *Un espace* $\mathcal{L}^\infty$ *jouissant de la propriété de Schur et de la propriété de Radon-Nikodym*, Seminaire d'Analyse Fonctionnelle 1978-1979, Ecole Polytechnique, Paris, Exposé IV.

J. Bourgain, D. H. Fremlin and M. Talagrand

[1] *Pointwise compact sets of Baire measurable functions*, American J. of Math. 100 (1978), pp. 845-886.

T. A. Chapman

[7] *Lectures on Hilbert cube manifolds*, Conference board of the mathematical sciences, Regional conference series in mathematics, Volume 28, American Math. Soc., Providence, R. I., 1977.

H. B. Cohen

[1] *A bounded to isomorphism between* $C(X)$ *Banach spaces*, Proc. Amer. Math. Soc. 50 (1975), pp. 215-217.

F. K. Dashiell

[1] *Isomorphism problems for the Baire Classes*, Pacific J. Math. 52 (1974), pp. 29-43.

F. K. Dashiell and J. Lindenstrauss

[1] *Some examples concerning strictly convex norms on* $C(K)$ *spaces*, Israel J. Math. 16 (1973), pp. 329-342.

J. Diestel

[1] *Geometry of Banach spaces - selected topics*, Lecture Notes in Math., vol. 485, Springer Verlag, Berlin-New York 1975.

J. Diestel and J. J. Uhl, Jr.

[1] *Vector measures*, Math. Surveys, Vol. 15, Amer. Math. Soc., Providence, R. I., 1977.

S. Ditor and R. Haydon

[1] *On absolute retracts*, $P(S)$, *and complemented subspaces of* $C(D^{\omega_1})$, Studia Math. 56 (1976), pp. 243-251.

L. Dor

[1] *On sequence spanning a complex* l^1 *space*, Proc. Amer. Math. Soc. 47 (1975), pp. 515-516.

P. Enflo, J. Lindenstrauss and G. Pisier

[1] *On the three space problem*, Math. Scand. 36 (1975), pp. 199-210.

P. Enflo and H. P. Rosenthal

[1] *Some results concerning* $L^p(\mu)$*-spaces*, J. Functional Analysis 14 (1973), pp. 325-348.

P. Enflo and T. W. Starbird

[1] *Subspaces of* L^1 *containing* L^1, Studia Math. 65, to appear.

A. Etcheberry

[1] *Isomorphism of spaces of bounded continuous functions*, Studia Math. 53 (1975), pp. 103-127.

T. Figiel

[6] *On the moduli of convexity and smoothness*, Studia Math. 56 (1976), pp. 121-155.

[7] *Uniformly convex norms on Banach lattices*, ibid. 68, to appear.

[8] *Lattice norms and the geometry of Banach spaces*, Proceedings of the Leipzig Conference on operator ideals (Leipzig 1977).

T. Figiel, S. Kwapień and A. Pełczyński

[1] *Sharp estimates for the constants of local unconditional structure of Minkowski spaces*, Bull. Acad. Polon. Sci. Sér. math., astr. et phys. 25 (1977), pp. 1221-1226.

T. Figiel, J. Lindenstrauss and V. Milman

[1] *The dimension of almost spherical sections of convex bodies*, Acta Math. 139 (1977), pp. 53-94.

J. L. B. Gamlen and R. J. Gaudet

[1] *On subsequences of the Haar system in* $L_p[0,1]$ $(1 < p < \infty)$, Israel J. Math. 15 (1973), pp. 404-413.

S. P. Gulko and A. V. Oskin

[1] *Isomorphic classification of spaces of continuous functions on totally ordered compact sets*, Funktional. Analiz i Priloz. 9 (1975), pp. 56-57 (Russian).

I. Hagler

[1] *On the structure of S and C(S) for S dyadic*, Trans. Amer. Math. Soc. 214 (1975), pp. 415-428.

[2] *Some more Banach spaces which contain* l^1, Studia Math. 46 (1973), pp. 35-42.

R. Haydon

[1] *On dual* L^1*-spaces and injective bidual Banach spaces*, to appear.

[2] *On Banach spaces which contain* $l^1(\tau)$ *and types of measures on compact spaces*, Israel J. Math. 28 (1977), pp. 313-324.

[3] *On a problem of Pełczyński: Milutin spaces, Dugundji spaces and AE*(0-dim), Studia Math. 52 (1974), pp. 23-31.

[4] *Embedding* D^τ *in Dugundji spaces, with an application to linear topological classification of spaces of continuous functions*, ibid. 56 (1976), pp. 229-242.

R. C. James

[14] *Nonreflexive spaces of type* 2, to appear.

W. B. Johnson

[3] unpublished.

W. B. Johnson, H. König, B. Maurey and J. R. Retherford

[1] *Eigenvalues of p-summing and l_p-type operators in Banach spaces*, J. Functional Analysis, to appear.

W. B. Johnson, B. Maurey, G. Schechtman and L. Tzafriri

[1] *Symmetric structures in Banach spaces*, Mem. Amer. Math. Soc., to appear.

W. B. Johnson and A. Szankowski

[1] *Complementably universal Banach spaces*, Studia Math. 58 (1976), pp. 91-97.

N. Kalton and N. T. Peck

[1] *Twisted sums of sequence spaces and the three space problem*, to appear.

S. V. Kislyakov

[1] *Classification of spaces of continuous functions of ordinals*, Siberian Math. J. 16 (1975), pp. 226-231 (Russian).

J. L Krivine

[2] *Sous-espaces de dimension fini des espaces de Banach reticulês*, Ann. of Math. 104 (1976), pp. 1-29.

W. Lusky

[1] *The Gurariĭ spaces are unique*, Arch. Math. 27 (1976), pp. 627-635.

B. Maurey et G. Pisier

[3] *Séries de variables aléatoires vectorielles indépendantes et propriétés géométriques des espaces de Banach*, Studia Math. 58 (1976), pp. 45-90.

B. Maurey and H. P. Rosenthal

[1] *Normalized weakly null sequences with no unconditional subsequences*, Studia Math. 61 (1977), pp. 77-98.

B. Maurey et L. Schwartz

[1] *Seminaire Maurey-Schwartz 1972-1973; 1973-1974; 1974-1975; 1975-1976*, École Polytechnique, Paris.

V. Z. Meshkov

[1] *Smoothness properties in Banach spaces*, Studia Math. 63 (1978), pp. 111-123.

V. Milman and H. Wolfson

[1] *Minkowski spaces with extremal distance from the Euclidean space*, Israel J. Math. 29 (1978), pp. 113-131.

T. Odell and H. P. Rosenthal

[1] *A double dual characterization of separable Banach spaces containing l^1*, Israel J. Math. 20 (1975), pp. 375-384.

R. I. Ovsepian and A. Pelczyński

[1] *The existence in every separable Banach space of a fundamental total and bounded biorthogonal sequence and related constructions of uniformly bounded orthonormal systems in L^2*, Studia Math. 54 (1975), pp. 149-159.

A. Pelczyński

[9] *All separable Banach spaces admit for every $\varepsilon > 0$ fundamental total biorthogonal system bounded by $1 + \varepsilon$*, Studia Math. 55 (1976), pp. 295-304.

G. Pisier

[2] *Martingales with values in uniformly convex spaces*, Israel J. Math. 20 (1975), pp. 326-350.

M. Ribe

[1] *On uniformly homeomorphic normed spaces*, Arkiv f. Math. 14 (1976), pp. 233-244.

H. P. Rosenthal

[9] *The Banach spaces C(K) and $L^p(\mu)$*, Bull. Amer. Math. Soc. 81 (1975), pp. 763-781.

[10] *On a theorem of J. L. Krivine concerning local finite representability of l^p in general Banach spaces*, J. Functional Analysis 28 (1978), pp. 197-225.

[11] *A characterization of Banach spaces containing l^1*, Proc. Nat. Acad. USA 71 (1974), pp. 2411-2413.

[12] *Some recent discoveries in the isomorphic theory of Banach spaces*, Bull. Amer. Math. Soc. 84 (1978), pp. 803-831.

G. Schechtman

[1] *Examples of $\mathcal{L}_p$ spaces* $(1 < p \neq 2 < \infty)$, Israel J. Math. 19 (1974), pp. 220-224.

[2] *On Pełczyński's paper "Universal bases"*, ibid. 22 (1975), pp. 181-184.

R. Schneider

[1] *Equivariant endomorphisms of the space of convex bodies*, Trans. Amer. Math. Soc. 194 (1974), pp. 53-78.

C. P. Stegall

[1] *Banach spaces whose duals contain $l^1(\Gamma)$ with applications to the study of dual $L_1(\mu)$ spaces*, Trans. Amer. Math. Soc. 176 (1976), pp. 463-477.

[2] *The Radon-Nikodym property in conjugate Banach spaces*, ibid. 206 (1975), pp. 213-223.

A. Szankowski

[2] *Subspaces without the approximation property*, Israel J. Math. 30 (1978), pp. 123-129.

[3] *A Banach lattice without the approximation property*, ibid. 24 (1976), pp. 329-337.

[4] *B(H) does not have the approximation property*, to appear.

M. Talagrand

[1] *Espaces de Banach Faiblement K-analytiques*, Séminaire sur la géométrie des espaces de Banach, Exposés XII-XIII, École Polytechnique, Palaiseau 1977-1978.

H. Toruńczyk

[6] *Characterizing a Hilbert space topology*, preprint 143, Institute of Math. Polish Acad. Sci.

W. A. Veech

[1] *Short proof of Sobczyk's theorem*, Proc. Amer. Math. Soc. 28 (1971), pp. 627-628.

J. Wolfe

[1] *Injective Banach spaces of type C(T)*, Israel J. Math. 18 (1974), pp. 133-140.

M. Zippin

[3] *The separable extension problem*, Israel J. Math. 26 (1977), pp. 372-387.